# HISTORY OF SCIENCE IN INDIA

## Analytical Database of Information Sources

SHAILENDRA KUMAR

GYAN

GYAN PUBLISHING HOUSE

New Delhi 110 002

History of Science in India
Analytical Database of Information Sources
*(Science, Bibliography, Information and Library Science)*

ISBN : 81-212-0473-9
Library of Congress Card No. 93-900323



Published 1994 in India by
Gyan Publishing House
5, Ansari Road
New Delhi 110 002


printed at Mehra Offset Press, New Delhi

TO

**Dr. B.S. GARG**

# FOREWORD

The History of Science has been internationally recognised as a major area for study. The First General Congress in History of Science was held in Paris in 1929 and the last Congress was in Zaragoza (Spain) in 1993. These Congress are held every four years under the auspices of International Union for History and Philosophy of Science. In India a symposium on History of Sciences in South East Asia, held in Delhi in 1950, first emphasized the importance for a systemic study. To meet this requirement Indian National Science Academy constituted a History of Science Board which was enlarged into a Commission for Compilation of History of Sciences in India in 1965 in which eminent scientists, historians, indologists were members. The name of the Commission was modified in 1989 to Indian National Commission for History of Science to give its coverage both from national and international perspective.

The study involves critical assessment of materials ranging from antiquity to modern period, and has wide dimension in terms of country, science, transmission, chronology, and varieties of source materials involved in the process. The contributions in the field mostly remain buried under the brittle pages of old journals and often in monographs long gone out of print. Proper documentation of these materials for critical assessment is a major problem in the field.

The author is a well known information specialist who has got initial training in the History of Science division of the Academy and has proved himself competent enough through his publications and Ph.D. dissertation. He has made in this book a data base of information sources for History of Science in India and the scientometric analysis of the sources, of course on a limited number of journals.

The genesis of the plan followed in the book has been explained in the author's introduction. His effort has been concentrated mostly in the selection of research journals in the History of Science in India with bibliographical information, list of contents. These are arranged chronologically with various indexes, author index, title index, keyword index and the journal index. The data base has also been stored in floppy disks and are available for speedy retrieval and analysis.

The book provides a systematic analysis of the research articles highlighting research trends, highly productive authors alongwith their contributions and the growth of literature in the field. The studies are analysed and presented with various graphs and titles.

To prepare a data base of this type with all necessary details such as have been attempted in this work is really a very difficult undertaking. However, the author deserves our special thanks for the able manner he has accomplished this task.

**Dr. A.K. BAG**
Head, History of Science Division,
Indian National Science Academy,
New Delhi.

# PREFACE

Research programmes on History of Science in India have been initiated at the Indian National Science Academy under the auspicious guidance of National Commission for the Compilation of History of Sciences in India. With the advent of National Institute of Science, Technology and Development Studies, these programmes have got tremendous boost. The main thrust of such studies has been to assimilate the coherent growth and development of science and technology in India since its inception and to extend it for the global perspective. There is no denying the fact that for such studies, a strong data base as well as documentation of published and unpublished literature on science and technology in respective region, is of utmost significance. Although some sporadic attempts have been initiated in this direction, nothing concrete has been achieved in this field except preparation of lists of manuscripts available in various libraries, archives and bibliographies. In order to be of some concrete help to the research scholars, scientists and information scientists, this study was proposed to be undertaken.

In this study, a data base on History of Science in India which includes the articles from the selected journals since their inception upto 1986, has been developed. The data base is prepared in a machine-readable form to meet the latest information and technological innovations; it is also available on a hard copy for ready reference. Since it could not have been humanely possible to carry out this study without the aid of personal computer as well as programmable scientific calculator, the data base has been created by using Personal Computer ESPL/PCL - XT/AT 286/386 at Indian National Scientific Documentation Centre, Delhi. The software used for this purpose is CDS/ISIS which is a UNESCO supported software package for the bibliographic information.

One of the requisite conditions for the growth and development of the subject, is the required analysis of the information sources. This has been done with the help of various scientometric techniques such as Lotka's Law, Zipf's Law, Citation Analysis so on and so forth which help to analyse the information in its proper perspectives.

Growth in the subject has been determined from the view points of literature, published sources, and research workers. Passive authors in the subject have been calculated and the pattern in the subject has been traced. The contribution of the authors in terms of number of papers published on the subject has been treated as an indicator of the productivity of the author. Various productivity constants are determined and the Lotkas's Law of authors' productivity has been applied in the subject. Highly productive authors are identified and studied to find out the productivity rank, patterns of contributions, subject scattering, and journal preference.

The analysis of the topical content makes better conclusion about the state and trends of research development in a particular field of science. Here, the Keyword Index from the data base on History of Science has been used which made it possible to define readily the research topics in a given field. Substantive words have been tabulated; frequency distribution of keywords has identified the high frequency occurrences as well as mapped the keywords. Rank-ordered frequency counts have been applied to identify the applicability of Zipf's Law in the subject. Authors' most prolific years of contributions and their respective ratios of number of pages to the number of references have been used to generate the idea of [illegible]xt value of the authors.

Citation analysis has been applied in the History of Science and citation ranks of the scientists have been determined. The continuity in the citation occurrences in terms of citation gaps has been studied. Papers with citations and their source details alongwith related citations have been listed to identify the citing items. Every attempt has been made to make this study useful to the concerned scholars in order to enable them to carry on their work in more significant and scientific manner. It is hoped that this study will go a long way for mirroring the prevailing conditions in respect of information analysis pertaining to the History of Science in India.

I shall always look back in gratitude and admiration to Dr. B.S. Garg, Head, Department of Library & Information Science, University of Rajasthan, Jaipur, my research supervisor whom I deem as my friend, for his keen interest, constant encouragement, constructive criticism and distinguished guidance throughout the course of this investigation.

I also owe a deep debt of gratitude to Prof. Krishan Kumar, Head, Department of Library and Information Science, University of Delhi for generously providing his thoughts, comments, and technical assistance from time to time. My thanks are also due to Prof. B. Guha of Library and Information Science and Prof. A.K. Bag of Indian National Science Academy for their suggestions and comments.

I wish to record my sincere thanks for Prof. T. Vishwanathan, Dr. B.K. Sen and other colleagues of mine at INSDOC for allowing me the use of computer facilities and that too, without any let or hindrance.

I am also thankful to the librarians as well as the other staff members of Central Science Library and Central Reference Library of University of Delhi; National Social Science Documentation Centre; Indira Gandhi National Centre for Arts; National Archives; Nehru Memorial Library; National Science Library of INSDOC plus the Asiatic Society of Bengal, for the help rendered by them in gathering the relevant data.

Above all, a deep debt of gratitude must indeed be due to my parents for being the main motivating force behind this academic endeavour. My wife 'Promila' and children 'Swati' and 'Prateek' also deserve special thanks for sparing me from household duties in order to help me to complete this academic undertaking efficiently and effectively. I would be failing in my duty if I do not express my sincere thanks to Sh. Prashant Srivastava, a colleague of mine for undertaking much pain for inputting the data and typing the manuscript. Last but not the least, I wish to acknowledge gratefulness to all those authors and publishers whose books/and articles I have consulted and/ quoted in my work.

**SHAILENDRA KUMAR**
**Assistant Professor**

**School of Social Sciences**
**Indira Gandhi National Open University**
**New Delhi**

## CONTENTS

## ABBREVIATIONS

| | | |
|---|---|---|
| ASB | - | Asiatic Society of Bengal |
| CB | - | Critical Bibliography |
| CDROM | - | Compact Disk Read Only Memory |
| CDS/ISIS | - | Computerised Documentation System/ Integrated Set of Information Systems |
| DSDD | - | Double Side Double Density |
| FDT | - | Field Definition Table |
| FST | - | Field Select Table |
| IHQ | - | Indian Historical Quarterly, Calcutta |
| IJHS | - | Indian Journal of History of Science, Delhi |
| ISIS | - | ISIS International Review Devoted to the History of Science and Civilization, Baltimore, USA |
| JAS | - | Journal of the Asiatic Society, Calcutta |
| JASBL | - | Journal of the Asiatic Society of Bengal, Letters, Calcutta |
| JASBSC | - | Journal of Asiatic Society of Bengal, Science, Calcutta |
| JPASB | - | Journal and Proceedings of the Asiatic Society of Bengal, Calcutta |
| JRAS | - | Journal of the Royal Asiatic Society of Great Britain and Ireland, London |
| PC/AT | - | Personal Computer/Advanced Technology |
| PC/XT | - | Personal Computer/Extended Technology |
| RAM | - | Random Access Memory |
| SCI | - | Science Citation Index |
| SSCI | - | Social Sciences Citation Index |
| TPI | - | Tracks Per Inch |
| VIJ | - | Vishveshvaranand Indological Journal, Hoshiarpur |

# LIST OF TABLES

# LIST OF FIGURES

# PART I
# Scientometric Analysis

# 1. INTRODUCTION

## 1.1 BACKGROUND OF HISTORY OF SCIENCE IN INDIA

The science as it is today is the greatest triumph of the human mind over nature. The story of its origin, development and achievements is one of the less known parts of history and has just initialised its way into general literature of science. The study of the History of Science has emerged as a relatively new discipline. The ancient/medieval science can be understood in the light of the current context, only by tracing back and correlating the then developments with the latest emerging scientific developments. With this end in view, several earlier works started being translated into English in the nineteenth century, whereas some sporadic attempts to study History of Science were started in the beginning of the twentieth century (1).

Although several books on the History of Science were published in India, the serials literature in it started appearing with the advent of the Asiatic Society of Bengal in 1784. The establishment of Indian Association for the Cultivation of Science at Calcutta in 1876 is a landmark in the history of symbiotic relationship of Indian and Western scientific ideas. During the twentieth century, the Indian science progressed with great momentum and 1920's can be regarded as the most glorious decade for Indian science, when Prof. M.N. Saha, S.N. Bose, J.C. Bose, C.V. Raman etc., appeared on the scene of Indian science (2).

However, organised activity for tracing correlation and development of science in India, started with the formation of 'The Research Institute of Ancient Scientific Studies', New Delhi in 1963 as well as by setting up of 'National Commission for the Compilation of History of Sciences in India' in 1965. In 1966, the commission started the journal titled 'Indian Journal of History of Science'. It was followed by establishment of 'National Institute of Science, Technology and Development Studies' in New Delhi, for undertaking research projects in the History of Science (3).

## 1.2 THE PROBLEM

A great wealth of literature generated in this field at a considerable cost of time and energy, is lying scattered almost unknown and unused. There is hardly any bibliographic control specially in the field of periodical literature. Unnecessary duplication of work occurs for want of proper bibliographic control. In addition, study of the History of Science has a wide time scale ranging from the ancient period to modern period, requiring the use of all types of information sources ranging from clay tablets to printed journals. However, the research scholars are not in a position to use the original literature effectively, mainly due to language problem. There is a large amount of literature dealing with the History of Science in India. Scientific periodical literature is however, highly significant for the study of the development of scientific ideas in as much as contribution of Indian scientists to the growth and development of scientific literature can be determined to a large measure on its basis.

### 1.2.1 Literature Survey

The literature in the area of measurement of bibliographic information related to its theoretical and application aspects was surveyed. The metric study in this field through various view points is very extensive; as a result, certain choices had to be made for this literature survey. Initially the search was confined to readily available material in open literature, and the very early of original papers in the field which introduced basic ideas were identified. Later on "Library Literature, and Library & Information Science Abstracts (LISA)" were scanned for the period 1970-86 to find out what has been done in the field of scientometrics and the studies undertaken in the History of Science in India. The literature survey in this area over the aforesaid period of time revealed the following.

The term 'Scientometrics' has Russian origin and was used for the application of quantitative methods to the History of Science (4) and overlaps with 'Bibliometrics' to a considerable extent. Alan Pritchard in 1969 had coined and defined the term 'Bibliometrics' as the application of mathematics and statistical methods to books and other media of communication (5). The field of bibliometrics has grown in close association with its applications. The most basic bibliometric data, a count and elementary categorization of publications by country and field, is still of interest today.

The first publication count appears to have been the paper by F.J. Cole and N.B. Eales in 1917. Cole and Eales analysed comparative anatomy papers by counting the number of publications by country, from 1543 to 1860; this count included books and journal articles. This count carefully constructed and well interpreted, could still serve as a model for the application of these techniques in the assessment of international scientific activity (6).

Since then there have been many studies related to this field in the area of growth of literature, bibliometric laws, citation pattern, evaluation of research output etc.

**1.2.1.1 Growth of Literature :** D.D. Solla Price published in 1963 some of the first observations of the exponential rate of growth in the number of scientific journals and the corresponding exponential rate of growth of abstract journal (7). More recently, in 1971, William Goffman developed an epidemic theory for the growth rate of a specific scientific area of activity (8).

In 1964, Richard H. Orr and Alice A. Leeds looked at the growth and other characteristics of the U.S. biomedical journal literature to see if bibliometric analyses could shed light on communications within the biomedical community (9). Kenneth O. May measured the growth and quality of the mathematical literature. The well-known hypothesis of exponential growth of the scientific literature is strongly confirmed but at a rate less than half that was found by D.D. Solla Price and other investigators. This discrepancy appears to be due to the failure of previous studies to take into account the titles published before the beginning of the time series used (10). I.N. Sengupta studied the growth of the literature of biochemistry and changes in ranking of periodicals, and suggested that during phases of rapid and vigorous growth of knowledge in a scientific discipline, articles of interest to that discipline appear in increasing numbers in periodicals distant from the field (11).

In 1978, G. Nigel Gilbert reviewed a number of indicators of the growth of science to assess their strength and weaknesses. The focus has been on the problems involved in measuring two aspects of scientific growth, growth in man power and growth in knowledge (12). T. Braun and E. Bujdoso investigated growth of a journal from the view points of distribution of papers, languages and topics, turnover of papers, distribution and growth of the number of papers from various countries, and doubling times (13). Recent study made by J. Tague, J. Beheshti, and L Rees-Patter stressed on the law of exponential growth, its evidence, implications and forecasts (14).

Though the growth of literature initially concentrated on the cumulation of literature subsequently resulting in the exponential growth, latter studies emphasised on the growth from the view points of number of papers, journal publications, research workers in the field, growth behaviour, etc.

**1.2.1.2 Bibliometric Laws :** Based on the correlations of the data, various laws have been formulated. The three basic laws in the field of bibliometrics have been formulated, related to the articles in the journals; authors of papers; and word counts in the text of the papers. Bradford's Law is based on the distribution of publications (usually journals) in a particular discipline or articles in a set of journals. Lotka's Law is based on the number of authors publishing in a discipline or other defined field. Zipf's Law is based on word-frequency rankings in a defined set of documents.

**Lotka's Law -** Alfred J. Lotka in 1926 formulated a law of Scientific Productivity based on correlation between the number of authors and their contributions (15). D.D. Solla Price, also interested in scientists' productivity, has defined a law developed from Lotka's stating that "half of the scientific papers are contributed by the square root of the total number of scientific authors" (7). The literature of information science has been examined between 1966 and 1970 by Henry Voos (16). It was determined that a new constant, $1/n^{3.5}$ fitted information science best. "If the constant $1/n^2$ holds for the hard science it was hypothesized [and tested] that other disciplines would have other constants". Other investigations (17-18) indicate that a similar pattern does indeed obtain today, at least approximately, whereas Russell C. Coile pointed out that the above studies have misinterpreted Lotka's Law and he proposed the Kolmogorov-Smirnov test for the applicability of the Lotka's Law (19). Allison and others while studying Lotka's Law stressed on the problem in its interpretation and application (20). Abraham Bookstein (21) has argued that Lotka's Law is invariant under the impact of society and time on pattern of scientific productivity. I.K.R. Rao proposed the negative binomial distribution, which describes a pattern on scientific productivity under a wide variety of social conditions (by rewards or threats) (22).

**Scientific Collaboration** - D.D.Solla Price apparently on the basis of a sampling of 'Chemical Abstracts' for the period 1910-60, mentioned about the trend towards multiple authorship (7). Beverly L. Clarke has made an hypothesis that the multiple authorship trend among biomedical workers during the war years was caused by heavy pressure to get the research done, with forced emphasis on the team approach; he has also pointed the qualifications as the factor for multiple research collaboration (23). D.Deb. Beaver and R. Rosen have presented and developed the first comprehensive theory of scientific collaboration, investigated a number of predictions of the theoretical view of scientific collaboration, and selected parameters of the growth of scientific collaboration as a response to the professionalisation of science (24-26). Z.M. Mulchenko has revealed the existence of a new collaborative form of research 'productional-line', an almost industrial procedure for the generation of new information by setting up so-called "ephemeron teams" - is headed by an eminent scientist who advances new ideas amongst the postgraduate students, junior research staff members, students writing papers (27).

**Zipf's Law** - In addition to papers, journals, authors and their collaborations, the vocabulary of science and technology is also an area of interest for bibliometric analysis. George K. Zipf working on the statistics of vocabularies, formulated his law in 1935 taking statistical technique of ranking of word frequency, beginning with the most productive (28). Psychologists have combined word frequency counts with factor analysis (29). Renats Tagliacozzo has pointed the way for language analysis to indicate various levels of technicality in scientific communication (30). B.C. Brookes has stated the relationship between the distribution pattern of the Zipf's Law and Bradford's Law, has also pointed the Gaussian and Zipfian empirical distributions (31). The statistics of such distributions have been disputed by theoreticians among whom Benoit Mandelbort (32) and Herbet A. Simon (33) made useful clarifications, Ye-sho Chen and Ferdinand F. Leimkuhler have made a rigorous analysis of Zipf's Law using an index for the sequence of observed values of the variables in a Zipf-type relationship. This approach has identified three important properties relating count, rank, and frequency (34). More recently, A. Bookstein has attempted to clarify and resolve the relationship among the Bradford, the Lotka, and the Zipf distributions (35). A.P. Trofimenko has also worked on content analysis and proposed a new method for its quantitative evaluation (a mathematical model). A nucleus of terms defining the most developed directions of research is found (36). Various studies dealing with the change in the number of significant words in the titles/ text of the papers being published in the primary journals and their importance have been carried out. A study has been made by A.B. Burton and A.J. Meadow (37) regarding the change in the relative information content of periodicals. P.R. Bird and M.A.Knight (38) have also shown that the titles of scientific papers have recently assumed greater importance.. J.S. Ghosh (39) has attempted to evaluate the effectiveness of keywords in the title, from the view point of document selection and information retrieval.

The recent applications of Zipf's Law include program complexity in software engineering, keywords distribution in bibliographic data base design, data compacting in computer networking, information retrieval, and statistical analysis (34).

**1.2.1.3 References and Citations :** The presence of references in scientific documents has proved to be a valuable source of data for the information scientists and bibliometricians. References and citations are the basic data for many of the techniques of bibliometrics. P.L.K. Gross and E.M. Gross seem to have originated the concept of counting the references in a central scientific journal as a means of identifying the key journals in a subject (40). The wide range of, field to field differences in the number and characteristics of references in scientific papers has been used by D.D.Solla Price in 1970 to characterise scientific field (41). P.F. Cole has also drawn relationship between journals' titles and their references (42). The basic parameters that may be studied using references vary widely from field to field; these include the number of references in a paper, the publication dates of the papers referred to, and the distribution of references to other journals and scientific areas, and the relationship between references and the text.

The most active area of modern bibliometrics is concerned with citations. The tools which have provided the impetus for the development of this branch of bibliometrics has been the computer, with its ability to store and manipulate large quantities of bibliometric data, and the citation indexes produced by the Institute for Scientific Information in 1969. Eugene Garfield in June 1955 published a paper entitled 'Citation Indexes for Science' (43), it was in part based on suggestions made by Adair (44), Hart (45), and Seidel (46) which followed a paper on the applicability of the citation system for patents. Lipetz first suggested improving the selectivity of citation indexes through an analysis of citation relationship (47). More recently, several studies have investigated the various functions and the quality of citations (48-49). M.C. Gomperts in 1968 proposed the Law of Constant Citation for scientific literature (50). L. Leydesdorff has stressed on the existence of theory of citation to put greater emphasis on practical perspective and theoretical basis of citations (51). Citation analyses have been used in the identification of the research trends (52), ranking of scientists (53), citation rate to compare scientists' work (54), and in the evaluation of research program (55).

Bibliometric research has developed a body of theoretical knowledge and a group of techniques and applications based on the distribution of bibliographic data elements. The wider application of bibliometric techniques is leading to the development of new and more precise techniques for greater economical and efficient management of the materials and services. It was also found that the significant study on History of Science literature using scientometric techniques is almost nil. The bibliometric techniques thus identified for the use in this research investigation are growth of literature, scientific productivity and collaboration, content analysis and citation analysis.

Hence, it is proposed to undertake a scientometric study of periodical literature in History of Science in India.

## 1.3 AIMS AND OBJECTIVES

The main aims and objectives of the study are given below:

1. To trace the growth of literature in the subject of History of Science in India, mainly periodical literature;
2. To study the journal publications in terms of articles published;
3. To determine scientific productivity in terms of
   (a) Authorship pattern
   (b) Collaborative pattern;
4. To identify the elites in the subject in terms of
   (a) Contribution
   (b) Subject scattering
   (c) Journal preference;
5. To do the content analysis on the basis of keywords;
6. To analyse the citation pattern, to get the measurement of effectiveness of the articles;
7. To prepare a bibliography on the History of Science; and
8. To create a machine readable data base on the subject.

## 1.4 SCOPE OF THE STUDY

The study is proposed to be confined to the articles published on the ancient, medieval and modern periods in the History of Science, in journals related to the History of Science. Here, the ancient period ranges from Harappa period i.e., advent of writing to the 1000 AD; it also includes the vedic, post-vedic (2BC-3AD), and classic (4AD-1000AD) periods. Medieval period ranges from 1001 AD to 1750 AD, whereas the modern period starts from 1751 AD onwards. The core journals proposed to be studied in this study are given as under:

1. (a) Journal and Proceedings of the Asiatic Society of Bengal, Calcutta
   (b) Journal of the Asiatic Society of Bengal, Science, Calcutta
   (c) Journal of the Asiatic Society of Bengal, Letters, Calcutta
   (d) Journal of the Asiatic Society, Calcutta
2. Journal of the Royal Asiatic Society of Great Britain and Ireland, London
3. ISIS International Review devoted to the History of Science and Civilization, Baltimore
4. Indian Historical Quarterly, Calcutta
5. Vishveshvaranand Indological Journal, Hoshiarpur
6. Indian Journal of History of Science, Delhi

Period covered in the study ranges from the inception of the journal till 1986.

## 1.5 HYPOTHESES

It is proposed to test a set of hypotheses related to various characteristics of History of Science literature in this study. These hypotheses are stated below:

(1) That the depression in the growth of literature in the subject takes place during the period of political crisis and war conditions in the country;

(2) That the authors' distribution of contributions to History of Science would approximate the Lotka's Law;

(3) That the authors in History of Science mostly contribute individually;

(4) That the contributions of the authors never remain steady;

(5) That the authors penetrate into the areas other than the areas of their normal pursuit;

(6) That the ratio of text length and the references listed in the article influences the use of the article in the research area;

(7) That the keywords distribution in History of Science would approximate the Zipf's Law distribution;

(8) That the citations in History of Science followed a phenomenon of continued citations.

## 1.6 METHODOLOGY

It is proposed to conduct the study using techniques for scientometric analysis such as growth of literature, scientific productivity, content analysis, citation analysis etc. The creation of machine readable data base on the subject History of Science facilitates in the analysis and helps in the retrieval of bibliographic information, using various access points. The methodology consists of the following steps :

### 1.6.1 Data Collection

Keeping in view the various facets involved in the analysis, exhaustive data collection sheets were prepared. The data collection sheets were used for data inputting into the computer.

For the citation analysis, separate data collection sheets were prepared, one for noting the citation details and another for source item details. Samples of the data collection sheets are shown in figures 1.1, 1.2, and 1.3.

**Figure 1**

**1.1 Data Collection Sheet: Bibliographic Information**

Journal Imprint Remarks Reference Class no. Period S. No.

---

Author

Title

Institution

---

Keywords in title

---

Contents 1

2

3

4

---

Inputted Edited Initials

---

**1.2 Data Collection Sheet: Citation Information**

Author

Year Journal Volume Page

Citation

Citation Details

Citation Index Year Author Journal Vol. Page Year

---

Inputted Edited Initials

---

**1.3 Data Collection Sheet: Source Information**

Author (s) Title of the Article. Journal Imprint

Address (es) No. of references

Related citations

Author Journal Vol. Year Page

---

Inputted Edited Initials

---

Each journal undertaken in this study was scanned and the bibliographic information from the relevant articles was jotted down on the data sheets. For contents, each article was perused and various subject headings from the article were jotted down on the data sheets. As all the journals and their various issues were not available at one place, therefore, different libraries were visited in Delhi and correspondences were made with the Asiatic Society, Calcutta. Citation data was noted from the CDROM Science Citation Index and the hard bound Social Sciences Citation Index. CDROM SCI search output was taken on the paper in computer format. This was further noted down on the data sheets prepared for this purpose, whereas from SSCI the citation information was jotted directly on the data sheets.

### 1.6.2 Data Organisation

The data sheets were grouped into their respective journals and arranged yearwise in each group. The total number of data sheets were noted to have a check on further handlings. Similarly, the citations and source information were arranged authorwise.

### 1.6.3 Data Inputting

Half of the data sheets were inputted into one computer whereas the other half were inputted into an another computer. The outputs were taken out on continuous sheets. After careful editing on continuous sheets, the relevant information was incorporated in the records and browsed for its presentation.

### 1.6.4 Information Processing

To convert the data into a machine readable data base and facilitate analysis, a great deal of information processing was done. Data from the machines, was down-loaded at a single place using export-import facilities in the software. The whole data base was made searchable using inverted file services. Main data base was arranged yearwise and later on printed as main text of the data base on History of Science. In addition, various outputs were brought out as indexes to the data base on History of Science. Many searches and sorts were performed to help in the analysis at later stage.

### 1.6.5 Machine-Readable Data Base And Its Facilitation In Analysis

Information sources in the History of Science were collected and subsequently arranged. Later on these were sorted, merged, edited, updated and various sequences were brought out with the help of a computer to find out various scientometric parameters. For it, a machine-readable data base was created.

#### 1.6.5.1 Machine-Readable Data Base

The details about the data base is given in the **Chapter 3 - Creation of a Data Base on History of Science.** This data base can be further used by researchers and research institutes dealing with History of Science. They can, with simple options run the data base to get relevant research papers in their research projects. Libraries in this area can further build up this data base to get bibliographies and other details in various topics pertaining to History of Science.

#### 1.6.5.2 Facilitation of The Data Base on History of Science

The use of machine-readable data base on History of Science is far more helpful in performing the following scientometric analyses on the subject:

(i) *Growth of literature* - In the beginning the data was inputted as and when it was collected from the journals and later on, the main text/main data base was exported and imported in a sorted sequence yearwise. Then the whole data base i.e., the main text was printed and manual counting was done for each years' output to get the growth rate of literature.

(ii) *Authors' productivity* - An author index was brought out using machine-readable data base and each author was counted for his occurrences in the entries to get the application of Lotka's Law in History of Science and from the same index highly productive authors were identified.

(iii) *Highly productive authors* - After the identification of highly productive authors from the author index, a separate search was made for each highly productive author and the search results were printed to get the following parameters :

(a) Co-authorship pattern

(b) Contributions

(c) Subject scattering

(d) Length of the text and references

(e) Journal preference

(iv) *Co-authorship pattern* - An authorwise sort list which included the name of first author and other authors was brought out using 'author field' in the print format. From the printed output, entries with single author, two authors, three authors, four authors and others were counted and co-authorship pattern was drawn.

(v) *Content analysis* - Title field, content field, and keyword field from the data base were managed into a single sort key so as to get the keyword index alphabetically. This index was later on used to count the record numbers printed against each word and a table was drawn for words and their frequencies.

(vi) *Journal scattering* - A computer output related to journal index was used to find out the number of articles that appeared in the various research journals taken in this study. A table related to the scattering of articles in various journals was drawn.

## 1.7 ANALYSIS

The following scientometric analyses were performed on the data base created in this study :

(i) Growth of the History of Science was determined using the yearwise listing of the records, along with authors' field;

(ii) Overview of research journals was undertaken;

(iii) Scientific productivity which traces the application of Lotka's Law in the subject and collaborative pattern was described with the help of an author index brought out from the data base ;

(iv) Highly productive authors were identified and their subject scattering, contribution pattern, journal preference were depicted ;

(v) Content analysis was described using the keyword index brought out from the data base ;

(vi) Citation analysis was undertaken and highly cited authors were listed along with their citations and source items ;

(vii) Printed data base of source materials on History of Science along with various indexes was prepared from the machine readable data base ;

(viii) A machine readable data base on History of Science was created with the help of various techniques available in the software package CDS/ISIS.

## 1.8 DEFINITION OF KEY TERMS

### 1.8.1 History of Science

It is the study of the development of science and technology (58).

### 1.8.2 Information Source

It is an item/entity which gives information. It can be in a recorded form or in an object form (59).

### 1.8.3 Scientometrics

It is the study relating to the measurement of science using bibliometric measurements for evaluation of scientific progress, level of scientific development, social relevance and the impact of application of science and technology, etc. (60-61).

### 1.8.4 Authors Productivity

Lotka's Law describes the productivity of authors. It deals with the relationship of quantity of publications of individuals and scientific organizations, identification of elites in science on the basis of their publication output as well as impact of social change on productivity (15).

### 1.8.5 Content Analysis

Analysis of the manifest and latent content of a body of communicated material (as a book or film) through classification, tabulation, and evaluation of its key symbols and themes in order to ascertain its meaning and probable effect (59).

### 1.8.6 Text Value

It is the ratio of length of the text to the number of references of the research paper. Higher the text value, greater will be the use of the paper. Hence, text value is directly proportional to the use of paper.

### 1.8.7 Citation Analysis

It is used as a measure of the impact of individual articles, journals, authors, etc. It describes various types of linkages between references on the basis of citation counts, co-citation, bibliographic coupling, etc (62).

### 1.8.8 Data Base

An organized collection of computer records, standardized in format and content; it is stored in any of a variety of computer-readable modes. It is the basic set of data from which computer-readable files are created. In the data base, all records are interrelated by some common denominator (59).

## REFERENCES


1. Svami, S.P.S., *Founders of Science in Ancient India Part 1* (New Delhi: Govindram Hasanand, 1965).

2. *Proceedings of the Seminar on Learned Periodical Publication in India Past, Present and Future* (New Delhi: India International Centre, 1989).

3. D. Chattopadhyay, *History of Science and Technology in Ancient India : The Beginnings* (New Delhi: CSIR/NISTADS, 1986).

4. G.M. Dobrov, and A.A. Korennoi, "The Information basis of Scientometrics" in *On Theoretical Problems of Informatics*. A.I. Mikhailov, and other (eds.). [Moscow: VINITI (FID), 1969], 165-91.

5. Allan Pritchard, "Statistical Bibliography of Bibliometrics?" *Journal of Documentation, 25,* No. 4 (1969), 348-49.

6. F.J. Cole, and Nellie B. Eales, "The History of Comparative Anatomy", *Science Progress,* 11 (1917), 578-96.

7. Derek Desolla Price, *Little Science, Big Science* [New Haven: Yale University Press, (1963)].

8. William Goffman, "A Mathematical Method for Analyzing the Growth of a Scientific Discipline," *Journal of the Association for Computing Machinery,* 18, No.2 (1971), 173-85.

9. Richard H. Orr, and Alice A. Leeds, "Biomedical Literature: Volume, Growth, and other Characteristics," *Federation Proceeding,* 23, No.6 (1964), 1310-31.

10. Kenneth O. May, "Growth and Quality of the Mathematical Literature," *Science,* (1966), 363-73.

11. I.N. Sengupta, "Recent Growth of the Literature of Biochemistry and Changes in Ranking of Periodicals," *Journal of Documentation,* 29, No.2 (1973), 192-211.

12. G. Nigel Gilber, "Measuring the Growth of Science: A Review of Indicators of Scientific Growth," *Scientometrics,* 19, No.1 (1978), 9-34.

13. T. Braun, and E. Biydoso, "Growth of a Journal Reflects Trends in the Literature of Nuclear Analytical Methods," *Journal of Radioanalytical Chemistry,* 50, No.1 (1979), 9-31.

14. Jean Tague, J. Beheshti, and L. Rees-Potter, "The Law of Exponential Growth : Evidence, Implications and Forecasts," *Library Trends,* (1981), 125-45.

15. Alfred J. Lotka, "The Frequency Distribution of Scientific Productivity," *Journal of the Washington Academy of Science,* 16, No. 12 (1926), 317-23.

16. Henry Voos, "Lotka and Information Science," *Journal of the American Society for Information Science,* (1974), 270-72.

17. Larry J. Murphy, "Lotka's Law in the Humanities?," *Journal of the American Society for Information Science,* (1973), 461-62.

18. Alan Edward Schoor, "Lotka's Law and Map Librarianship," *Journal of the American Society for Information Science,* (1975), 189-90.

19. Paul D. Allison, and others, "Lotka's Law: A Problem in the Interpretation and Applications," *Social Studies of Science,* 6, No.2 (1976), 269-76.

20. Russell C. Coile, "Lotka's Frequency Distribution of Scientific Productivity," *Journal of the American Society for Information Science,* (1977), 366-70.

21. Abraham Bookstein, "Patterns of Scientific Productivity and Social Change : A Discussion of Lotka's Law and Bibliometric Symmetry," *Journal of the American Society for Information Science*, (1977), 206-10.

22. I. K. Ravichandra Rao, "The Distribution of Scientific Productivity and Social Change," *Journal of the American Society for Information Science*, (1980), 111-20.

23. Beverly L. Clarke, "Multiple Authorship Trends in Scientific Papers," *Science*, 143 (1964), 822-24.

24. D. Deb Beaver, and R. Rosess, "Studies in Scientific Collaboration : Part I The Professional Origins of Scientific Co-Authorship," *Scientometrics*, 1, No.1 (1978), 65-84."

25. ______________, "Studies in Scientific Collaboration : Part II Scientific Co-authorship, Research Productivity and Visibility in the Fresh Scientific Elite, 1799-1830," *Scientometrics*, 1, No.2 (1979), 133-49.

26. ______________, "Studies in Scientific Collaboration : Part III Professionalization and the Natural History of Modern Scientific Co-authorship," *Scientometrics*, 1, No.3 (1979), 231-45.

27. Z.M. Mulchenko, and other, "On Scientometrical Characteristics on Information Activities of Leading Scientists," *Scientometrics* 1, No.4 (1979), 307-25.

28. George K. Zipf, *Psycho-biology of language: An Introduction to Dynamic Biology* (Cambridge, 1935).

29. Ross Hartsough, and Julius Laffal, "Content Analysis of Scientific Writings," *The Journal of General Psychology*, 83 (1970), 193-206.

30. Renata Tagliacozzo, "Levels of Technicality in Scientific Communication," *Information Processing and Management*, 2, No.2 (1976), 95-110.

31. B.C. Brookes, "Towards Informatics : Haitun, Laplace, Zipf, Bradford and the Alvery Progress," *Journal of Documentation*, 40, No.2 (1984), 120-43.

32. Benoit Mandelbrot, On the Language of Taxonomy, *In Information Theory Third London Symposium* (1955), 135-45.

33. Herbet A. Simon, "On a class of skew Distribution Functions," *Biometrika*, 42 (1955), 425-40.

34. Ye-sho Chen, and Ferdinand F. Leimkuhler, "Analysis of Zipf's Law : An Index Approach," *Information Processing and Management*, 23, No.3 (1987), 171-82.

35. Abraham Bookstein, "The Bibliometric Distributions," *Library Quarterly*, 46, No.4 (1976), 416-23.

36. A.P. Trofimenko, "Scientometric Analysis of the Topical Content of Scientific Research and Particularities," *Scientometrics*, 18, No. 5-6 (1990), 409-35.

37. A.B. Buxton, and A.J. Meadow, "The Variation in the Information Content of Titles of Research Papers with Time and Discipline," *Journal of Documentation*, 33, No.1 (1977), 46-52.

38. P.R. Bird, and M.A. Knight, "Word Count Statistics of the Titles of Scientific Papers," *Information Scientists*, 9, No.2 (1965), 67-69.

39. J.S. Ghosh, "The Information Content of Title in Contraception Literature," *Journal of Chemical Information and Computer Sciences*, 17, No.1 (1977), 36-40.

40. P.L.K. Gross, and E.M. Gross, "College Libraries and Chemical Education," *Science*, 66 (1927), 1229-34.

41. D.D. Solla Price, "Citation Measure of Hard Science, Soft Science, Technology, and Nonscience," in *Communication Among Scientists and Engineers*, Carnot E. Nelson, and Donald K. Pollock (eds.).(Lexington:D.C. Heath and Co., 1970).

42. P.F. Cole, "A New Look at Reference Scattering," *Journal of Documentation*, 18, No. 2 (1962), 58-64.

43. Eugene Garfield, "Citation Indexes for Science," *Science*, 122, No.3159 (1955), 108-11.

44. W.C. Adair, "Citation Indexes for Scientific Literature," *American Documentation*, 6, No. 31 (1955).

45. H.C. Hart, "Re: Citation System for Patent Office," *Journal of Patent Office Society*, 31, No. 714 (1949).

46. A.H. Seidel, "Citation System for Patent Office," *Journal of Patent Office Society*, 31, No.554 (1949).

47. Ben-Ami Lipetz, "Improvement of the Selectivity of Citation Indexes to Science Literature Through Inclusion of Citation Relationship Indicators," *American Documentation*, 16, No.2 (1965), 81-90.

48. Michael J. Moravcsik, and Poovanalingam Murugesan, "Some Results on the Function and Quality of Citations," *Social Studies of Science*, 5, No.1 (1975), 86-92.

49. Henry Voos, and Katherine S. Dagaev, "Are All Citations Equal: Or, Did We Op. Cit. Your Idem?," *Academic Librarianship*, 1, No.6 (1976), 19-21.

50. M.C. Gomperts, "The Law of Constant Citation for Scientific Literature," *Journal of Documentation*, 24, No.2 (1968), 113-19.

51. L. Leydesdorff, "Towards a Theory of Citation?" *Scientometrics*, 12, No 5-6 (1987), 305-09.

52. Harry Rothman, and Michael Woodhead, "The Use of Citation Counting to Indentify Research Trends." *Journal of Documentation*, 27, No. 4 (1971), 287-94.

53. G. Folly, and others, "Some Methodological Problems in Ranking Scientists by Citation Analysis," *Scientometrics*, 3, No.2 (1981), 135-47.

54. Nancy L. Geller, and John S.D. Cani, "Lifetime-Citation Rates: A Mathematical Model to Compare Scientists' Work," *Journal of the American Society for Information Science*, (1981), 4-15.

55. Francis Narin, "Citation Analysis: Bibliometric Techniques in the Evaluation of Research Programs," *Science and Public Policy*, 14, No.2 (1987), 99-106.

56. R.K. Verma and K. Shailendra, "An Analysis and Review of Bibliometrics Literature," *International Conference on Bibliometrics and Theoretical Aspects of Information Retrieval* (Belgium, 1987).

57. B.K. Sen and N. Kumar, "Indian Contributions in Bibliometrics 1958-1984 : A Review," *Annals of Library Science and Documentation*, 33, No.3 (1986), 85-103.

58. D.M. Bose, and others (eds.), *A Concise History of Science in India*. (New Delhi : Indian National Science Academy, 1983).

59. S. Parathasarathy, "Studies in Information Management : Application of Statistical Analysis," *IASLIC Bulletin*, 30, No. 4 (1985), 131-37.

60. T.N. Rajan and B.K. Sen, "An Essay on Informetrics: A Study on Growth and Development," *Annals of Library Science and Documentation*, 33, No. 1-2 (1986), 1-12.

61. *The ALA Glossary of Library and Information Science*. (ALA, 1983).

62. E. Garfield, *Science Citation Index : Journal Citation Reports* (Philadelphia : ISI, 1984).

# 2. RESEARCH JOURNALS - AN OVERVIEW

Research in the History of Science in India started getting published in various research journals in the nineteenth century. It was organised in its present form in 1966 with the publication of 'Indian Journal of History of Science'. Research journals selected for this study on the basis of advice of subject experts are listed below:

1. Journal and Proceedings of the Asiatic Society of Bengal, Calcutta
2. Journal of the Asiatic Society of Bengal, Calcutta
3. Journal of the Asiatic Society, Calcutta
4. Journal of the Royal Asiatic Society of Great Britain and Ireland, London
5. ISIS International Review devoted to the History of Science and Civilization, Baltimore
6. Indian Historical Quarterly, Calcutta
7. Vishveshvaranand Indological Journal, Hoshiarpur
8. Indian Journal of History of Science, Delhi

## 2.1 JOURNAL AND PROCEEDINGS OF THE ASIATIC SOCIETY OF BENGAL, CALCUTTA

The first volume of this journal was started in 1905 from Calcutta, printed at the Baptist Press, and published by the Asiatic Society, 57, Park Street. It was continued from the 'Journal of the Asiatic Society of Bengal (1-73; 1832 - 1904)' and continued to be published upto 1934. The title of the volume of 1909 as given on title page was 'Journal of the Asiatic Society of Bengal, New Series, Calcutta.' The title of the volume of 1913 as printed on its cover page was 'Journal and Proceedings of the Asiatic Society of Bengal, New Series,' whereas published in 1918, the title page of this journal printed the title as 'Journal of the Asiatic Society of Bengal, New Series.'

## 2.2 JOURNAL OF THE ASIATIC SOCIETY OF BENGAL, CALCUTTA

Volume 1, 1935 of the journal was printed at Baptist Mission Press, and was published by the Royal Asiatic Society of Bengal. For the convenience of the readers, to enable them to preserve and to bind separately the literary and scientific material, the 'Journal and Proceedings of the Asiatic Society of Bengal, Calcutta' was issued from 1935 in three parts, each with its own title, page numbering and index. The titles of the three parts were 'Journal of the Asiatic Society of Bengal. Letters'; 'Journal of theAsiatic Society of Bengal. Science'; 'Year-book of the Asiatic Society of Bengal'; The year-book contains the matter hitherto published in the proceedings of the society. The year and volume numbers were the same for the three parts of the journal.

The new form of issue constituted a third series of the journal of which the history is as follows:

| | | |
|---|---|---|
| First series | (1) Journal | Vols. 1-73, 1832-1904 |
| | (2) Proceedings | Vols. 1-40, 1865-1904 |

New (i.e. second) Series Journal and Proceedings Vols. 1-30, 1905-1934

| | | |
|---|---|---|
| Third series | (1) Journal. Letters | Vol. 1-, 1935- |
| | (2) Journal. Science | Vol. 1-, 1935- |
| | (3) Year-book | Vol. 1-, 1935- |

Title of Third series appeared as

Journal of the Asiatic Society of Bengal. Letters.

Journal of the Asiatic Society of Bengal. Science.

Year-book of the Asiatic Society of Bengal.

The titles of 1936 volume of this journal were however, changed as follows:

Journal of the Royal Asiatic Society of Bengal. Letters.

Journal of the Royal Asiatic Society of Bengal. Science.

Year-book of the Royal Asiatic Society of Bengal.

Cross reference of second, and third authors to that of the first author appears in an alphabetical sequence in the journal.

## 2.3 JOURNAL OF THE ASIATIC SOCIETY, CALCUTTA

The journal was started in 1959 and is still continuing. It was continued from the 'Journal of the Asiatic Society. Science'; 'Journal of the Asiatic Society. Letters'; and 'Year-book of the Asiatic Society'. Each volume of the journal contains four issues. Journal includes not only articles, but also reprints, book reviews, and bibliographical supplements.

## 2.4 JOURNAL OF THE ASIATIC SOCIETY OF GREAT BRITAIN AND IRELAND, LONDON

The journal was started in 1843+; it continued from the 'Transactions of the Royal Asiatic Society of Great Britain and Ireland'. The journal has been published by the Royal Asiatic Society of Great Britain and Ireland with which is incorporated the 'Society of Biblical Archaeology'. Each volume of the journal has a table of contents, which includes Notes, Articles, Reviews of Books, Obituary, The Library, Index and Alphabetical list of authors along with articles' titles etc.

Articles in different volumes of this journal range over a wide scale of subjects/topics amply borne by a few examples which are given below:-

The volume of 1943 contained most of the articles on magic. The volume of 1946 has many articles on language study and alphabets studies. The volume of 1951 listed "Catalogue of oriental manuscripts in the library of the Royal College of Physicians" (It listed the sixty three manuscripts giving size, author, title, subject etc.). The volume of 1981 listed manuscripts in United Kingdom, which are available in Library of the Wellcome Institute for the History of Medicine.

## 2.5 ISIS INTERNATIONAL REVIEW DEVOTED TO THE HISTORY OF SCIENCE AND CIVILIZATION, BALTIMORE

The journal is a quarterly organ of the 'History of Science Society (Constituent of the American Council of Learned Societies) and of the International Committee of the History of Science'. ISIS was founded and edited in Wondelgem-Lez-Gand, Belgium in 1912 by George Sarton. The first issue of ISIS was published from 1913 onwards. But the journal was not published from June 1914 to September 1919 during the period of first world war. The journal has an Index vol 1-20, 1913-34. The volume of 1920 has an article on "Synthetic Bibliography with special reference to the History of Science". The volume of 1936 includes an article on library and information science giving bibliography on a subject as well as Notes and Correspondence section, which describes notes, correspondence of works and queries which are made to seek solutions to some problems e.g. ISIS, 1936, 24 (68), 432

Query no. 57

Sarton G. "Early decimal instrument. Who was the instrument maker ? Has any of his instruments and especially of the one decimally graduated survived?.... If such instruments are known, I would be glad to publish description of them in ISIS".

Works on History of Science which appeared in other journals are given in its notes and correspondence section.

e.g. ISIS. 1936, 24 (68), 433

Sarton G. "A special number of *Nature* (No. 3418, vol.155, 699-755, May 4, 1935) is devoted to such retrospective reviews prepared by leading scientists of England. It is a very useful and instructive account. Historians of science will do well to make a note of it for future reference".

The volume of 1937 mentioned about OSIRIS - a publication. Every volume of OSIRIS is dedicated to a living historian of science, and contains biographical data concerning him as well as full bibliographical details and a portrait. For instances Vol. 1 was dedicated to D.E. Smith, Historian of Mathematics; Vol. 2 was dedicated to T.L. Heath, Historian of Greek Mathematics, and Vol. 3 was dedicated to E.O.V. Lippmann, Historian of Chemistry. Noticeable language articles have also appeared in the ISIS 1937.

The volume of 1947, 32 (85) scheduled for publication in July 1940, was published in January 1947.

The volume of 1942 contains contents page as follows: Main Article, Queries and Answers, Notes and Correspondence, Reviews, XIth Century, XVIIIth Century, XXth Century, Islam, Chemistry, Botany, Zoology, Sociology, Religion, Medicine, Education, Critical Bibliography. This volume has six issues (93-98) - 1942, 34(93) (94); 1943, 34(95-98). Volume number 35 has four issues. The volume of 1947 contains articles on "Teaching of the History of Science." Since 1953 editorial responsibility of the journal has been shifted from single editor to an Editorial Board, of which Prof. I. Bernard Cohen was the Chairman, besides being the Managing Editor.

The volume of 1957 was devoted to George Sarton as a memorial issue.

The volume of 1959 printed cover-pages of the S&T Books, History of Science books in the beginning of ISIS issue.

The volume of 1986 has Critical Bibliography, which includes classified arrangement, book reviews, index etc. Covering works, by Indian/on India of seventeenth century to twentieth century.

**ISIS Critical Bibliography**

One of the major feature of ISIS journal is its Critical Bibliography (CB). The bibliography (Called "analytical" at first, later "Critical") appeared at frequent but irregular intervals, at least once a year. It recorded all material of interest for the History of Science, including the History of Civilization and the Philosophy of Science. The editor's aim was to establish the History of Science as an independent discipline and to have the journal (ISIS) serve as a rallying ground for scholars engaged in these studies.

The first seventy-nine issues of the CB were edited by Sarton himself from 1913 to 1952, a period which witnessed two world wars, during which time Sarton moved from Belgium to United States, ISIS become the official journal of the History of Science Society (Founded in 1924), and the History of Science was established as an academic discipline. In 1951 Sarton gave up editing ISIS, and the next year he likewise retired from editing the CB in order to devote his energies to writing. Since CB 80 the coverage was somewhat restricted and the classification scheme modified.

CB was selective, mainly confined to publications in English, French, German, Italian, and Latin. Coverage of Publications in other languages was sporadic. From the point of view of the historian of science, its scope was encyclopaedic. It included the Philosophy of Science, and until CB 80, the History of Civilization. Until CB 76 it even included catalogues of antiquarian book dealers. No item was entered more than once. There was no subject index. CB 1-90 contains some 80,000 entries.

The bibliography was not primarily intended for historians of special sciences; they were expected to supplement it with other sources. During the period 1913-1965 a number of specialized bibliographies in the History of Science and related subjects, prepared independently of the CB, were published.

The year 1972 witnessed a publication titled "ISIS Cumulative Bibliography : A bibliography of the History of Science formed from ISIS Critical Bibliography 1-90; 1913-65", with Whitrow as its Editor. This cumulative bibliography was published from London in conjunction with the History of Science Society. This publication was intended to be published in four volumes. The first two volumes of the Cumulated Bibliography were published in 1972, the third volume was published in 1976, whereas the fourth volume was published later on.

### 2.6 INDIAN HISTORICAL QUARTERLY, CALCUTTA

The journal was published from March 1925 to June 1963 by Calcutta Oriental Press, Calcutta. This journal includes Articles, Miscellany, Review, Selected contents of Oriental Journals, and Bibliographical Notes. After the first two issues of 1963 volume, publication of the journal was suspended. Cumulated index to the Indian Historical Quarterly : 1925 - 1963 was also brought out separately later on.

### 2.7 VISHVESHVARANAND INDOLOGICAL JOURNAL, HOSHIARPUR

The journal started its publication in 1963 under the sponsorship of Vishveshvaranand Vedic Research Institute (VVRI). Each volume of the journal has two issues. This journal includes Papers, Reviews, Obituary Notices, Literary Notices, News and Notes (Internal and External), Index to Contributors (Alphabetical authorwise), and Supplements. This journal has published a few memorial volumes in honour of contributions made by Dr. Siddheshwar Varma, Dr. Vishva Bandhu and Prof. K.V. Sarma. Prof. K.V. Sarma alone has to his credit more than sixty books, and one hundred and forty five research papers, besides a large quantum of other academic writings relating to Sanskrit and Indology. Prof. K.V. Sarma published several series on scientific works as supplements in the various issues of the Vishveshvaranand Indological Journal.

### 2.8 INDIAN JOURNAL OF HISTORY OF SCIENCE, DELHI

The journal started in 1966, published under the general guidance of the **National Commission for the Compilation of History of Sciences in India,** by the Indian National Science Academy, New Delhi, and printed at its publication office, Calcutta. From 1966 to 1983 the journal was semi-annual and since 1984 the journal has become quarterly. The journal presently contains not only articles, but also books' reviews as well as notes and news.

The volume of 1982 has published proceedings of the Seminar on Science and Technology in India from eighteenth century to nineteenth century which was held from 11th to 13th February 1980 at New Delhi. The volume of 1985 published articles on Astronomy, as the publication was specially planned to coincide with the inauguration of the International Astronomical Union colloquium on History of Oriental Astronomy held from 13th to 16th November, 1985 at the Indian National Science Academy, New Delhi. Notes and references of each article are given at the end. The volume has special Bibliography with keyword index on Primary Sources and Secondary Sources in Sanskrit, Arabic and Persian.

# 3. CREATION OF A DATA BASE ON HISTORY OF SCIENCE

Data base is a collection of data stored in a computer and coded in such a way that it can be retrieved in various ways, under various headings (1). A machine readable data base was generated for the History of Science. The data collected for the study was inputted in the microcomputer PC-XT/AT 286/386 using a software package for such bibliographic information. The data was so structured that useful applications could be made. The data in the computer was arranged yearwise, alongwith its outputs, such as main text, title index, author index, keyword index and journal index brought for ready reference. The data base is also available on floppy diskettes DSDD 48 TPI. The data base from the floppy diskettes was downloaded in the computer, using specified system requirements and a software package, making it machine readable.

As the data base is expected to be used and updated by the users with little computer exposure, the data base has a very easy means of updating and retrieving information. Keeping in view the requirements for the proposed data base, CDS/ISIS software was chosen for developing the data base on the basis of its merits given below:-

1. Among all the popular DBMS packages (dBASE IV PLUS, Fox Base,) only CDS/ISIS supports variable length fields;
2. In CDS/ISIS even if some fields are not mandatory, there will be no waste of space;
3. CDS/ISIS provides a number of indexing techniques, which are essential for a data base like the proposed one;
4. CDS/ISIS provides user defined print formate; so it is easy to get outputs according to the user's requirements; and
5. CDS/ISIS provides excellent exporting and importing facilities which facilitate exchange of information among various systems and organisations.

While designing the data base, a study was made to identify the data element fields, which were to be included in the data base to make it fairly exhaustive and an attempt was also made to identify all possible search and print outputs (2-6). At this time all the bibliographic data bases available at INSDOC using CDS/ISIS software package and data bases available in CDROM such as Science Citation Index, Applied Science and Technology Index, General Science Index, Agricultural and Biological Index etc., were deeply studied and advice of experts in this area was also sought and taken care of. After the data base development, some test runs were conducted, resulting successfully into compilation of a single, coherent machine readable source from widely scattered sources, with facility of easy updation. CDS/ISIS software is mainly used in India for collating bibliographic data bases, whereas this work also explored the capabilities of CDS/ISIS in developing the data base.

## 3.1 SYSTEM INSTALLATION

The minimum hardware requirements for running the data base are as follows : IBM Personal Computer PC-XT/AT 286/386, 640 K RAM, one floppy disk drive and 40 MB hard disk. A software CDS/ISIS version 2.32 was used to develop the data base on History of Science. This software is available on three diskettes labelled as ISIS01, ISIS02, and Samples. A proper installation of the software was needed to facilitate the smooth functioning of the system.

### 3.1.1 Downloading the Data Base on History of Science

The data base on History of Science available on the floppy diskettes kept was downloaded with a software CDS/ISIS as follows:

(1) C:\ISIS>CD DATA **OR** C:\>CD\ISIS\DATA

C:\ISIS\DATA>

(2) Copied the files from the floppy containing data base on History of Science to subdirectory DATA

C:\ISIS\DATA>Copy a:*.*

(3) Copied the link files and hit files of the data base from the floppy to main directory ISIS

C:\ISIS>Copy a:*.*

## 3.2 SOFTWARE

A software package named 'CDS/ISIS', Ver 2.32 was used, to develop the data base on History of Science. CDS\ISIS (Computerised Documentation System/Integrated Set of Information Systems), software was prepared by UNESCO, in 1969 under the guidance of Dr. Del Bigio the Chief Programmer. Initially it was a program for the UNESCO Library automation. Later on this program was developed and released as a software package CDS\ISIS version 1.0 in 1985. It is a menu driven system; for each main function, there is a submenu and a relevant action can be executed by using the given option(s). The software is written in PASCAL language. CDS\ISIS has allowed to build and manage structured nonnumerical data bases, i.e., data base whose major constituent is text. Here the information was structured into data elements. Data base on History of Science consisted of such data elements such as journal, imprint, author, title, keywords, contents etc. Data elements were stored in various fields, and each of which was assigned a numeric tag indicative of its contents. CDS\ISIS identified the name of the field as *numeric tag*.

A field may contain a single data element, or two or more variable length data elements. In the latter case, the field contains further subfields. Moreover, a field may be repeatable, i.e., any given record may contain more than one occurrences in the same field (7).

### 3.2.1 System Functions

The major functions provided by CDS\ISIS allowed:

- to define a data base containing the required data elements
- to enter new record(s) into a given data base
- to modify, correct or delete the existing record(s)
- to retrieve record(s) in any sequence desired
- to print, the range of records, sort sequence and indexes
- to develop special applications using integrated programming facilities.

These facilities are provided through a set of eight major services given in main menu of CDS\ISIS as follows:

**Main Menu CDS\ISIS**

Micro CDS\ISIS - Version 2.32

C - Change data base

L - Change dialogue language

E - ISISENT - Data entry services

S - ISISRET - Information retrieval services

P - ISISPRT - Sorting and printing services

I - ISISINV - Inverted file services

D - ISISDEF - Data base definition services

M - ISISXCH - Master file services

U - ISISUTL - System utility services

A - ISISPAS - Advanced programming services

X - Exit (to MSDOS)

### 3.2.2 Selecting Menu Option

A menu option was selected simply by pressing the relevant alphabetic key associated with the option.

### 3.2.3 Submenus

If any of the other options given in the main menu was chosen, other than X-Exit and C-Change Data Base, the system would have displayed another set of options in the submenu. For submenu the choice was made by pressing the relevant alphabetic key associated with the option.

### 3.2.4 System Restrictions

The following system restrictions are available in the version 2.32 of CDS/ISIS :-

Maximum number of data bases - unlimited

Maximum number of records in a data base - 16 millions (500MB)

Maximum size of a record in a data base - 8000 characters

Maximum number of fields of a record in a data base - 200

Maximum number of FST lines of a record in a data base - 200

Maximum number of field size of a record in a data base - 8000 character

Maximum number of fields in a worksheet page of record in a data base - 19

Maximum number of pages in a worksheet of a record in a data base - 20

Maximum size of a display format of a record in a data base - 4000 characters

### 3.2.5 Data Base Structure

A CDS\ISIS data base consists of a number of logically related but physically distinct computer files. The management of the physical files is the responsibility of CDS\ISIS, and the functions of some of the data base files are as follows:

| File | Size | Date of Creation | Time | Function |
|---|---|---|---|---|
| IHST MST | 767488 | 09-02-90 | 2:17p | To contain Master file records |
| IHST XRF | 6656 | 09-02-90 | 2:17p | To contain Cross-reference file |
| IHST IFP | 243200 | 08-11-90 | 12:10p | To search link files |
| IHST NO1 | 5920 | 08-11-90 | 12:10p | |
| IHST NO2 | 15312 | 08-11-90 | 12:10p | |
| IHST LO1 | 57984 | 08-11-90 | 12:10p | |

| File | Size | Date | Time | Purpose |
|---|---|---|---|---|
| IHST LO2 | 125832 | 08-11-90 | 12:10p | To search link files |
| IHST LN1 | 183062 | 08-10-90 | 5:29p | |
| IHST LN2 | 128167 | 08-10-90 | 5:29p | |
| IHST LK1 | 183062 | 08-10-90 | 5:44p | |
| IHST LK2 | 128167 | 08-10-90 | 5:51p | |
| IHST CNT | 53 | 08-11-90 | 12:10p | |
| IHST FDT | 702 | 12-12-90 | 6:06p | To contain information on various fields |
| IHST FST | 100 | 08-12-90 | 4:33p | To select field for search |
| IHST PFT | 128 | 12-12-90 | 6:06p | To provide print format for record(s) |
| AIHST FMT | 531 | 08-12-90 | 4:34p | To design worksheets |
| BIHST FMT | 85 | 08-12-90 | 4:34p | |
| IHST HIT | 451072 | 09-02-90 | 2:27p | Sort files |
| IHST HXF | 10566 | 09-02-90 | 2:30p | |

## 3.3 DATA BASE DESIGN

A data base structure in CDS/ISIS involved the preparation of Field Definition Table, Worksheet, Format, and Field Select Table.

### 3.3.1 Field Definition Table (FDT)

The FDT provided information on the contents of the record(s) in a given data base. In particular it defined the various fields which were present and a number of parameters for each field. Various parameters such as length of the field, type of the field (A-Alphabetic, N- Numeric and X-Alphanumeric), subfields (Delimiters) and repeatable (Occurrences of the data element) fields for each field were analysed and presented in the Table 1.

**Table 1**

**Field Definition Table**

Field Definition Table (FDT) Data Base: IHST

| | Tag | Name | Len | Typ | Rep | Delimiters/Pattern |
|---|---|---|---|---|---|---|
| - | 1 | Journal | 25 | X | | |
| - | 2 | Imprint | 30 | X | | YVIP |
| - | 3 | Reference | 3 | N | | |
| - | 4 | Class No. | 25 | X | | |
| - | 5 | Period | 25 | X | R | |

| | | | | | | |
|---|---|---|---|---|---|---|
| - | 6 | Author | 150 | X | R | |
| - | 7 | Title | 230 | X | | |
| - | 8 | Institution | 380 | X | R | PA |
| - | 9 | Keyword | 230 | X | | |
| - | 10 | Information Source | 55 | X | | |
| - | 11 | Contents | 1500 | X | R | |

### 3.3.2 Worksheet (FMT)

The FMT defined the positioning, attributes of data line (normal, blinking, invisible etc.) and default value of each field on the screen. The History of Science data base used the following worksheet.

**Figure 2**

**Data Base Worksheet**

Journal ____________________ Imprint ____________________

Reference ___ Class No. ____________________ Period ____________________

Author ____________________

____________________

Title ____________________

____________________

Institution ____________________

____________________

____________________

____________________

____________________

Keyword ____________________

____________________

____________________

Information Source ____________________

More...

Contents ____________________

____________________

____________________

____________________

____________________

____________________

____________________

### 3.3.3 Field Select Table (FST)

The FST provided information about the search elements in the fields and how the data should be extracted from the fields. The History of Science data base used the following FST.

**Table 2**

**Field Select Table**

Data Base Name:IHSTFST for Inverted File FST name:IHST

| | ID | IT | Data extraction format |
|---|---|---|---|
| - | 1 | 0 | V1 |
| - | 2 | 1 | V2 |
| - | 4 | 0 | V4 |
| - | 5 | 0 | (V5/) |
| - | 6 | 0 | V6 |
| - | 7 | 0 | V7 |
| - | 8 | 1 | V8 |
| - | 9 | 2 | (V9) |
| - | 10 | 0 | V10 |
| - | 11 | 2 | V11 |

### 3.3.4 Format

The format is a collection of commands used to display print data. Various formats were used depending upon the nature of the outputs. The History of Science data base used the following format for the Main Text :

**Main Text Format**

Data Base Name: IHST

Format name: IHST

```
f(MFN,1,0)/MHU,(V6+¦;¦)”. “,MHL,V7”, "V1". “,V2^Y.”, “V2^V,”;”V2^I,” : “V2^P”. “,”Ref
“V3/MHL, (V8/)/MHU, “Keyword: “V9/# ‘CONTENTS: “(V11/)/##
```

## 3.4 DATA ENTRY

Data entry in CDS/ISIS was done through data entry worksheets, which have been defined previously. Data entry services menu was called by using E-option from the main menu of CDS/ISIS. A worksheet similar to a printed form with spaces to be filled in, was displayed on the screen. A keyboard for data entry was used almost like a typewriter. A sample data entry worksheet page is shown in Figure 3. After filling the information in a record, a new record was created by selecting the N-option and whenever editing was required M-Modify option was used, which are present in the worksheet itself.

**Figure 3**

**Data Entry Worksheet**

Journal IJHS ________________________ Imprint^y1985^v20^i1-4^p122-30________

Refernece 58_Class No.________________________ Period ____________

Author Bag AK ________________________________

**Title** <Astronomy> in <Indus Civilization> and during vedic times ________________

Institution ^pNew Delhi^aIndian National Science Academy ________________

Keyword <Time><Seasons> ________________

**Information Source** ________________

| | | | |
|---|---|---|---|
| Y-Next page | B-Previous page | M-Modify | N-New Record |
| X-Exit | D-Delete | C-Cancel | T-End revise |
| | More... | | MFN= 1510 |

Contents 1. _____ New Year Festival%2. Lunar Mansion and Lunar Months%3. Unit of Time%4. Lunar Months, Seasons & Solar Months%5. Solstices (Ayanas) and Equinoxes ________________

## 3.5 SEARCH

The search in CDS/ISIS is based on Boolean algebra, which provides a convenient way of expressing logical operations between classes. To perform search in CDS/ISIS the data base was made searchable by creating an inverted file (using F-option from the inverted file services menu). In formulating the search expression, three types of terms were used :-

Precise terms,

Right-Truncated terms and

ANY terms

### 3.5.1 Precise Terms

A precise term is, any searchable element defined for a given data base, such as subject descriptors, keywords, words in the titles, author names, etc. For precise search term, the term was specified in exactly the same way as it was presented in the data base, as shown in the following example:-

Set 5: AL-HAKIM

P= 1 AL-HAKIM

T= 1 - #5: AL-HAKIM

Set 6: FISH

P= 23 FISH

T= 23 - #6: FISH

Set 8: ASTRONOMY

P= 16 ASTRONOMY

T= 16 - #8: ASTRONOMY

Term Dictionary (T - Option) created by the system was also used for Precise Terms.

### 3.5.2 Right-Truncated Terms

This allows to search on leading sequences of characters. Right-truncation was indicated by placing a dollar sign '$' immediately after the last root character. Example:

Search Expression?

Fish$

Set 2: Fish $

P= 23 Fish

P= 2 Fish - Respiration

P= 1 Fish - Symbiotic

P= 1 Fish And Fishery Biology Labor

P= 1 Fish-Classification

P= 1 Fish Epidemic

P= 1 Fish-Fossils

P= 1 Fish-Live

P= 1 Fish-Trap

P= 1 Fisher C

P= 1 Fisheries

P= 1 Fisheries - Marine

P= 1 Fishery

P= 2 Fishes

P= 1 Fishes - Abnormalities

P= 1 Fishes of Northern <Bengal>

P= 1 Fishing

T= 39 - #24: Fish$

T= 39-#7: #24

Search Expression?

Astro$

Set 9: Astro$

P= 2 Astrology

P= 1 Astronomic Constants

P= 1 Astronomical <Observatories>

P= 1 Astronomical Basis for Fixing

P= 1 Astronomical Chapter of the BU

P= 1 Astronomical Evidence on the A

P= 1 Astronomical References From T

P= 16 Astronomy

P= 2 Astronomy - Source Materials

P= 1 Astronomy - Spherical

P= 1 Astronomy <Astronomy> In Ancie

P= 1 Astronomy Jaina

P= 1 Astronomy of the Mamluks

P= 1 Astronomy-Development

T= 29 - #23: ASTRO$

T= 29 - #9:#23

--------->

### 3.5.3 ANY Terms

An 'ANY' term is a collective term standing for a pre-defined set of search terms. An 'ANY' file was prepared and the search was performed by prefixing the word 'ANY' followed by search term. Example:

Preparation of 'ANY' file - 'Edlin' text editor was used for the preparation of 'ANY' file as given below:

Edlin IHST. ANY <CR>

| Ist column | 31st column |
|---|---|
| **ANY ASIA** | **BANGLADESH** |
| **ANY ASIA** | **INDIA** |
| **ANY ASIA** | **PAKISTAN** |
| **ANY ASIA** | **SRI LANKA** |
| **ANY CHEMISTRY** | **ALCHEMY** |
| **ANY CHEMISTRY** | **CHEMISTRY** |
| **ANY CHEMISTRY** | **INORGANIC** |
| **ANY CHEMISTRY** | **ORGANIC** |

Search expression

**ANY Asia <CR>** produced all the records containing the words Bangladesh, India, Pakistan, Sri Lanka.

### 3.5.4 Boolean Search

It uses the combination of two or more search terms in a search expression by using search operators which indicate the intended relationship between the terms. The logical search operators are :-

Logical OR + 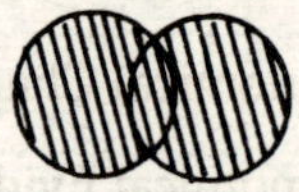

Logical AND * 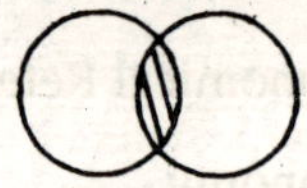

Logical NOT ^ 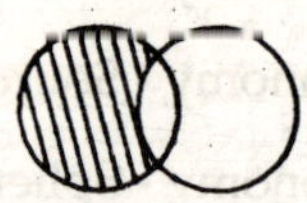

Various Boolean searches were performed on the data base; search expressions, along with the system execution are presented below :-

Search expression?

JAIN$+ASTRO$

SET 24: JAIN$+ASTRO$

P= 1 Jain - Calendar

P= 1 Jain - Mathematics

P= 1 Jain - Source Materials, Agric

P= 1 Jain BL

P= 2 Jain BS

P= 1 Jain JC

P= 1 Jain KK

P= 7 Jain LC

P= 1 Jain-Mysticism

P= 6 Jaina

P= 1 Jaina <Calendar>

P= 1 Jaina Astronomy

P= 2 Jaina School of <Astronomy>

T= 20 - #38:Jain

P= 2 Astrology

P= 1 Astronomic Constants

P= 1 Astronomical <Observatories>

P= 1 Astronomical Basis for Fixing

P= 1 Astronomical Chapter of the BU

P= 1 Astronomical Evidence on the A

P= 1 Astronomical References from T

P= 16 Astronomy

P= 2 Astronomy - Source Materials

P= 1 Astronomy - Spherical

P= 1 Astronomy <Astronomy> in Ancie

P= 1 Astronomy Jaina

P= 1 Astronomy of the Mamluks

P= 1 Astronomy-Development

T= 29 - #52: Astro$

T= 48 - #53: #38 + #52

T= 48 - #25: #53

——>

Search expression?

Jain$ * Astro$

Set 24: Jain$ * Astro$

P= 1 Jain - Calendar

P= 1 Jain - Mathematics

P= 1 Jain - Source Materials, Agric

P= 1 Jain BL

P= 2 Jain BS

P= 1 Jain JC

P= 1 Jain KK

P= 7 Jain LC

P= 1 Jain-Mysticism

P= 6 Jaina

P= 1 Jaina <Calendar>

P= 1 Jaina Astronomy

P= 2 Jaina School of <Astronomy>

T= 20 - #36:Jain$

P= 2 Astrology

P= 1 Astronomic Constants

P= 1 Astronomical <Observatories>

P= 1 Astronomical Basis for Fixing

P= 1 Astronomical Chapter of the BU

P= 1 Astronomical Evidence on the A

P= 1 Astronomical References from T

P= 16 Astronomy

P= 2 Astronomy - Source Materials

P= 1 Astronomy - Spherical

P= 1 Astronomy <Astronomy> In Ancie

P= 1 Astronomy Jaina

P= 1 Astronomy of the Mamluks

P= 1 Astronomy-Development

T= 29 - #50: Astro$

T= 1 - #51: #36 + #50

T= 1 - #23: #51

——>

Search expression?

Jain$ ^ Astro$

Set 26: Jain$ ^ Astro$

P= 1 Jain - Calendar

P= 1 Jain - Mathematics

P= 1 Jain - Source Materials, Agric

P= 1 Jain BL

P= 2 Jain BS

P= 1 Jain JC

P= 1 Jain KK

P= 7 Jain LC

P= 1 Jain-Mysticism

P= 6 Jaina

P= 1 Jaina <Calendar>

P= 1 Jaina Astronomy

P= 2 Jaina School of <Astronomy>

T= 20 - #39:Jain$

P= 2 Astrology

P= 1 Astronomic Constants

P= 1 Astronomical <Observatories>

P= 1 Astronomical Basis for Fixing

P= 1 Astronomical Chapter of the BU

P= 1 Astronomical Evidence on the A

P= 1 Astronomical References from T

| | | |
|---|---|---|
| P= | 16 | Astronomy |
| P= | 2 | Astronomy - Source Materials |
| P= | 1 | Astronomy - Spherical |
| P= | 1 | Astronomy <Astronomy> in Ancie |
| P= | 1 | Astronomy Jaina |
| P= | 1 | Astronomy of the Mamluks |
| P= | 1 | Astronomy-Development |
| T= | 29 - #53: Astro$ | |
| T= | 19 - #54: #39 + #53 | |
| T= | 19 - #26: #54 | |

——>

### 3.5.5 Proximity Search Operators

These operators are more restrictive types of logical 'AND'. Here the field on which search was performed, was first indexed using indexing technique '4' (words) and later on search expressions were made using proximity operators as follows :-

(i) '.' was used when both the terms occur in the same field, and

(ii) '$' was used when both the terms occur in the same field or the same occurrence of a repeatable field.

### 3.5.6 Free Text Search

This technique allows to specify search requirements on fields which have not been inverted. In order to distinguish it from a normal search expression, it was preceded by a question mark, as follows :

? Boolean expression **or**

Set 10: ? v6 : 'HORA SL' AND VAL (v2^y)>=1950

Figure 4 shows the screen layout during the execution of a free text search.

**Figure 4**

**Execution of a Free Text Search**

| MFN | HITS | % | RECS |
|---|---|---|---|
| 821 | 5 | 0.32 | 1569 |

MFN - Shows the MFN of the record being processed;

Hits - Shows the current number of records retrieved;

% - Shows the current hit rate; and

Recs - Shows the total number of records to be processed.

*Search Results*

With the press of enter/execute key the search expressions were executed displaying the searched postings and terms on the screen.

After this the enter key was pressed again and the 'Information Retrieval Services' menu was displayed and the search results were browsed and saved using D- and P- options respectively. The saved search results were further printed using P-option from the main menu of CDS/ISIS.

### 3.6 SORTING AND PRINTING

The sorting and printing option in CDS/ISIS was used, to print a given search (using search option), a whole data base and a specific range of records. Various print parameters were filled, in the **system print worksheet**. For sorting the 'Sort-Y' option in print worksheet was used, which displayed the **system sort worksheet.** Accordingly various sort parameters were filled in. Various samples of print and sort worksheets are given below:

**Main Text**

AEYPRT. FMT (System Print Worksheet)

First Title History of Science

Second Title *Complete Records in Yearwise Order

Print format @IHST

Line width 35

Number of columns 2

Column width 40

Print file name **IHST**

**Title Index**

AEYPRT.FMT

First Title History of Science

Second Title Title Index

Print format*

Line width 60

Number of columns 1

Column width 60

Sort Y

Print file name **TITLI**

AEYSRT. FMT (Sort Worksheet)

Number of headings 1

Heading format : MHU,#v1

Length of first sort key 254

Heading processing indicator 0

FST for first sort key 1 0 MHU, V7

Length of second sort key 5

Heading processing indicator 0

FST for second sort key 2 0 '<', f(mfn,1,0),'=',mfn(5),'>'

**Author Index**

AEYPRT. FMT

First Title History of Science

Second Title* Author Index

Print format *

Line width 33

Number of columns 2

Column width 35

Sort Y

Print file name **AUTHIN**

AEYSRT. FMT

Number of headings 1

Heading format: MHU, #V1

Length of first sort key 100

Heading processing indicator 1

FST for first sort key 1 0 (V6/)

Length of second sort key 5

Heading processing indicator 0

FST for second sort key 2 0 '<', f(mfn,1,0),'=',mfn(5),'>'

**Keyword Index**

AEYPRT. FMT

First Title History of Science

Second Title Keyword Index

Print format*

Line width 33

Number of columns 2

Column width 35

Sort Y

Print file name **KINDE**

AEYSRT. FMT

Number of headings 1

Heading format: MHU, #V1

Length of first sort key 100

Heading processing indicator 1

FST for first sort key 1 2 V7 + 1 2 V9 + 1 2 V11

Length of second sort key 5

Heading processing indicator 0

FST for second sort key 2 0 '<',f(mfn,1,0),'=',mfn(5),'>'

**Journal Index**

AEYPRT. FMT

First title History of Science

Second title Journal Index

Print format*

Line width 33

Number of columns 2

Column width 35

Sort Y

Print file name **JOUIN**

AEYSRT. FMT

Number of headings 1

Heading format: MHU, #V1

Length of first sort key 10

Heading processing indicator 0

FST for first sort key 1 0 V1

Length of second sort key 5

Heading processing indicator 0

FST for second sort key 2 0 '<',f(mfn,1, 0),'=', mfn(5),'>'

## REFERENCES


1. A. Chandor, *The Penguin Dictionary of Computer.* (3rd ed; Penguin, 1985).
2. K. Shailendra, R.N. Neogi, and K.L. Gogia, "Use of CDS/ISIS for creating India's data base of serials," *Program*, 23, No. 3(1989), 287-94.
3. K. Shailendra, "Networking of data bases created by CDS/ISIS using common communication format tags," *Library Herald*, 27, No. 1-2(1988), 9-19.
4. K. Shailendra, V.V.Lakshmi, and S.R. Kundu, "Creation of the directory of participating institutions for NUCSSI using mini-micro CDS/ISIS, "*Annals of Library Science and Documentation*, 35, No. 4(1988), 178-88.
5. P. Nieuwenhuysen, "Computerised storage and retrieval of structured text information: CDS/ISIS Version 2.3," *Program*, 25, No. 1 (1991), 1-12.
6. A.A. Chaudhry, "Interfacing CDS/ISIS with Wordstar: ICIMOD's experience," *Program*, 23, No. 1 (1989), 65-72.
7. *Mini-Micro CDS/ISIS 2.3 Reference Manual and CDS/ISIS Pascal Manual* (Paris: Unesco, 1989), p.362.

# 4. GROWTH OF HISTROY OF SCIENCE

History of Science is a subject, relatively of recent origin. The amount of literature produced in it is not as high as compared to the subjects in the basic sciences. Though the articles in this had started coming up in the beginning of the 20th Century but many of the articles were initially related to the research in basic sciences.

The contributions in the History of Science were scattered in the beginning, but later on, these were concentrated in specific publications. There is a gradual increase in the research workers with the advent of various institutions in the subject in the country (1-2).

The present study traces the growth in History of Science in the journals included in this study from the view points of the following :

- **Literature**
- **Publication Sources**
- **Research Workers**

## 4.1 LITERATURE

Literature in History of Science had steady growth from 1905 to 1969 as shown in Table 3. But since 1970 the amount of literature produced in the subject was increased considerably. In year 1955 there were only 4 contributions, whereas in the year 1970, the number of contributions rose to 43. Out of the period and the journals studied, 4 happen to be the minimum number of contributions in a year, whereas 43 happen to be the maximum number of contributions in a year. The individual year contribution is shown in Table 5.

Table 3 shows the growth rate in various blocks of periods. The growth rate of literature in the later half of the first decade in the twentieth century was 25.6 contributions/year. This growth rate then had a sharp decrease; this may be due to the start of first world war. It has been observed that the general tendency in the growth rate of literature has apparent increase. Various clinks in the growth rate are due to the country's involvement in the war at various stages. The decade of 1940s' witnessed the country's war for independence in full swing; whereas in the 1960s', the country was involved in two wars - one with China and another with Pakistan. During the 1980s', the rate of growth of literature was growing steadily.

**Table 3**

**Growth Rate in Different Periods**

| Decade | | # Contributions C/10 | Cum. Contributions (ΣC) | Cum. Years(t) | $Gr = \frac{\Sigma C}{t}$ |
|---|---|---|---|---|---|
| 1905-1909* | 128 | 25.6 | 128 | 5 | 25.6 |
| 1910-1919 | 165 | 16.5 | 293 | 15 | 19.53 |
| 1920-1929 | 159 | 15.9 | 452 | 25 | 18.08 |
| 1930-1939 | 176 | 17.6 | 628 | 35 | 17.94 |
| 1940-1949 | 149 | 14.9 | 777 | 45 | 17.26 |

| | | | | | |
|---|---|---|---|---|---|
| 1950-1959 | 159 | 15.9 | 936 | 55 | 17.02 |
| 1960-1969 | 135 | 13.5 | 1071 | 65 | 16.48 |
| 1970-1979 | 277 | 27.7 | 1348 | 75 | 17.97 |
| 1980-1985* | 178 | 35.5 | 1526 | 80 | 19.1 |
| 1986.... | 41 | 41.0 | 1567 | 81 | 19.34 |

*Block of 5 Years

## 4.2 PUBLICATION SOURCES

The study covered six research journals to get contributions on various subjects in the History of Science. The Asiatic Society of Bengal, one of the founders of research journals in the country started publishing the Journal of the Asiatic Society of Bengal in 1832. Since then the journal has undergone various changes and has continued till now as Journal of the Asiatic Society, Calcutta (3). Other journals taken into account are listed below:-

ISIS, 1913+;

Indian Historical Quarterly, 1925-1963;

Journal of the Asiatic Society of Great Britain and Ireland ; 1927+

Vishveshvaranand Indological Journal, 1963+;

Indian Journal of History of Science, 1966+ .

While scanning the twentieth century publications on History of Science in India, the total number of journals available during a particular decade, have been at variance as shown in Table 4.

**Table 4**

**Number of Journals in Different Periods**

| Decade | Number of Journal (s) | Journal (s) |
|---|---|---|
| 1905-1909* | 1 | ASB |
| 1910-1919 | 2 | ASB+ISIS |
| 1920-1929 | 4 | ASB+ISIS+JRAS+IHQ |
| 1930-1939 | 4 | ASB+ISIS+JRAS+IHQ |
| 1940-1949 | 4 | ASB+ISIS+JRAS+IHQ |
| 1950-1959 | 4 | ASB+ISIS+JRAS+IHQ |
| 1960-1969 | 6 | ASB+ISIS+JRAS+IHQ+VIJ IJHS |
| 1970-1979 | 5 | ASB+ISIS+JRAS+VIJ+IJHS |
| 1980-1986** | 5 | ASB+ISIS+JRAS+VIJ+IJHS |

* Block of 5 years

** Block of 7 years

Detailed breakup of contributions in various journals is given in Table 6 and Figure 5. Maximum contributions during a year, recorded in the journals individually is given as follows :

**Table 5**
**Maximum contributions in the Journals Individually During a Year**

| Journal | | Maximum Contributions | Year(s) |
|---|---|---|---|
| ASB | JPASB | 41 | 1907 |
| | JASBSC | 17 | 1954 |
| | JASBL | 5 | 1952 |
| | JAS | 11 | 1959 |
| ISIS | | 7 | 1947 |
| JRAS | | 7 | 1933 |
| IHQ | | 11,11 | 1932,1933 |
| VIJ | | 7 | 1977 |
| IJHS | | 34,34 | 1984,1986 |

### 4.3 RESEARCH WORKERS

The growth of the authors in History of Science during 1905 to 1986 had a constant increase in terms of number of authors as shown in Tables 7-8 and Figure 6. Authors contributed during all these years ranging from minimum 4 authors in each of the two years 1955 & 1960, to maximum 47 authors in the year 1984. New authors added during these years had a range from minimum 1 author in the year 1934 to maximum 35 authors in the year 1970. The authors who continued their contributions were minimum zero in 1925 to maximum 19 authors in 1984.

With the cumulative increase in the number of authors, each year the number of passive authors also increased with the exceptions of being constant twice and three falls therein. The passive authors remain steady in the years 1916 and 1917 with 125 authors each year; in the years 1928 and 1929 with 209 authors each year. Number of passive authors decreased during the period 1921 to 1922 from 153 to 149; during the period 1925 to 1926 from 193 to 190 ; during the period 1951 to 1952 from 434 to 428.

The growth rate of the authors during 1905 to 1986 varied from 7.0 to 19.83. In the year 1907 the growth rate was observed as 13.7 maximum in the individual year, whereas growth rate of the authors in decades shown in Table 8 and Figure 6.3, observed a downfall in the beginning though it improved in later decades (4-5). From 1980 to 1986 alone, the growth rate was recorded as 19.83. Barring two decades (i.e., 1920-1929 and 1950-1959) when slight fall was found in the number of authors, these increased upward decadewise as shown in the Figure 6.1. The cumulative number of authors decadewise had almost a straight upward line as is evident from the Figure 6.2.

## Table 6

### Breakup Of Contributions In Different Journals Upto A Particular Year

| Year of Contri-bution | # Contri-butions | Journals | Cum. Year | Cum. Contri-butions |
|---|---|---|---|---|
| 1905 | 10 | JPASB | 1 | 10 |
| 1906 | 39 | JPASB | 2 | 49 |
| 1907 | 41 | JPASB | 3 | 90 |
| 1908 | 26 | JPASB | 4 | 116 |
| 1909 | 12 | JPASB | 5 | 128 |
| 1910 | 17 | JPASB | 6 | 145 |
| 1911 | 18 | JPASB | 7 | 163 |
| 1912 | 21 | $JPASB_{2}$ $JASB_{19}$ | 8 | 184 |
| 1913 | 32 | JPASB | 9 | 216 |
| 1914 | 22 | JPASB | 10 | 238 |
| 1915 | 12 | JPASB | 11 | 250 |
| 1916 | 5 | JPASB | 12 | 255 |
| 1917 | 10 | JPASB | 13 | 265 |
| 1918 | 11 | JPASB | 14 | 276 |
| 1919 | 17 | $JPASB_{15}$ $ISIS_{2}$ | 15 | 293 |
| 1920 | 14 | $JPASB_{13}$ $ISIS_{1}$ | 16 | 307 |
| 1921 | 5 | JPASB | 17 | 312 |
| 1922 | 17 | JPASB | 18 | 329 |
| 1923 | 20 | JPASB | 19 | 349 |
| 1924 | 22 | JPASB | 20 | 371 |
| 1925 | 6 | $JPASB_{2}$ $IHQ_{4}$ | 21 | 377 |
| 1926 | 23 | $JPASB_{19}$ $IHQ_{4}$ | 22 | 400 |
| 1927 | 26 | $JPASB_{15}$ $IHQ_{10}$ $JRAS_{1}$ | 23 | 426 |
| 1928 | 11 | $JPASB_{6}$ $IHQ_{5}$ | 24 | 437 |
| 1929 | 15 | $JPASB_{7}$ $IHQ_{7}$ $ISIS_{1}$ | 25 | 452 |
| 1930 | 18 | $JPASB_{11}$ $IHQ_{5}$ $ISIS_{2}$ | 26 | 470 |
| 1931 | 26 | $JPASB_{12}$ $IHQ_{8}$ $JRAS_{6}$ | 27 | 496 |
| 1932 | 22 | $JPASB_{7}$ $IHQ_{11}$ $JRAS_{3}$ $ISIS_{1}$ | 28 | 518 |
| 1933 | 34 | $JPASB_{13}$ $IHQ_{11}$ $JRAS_{7}$ $ISIS_{3}$ | 29 | 552 |
| 1934 | 10 | $JPASB_{3}$ $IHQ_{7}$ | 30 | 562 |
| 1935 | 14 | $IHQ_{4}$ $ISIS_{6}$ $JASBSC_{4}$ | 31 | 576 |
| 1936 | 17 | $IHQ_{5}$ $ISIS_{4}$ $JASBSC_{8}$ | 32 | 593 |
| 1937 | 10 | $IHQ_{4}$ $ISIS_{4}$ $JASBSC_{1}$ $JASBL_{1}$ | 33 | 603 |
| 1938 | 13 | $IHQ_{2}$ $JRAS_{2}$ $ISIS_{2}$ $JASBSC_{7}$ | 34 | 616 |
| 1939 | 12 | $IHQ_{7}$ $JRAS_{3}$ $JASBSC_{2}$ | 35 | 628 |
| 1940 | 13 | $IHQ_{5}$ $JASBSC_{6}$ $JASBL_{2}$ | 36 | 641 |
| 1941 | 7 | $JASBSC_{5}$ $JASBL_{2}$ | 37 | 648 |
| 1942 | 13 | $IHQ_{5}$ $JRAS_{3}$ $JASBSC_{4}$ $JASBL_{1}$ | 38 | 661 |
| 1943 | 22 | $IHQ_{7}$ $JRAS_{3}$ $ISIS_{1}$ $JASBSC_{11}$ | 39 | 683 |
| 1944 | 15 | $IHQ_{1}$ $JRAS_{2}$ $ISIS_{4}$ $JASBSC_{8}$ | 40 | 698 |
| 1945 | 21 | $IHQ_{2}$ $JRAS_{3}$ $JASBSC_{16}$ | 41 | 719 |
| 1946 | 15 | $IHQ_{2}$ $JRAS_{2}$ $ISIS_{3}$ $JASBSC_{8}$ | 42 | 734 |
| 1947 | 16 | $JRAS_{2}$ $ISIS_{7}$ $JASBSC_{7}$ | 43 | 750 |

| | | | | | | | | | | | | |
|---|---|---|---|---|---|---|---|---|---|---|---|---|
| 1948 | 12 | | IHQ$_4$ | JRAS$_3$ | | JASBSC$_5$ | | | | | 44 | 762 |
| 1949 | 15 | | IHQ$_4$ | JRAS$_1$ | ISIS$_2$ | JASBSC$_8$ | | | | | 45 | 777 |
| 1950 | 24 | JPASB$_8$ | IHQ$_1$ | JRAS$_1$ | ISIS$_2$ | JASBSC$_8$ | JASBL$_4$ | | | | 46 | 801 |
| 1951 | 7 | | IHQ$_2$ | JRAS$_1$ | ISIS$_2$ | | JASBL$_2$ | | | | 47 | 808 |
| 1952 | 28 | | IHQ$_7$ | JRAS$_2$ | ISIS$_1$ | JASBSC$_{13}$ | JASBL$_5$ | | | | 48 | 836 |
| 1953 | 21 | | IHQ$_5$ | | ISIS$_2$ | JASBSC$_{12}$ | JASBL$_2$ | | | | 49 | 857 |
| 1954 | 25 | | IHQ$_5$ | JRAS$_2$ | | JASBSC$_{17}$ | JASBL$_1$ | | | | 50 | 882 |
| 1955 | 4 | | IHQ$_4$ | | | JASBSC$_2$ | | | | | 51 | 886 |
| 1956 | 9 | | | JRAS$_2$ | ISIS$_2$ | JASBSC$_3$ | JASBL$_1$ | JAS$_1$ | | | 52 | 895 |
| 1957 | 12 | | IHQ$_3$ | JRAS$_1$ | ISIS$_1$ | JASBSC$_3$ | JASBL$_4$ | | | | 53 | 907 |
| 1958 | 9 | | IHQ$_1$ | JRAS$_1$ | ISIS$_4$ | | JASBL$_3$ | | | | 54 | 916 |
| 1959 | 20 | | IHQ$_3$ | JRAS$_3$ | ISIS$_3$ | | | JAS$_{11}$ | | | 55 | 936 |
| 1960 | 5 | | IHQ$_1$ | JRAS$_1$ | ISIS$_3$ | | | | | | 56 | 941 |
| 1961 | 9 | | IHQ$_4$ | JRAS$_1$ | ISIS$_1$ | | | JAS$_3$ | | | 57 | 950 |
| 1962 | 17 | | IHQ$_4$ | JRAS$_2$ | ISIS$_6$ | | | JAS$_5$ | | | 58 | 967 |
| 1963 | 11 | | | JRAS$_3$ | ISIS$_3$ | | | JAS$_2$ | VIJ$_3$ | | 59 | 978 |
| 1964 | 8 | | | JRAS$_2$ | ISIS$_1$ | | | JAS$_2$ | VIJ$_3$ | | 60 | 986 |
| 1965 | 6 | | | JRAS$_1$ | | | | JAS$_3$ | VIJ$_2$ | | 61 | 992 |
| 1966 | 27 | | | | ISIS$_2$ | | | JAS$_3$ | VIJ$_4$ | IJHS$_{18}$ | 62 | 1019 |
| 1967 | 14 | | | JRAS$_1$ | ISIS$_1$ | | | JAS$_2$ | VIJ$_1$ | IJHS$_9$ | 63 | 1033 |
| 1968 | 12 | | | JRAS$_1$ | ISIS$_2$ | | | JAS$_1$ | VIJ$_1$ | IJHS$_7$ | 64 | 1045 |
| 1969 | 26 | | | | ISIS$_2$ | | | JAS$_6$ | VIJ$_3$ | IJHS$_{15}$ | 65 | 1071 |
| 1970 | 43 | | | JRAS$_3$ | ISIS$_1$ | | | JAS$_2$ | VIJ$_4$ | IJHS$_{33}$ | 66 | 1114 |
| 1971 | 19 | | | JRAS$_2$ | | | | JAS$_2$ | | IJHS$_{15}$ | 67 | 1133 |
| 1972 | 24 | | | JRAS$_3$ | | | | JAS$_5$ | VIJ$_4$ | IJHS$_{12}$ | 68 | 1157 |
| 1973 | 13 | | | | | | | JAS$_2$ | VIJ$_4$ | IJHS$_7$ | 69 | 1170 |
| 1974 | 32 | | | JRAS$_1$ | | | | | VIJ$_6$ | IJHS$_{25}$ | 70 | 1202 |
| 1975 | 38 | | | JRAS$_1$ | ISIS$_2$ | | | JAS$_5$ | VIJ$_6$ | IJHS$_{24}$ | 71 | 1240 |
| 1976 | 21 | | | | ISIS$_1$ | | | JAS$_2$ | VIJ$_5$ | IJHS$_{13}$ | 72 | 1261 |
| 1977 | 38 | | | JRAS$_1$ | ISIS$_1$ | | | JAS$_3$ | VIJ$_7$ | IJHS$_{26}$ | 73 | 1299 |
| 1978 | 25 | | | JRAS$_2$ | | | | | VIJ$_5$ | IJHS$_{18}$ | 74 | 1324 |
| 1979 | 24 | | | JRAS$_1$ | ISIS$_2$ | | | JAS$_1$ | VIJ$_4$ | IJHS$_{16}$ | 75 | 1348 |
| 1980 | 31 | | | JRAS$_3$ | ISIS$_1$ | | | JAS$_3$ | VIJ$_5$ | IJHS$_{19}$ | 76 | 1379 |
| 1981 | 34 | | | JRAS$_1$ | ISIS$_2$ | | | JAS$_2$ | | IJHS$_{29}$ | 77 | 1413 |
| 1982 | 35 | | | | ISIS$_4$ | | | JAS$_1$ | | IJHS$_{30}$ | 78 | 1448 |
| 1983 | 20 | | | JRAS$_1$ | ISIS$_2$ | | | JAS$_1$ | | IJHS$_{16}$ | 79 | 1468 |
| 1984 | 39 | | | JRAS$_2$ | ISIS$_1$ | | | JAS$_2$ | | IJHS$_{34}$ | 80 | 1507 |
| 1985 | 19 | | | JRAS$_1$ | ISIS$_3$ | | | JAS$_2$ | | IJHS$_{13}$ | 81 | 1526 |
| 1986 | 41 | | | JRAS$_2$ | | | | JAS$_5$ | | IJHS$_{34}$ | 82 | 1567 |

Maximum contribution was 43 in the year 1970
Minimum Contribution was 4 in the year 1955

**Figure 5**

DISTRIBUTION OF ARTICLES ON THE HISTORY OF SCIENCE IN THE JOURNALS

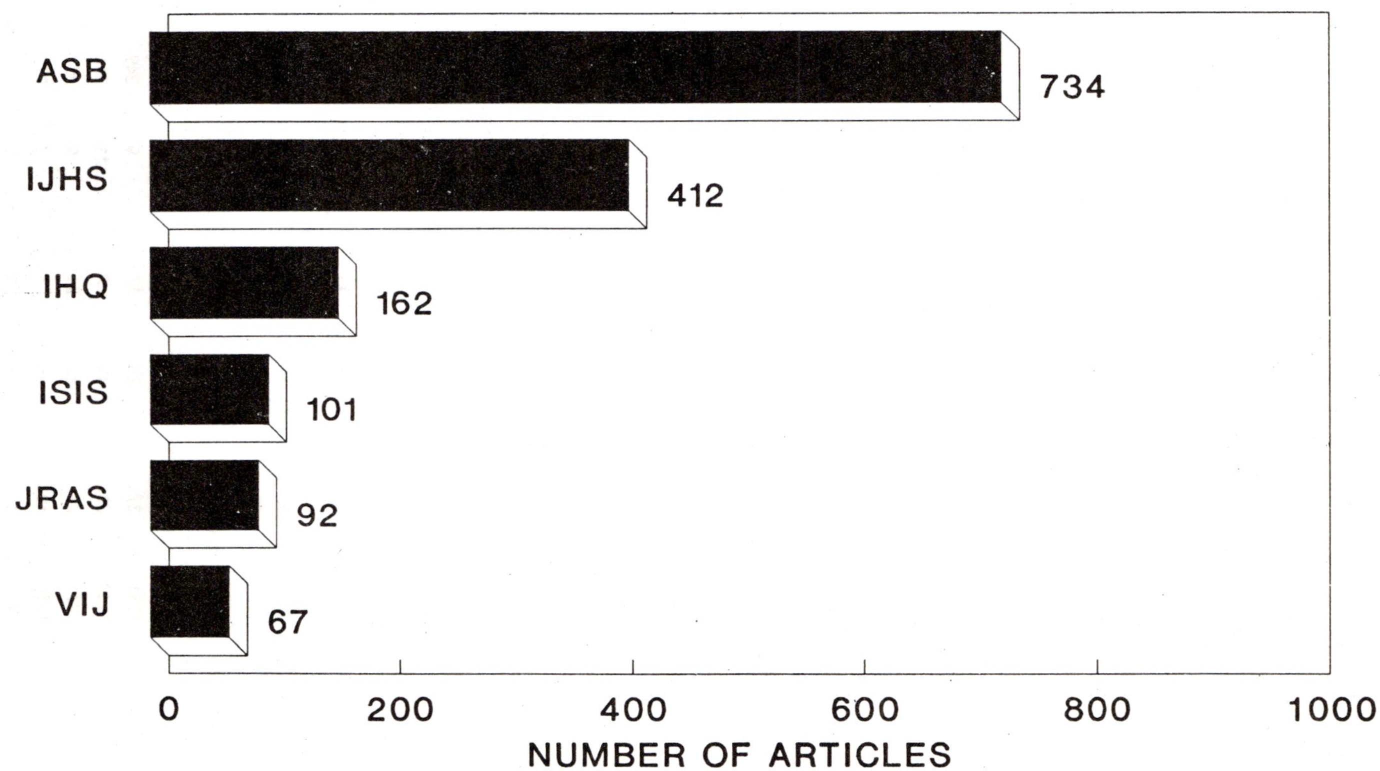

## Table 7

Growth of Authors in History of Science

| Year 'Y' | Σ Y | No. of Authors 'A1' | Authors Added 'A2' | Authors Continued (A1-A2)=A3 | Σ A2 'A4' | Passive Authors (A4-A2)-A3 'A5' | Growth of Authors A4/Y |
|---|---|---|---|---|---|---|---|
| 1905 | 1 | 7 | 7 | 0 | 7 | 0 | 7.00 |
| 1906 | 2 | 20 | 18 | 2 | 25 | 5 | 12.50 |
| 1907 | 3 | 22 | 16 | 6 | 41 | 19 | 13.70 |
| 1908 | 4 | 18 | 8 | 10 | 49 | 31 | 12.20 |
| 1909 | 5 | 10 | 4 | 6 | 53 | 43 | 10.60 |
| 1910 | 6 | 15 | 10 | 5 | 63 | 48 | 10.50 |
| 1911 | 7 | 15 | 8 | 7 | 71 | 56 | 10.14 |
| 1912 | 8 | 20 | 9 | 11 | 80 | 60 | 10.00 |
| 1913 | 9 | 26 | 19 | 7 | 99 | 73 | 11.00 |
| 1914 | 10 | 24 | 18 | 6 | 117 | 93 | 11.70 |
| 1915 | 11 | 12 | 10 | 2 | 127 | 115 | 11.54 |
| 1916 | 12 | 5 | 3 | 2 | 130 | 125 | 10.83 |
| 1917 | 13 | 9 | 4 | 5 | 134 | 125 | 10.31 |
| 1918 | 14 | 11 | 7 | 4 | 141 | 130 | 10.07 |
| 1919 | 15 | 16 | 9 | 7 | 150 | 134 | 10.00 |
| 1920 | 16 | 13 | 5 | 8 | 155 | 142 | 9.70 |
| 1921 | 17 | 5 | 3 | 2 | 158 | 153 | 9.30 |
| 1922 | 18 | 16 | 7 | 9 | 165 | 149 | 9.16 |
| 1923 | 19 | 18 | 13 | 5 | 178 | 160 | 9.37 |
| 1924 | 20 | 19 | 15 | 4 | 193 | 174 | 9.65 |
| 1925 | 21 | 5 | 5 | 0 | 198 | 193 | 9.43 |
| 1926 | 22 | 12 | 4 | 8 | 202 | 190 | 9.18 |
| 1927 | 23 | 20 | 12 | 8 | 214 | 194 | 9.30 |
| 1928 | 24 | 7 | 2 | 5 | 216 | 209 | 9.00 |
| 1929 | 25 | 17 | 10 | 7 | 226 | 209 | 9.04 |
| 1930 | 26 | 18 | 6 | 12 | 232 | 214 | 8.92 |
| 1931 | 27 | 26 | 17 | 9 | 249 | 223 | 9.22 |
| 1932 | 28 | 18 | 7 | 11 | 256 | 238 | 9.14 |
| 1933 | 29 | 30 | 21 | 9 | 277 | 247 | 9.55 |
| 1934 | 30 | 8 | 1 | 7 | 278 | 270 | 9.27 |
| 1935 | 31 | 13 | 10 | 3 | 288 | 275 | 9.29 |
| 1936 | 32 | 18 | 15 | 3 | 303 | 285 | 9.47 |
| 1937 | 33 | 11 | 8 | 3 | 311 | 300 | 9.42 |
| 1938 | 34 | 10 | 7 | 3 | 318 | 308 | 9.35 |
| 1939 | 35 | 11 | 6 | 5 | 324 | 313 | 9.26 |
| 1940 | 36 | 14 | 13 | 1 | 337 | 323 | 9.36 |
| 1941 | 37 | 5 | 3 | 2 | 340 | 335 | 9.19 |
| 1942 | 38 | 13 | 9 | 4 | 349 | 336 | 9.18 |
| 1943 | 39 | 22 | 12 | 10 | 361 | 339 | 9.26 |
| 1944 | 40 | 15 | 10 | 5 | 371 | 356 | 9.27 |
| 1945 | 41 | 18 | 9 | 9 | 380 | 362 | 9.27 |
| 1946 | 42 | 17 | 12 | 5 | 392 | 375 | 9.33 |
| 1947 | 43 | 14 | 10 | 4 | 402 | 388 | 9.35 |
| 1948 | 44 | 13 | 11 | 2 | 413 | 400 | 9.39 |
| 1949 | 45 | 13 | 11 | 2 | 424 | 411 | 9.42 |

| | | | | | | | |
|---|---|---|---|---|---|---|---|
| 1950 | 46 | 22 | 13 | 9 | 437 | 415 | 9.50 |
| 1951 | 47 | 6 | 3 | 3 | 440 | 434 | 9.36 |
| 1952 | 48 | 28 | 16 | 12 | 456 | 428 | 9.50 |
| 1953 | 49 | 21 | 11 | 10 | 467 | 446 | 9.53 |
| 1954 | 50 | 24 | 17 | 7 | 484 | 460 | 9.68 |
| 1955 | 51 | 4 | 3 | 1 | 487 | 483 | 9.55 |
| 1956 | 52 | 9 | 7 | 2 | 494 | 485 | 9.50 |
| 1957 | 53 | 14 | 10 | 4 | 504 | 490 | 9.24 |
| 1958 | 54 | 6 | 3 | 3 | 507 | 501 | 9.39 |
| 1959 | 55 | 20 | 16 | 4 | 523 | 503 | 9.51 |
| 1960 | 56 | 4 | 3 | 1 | 526 | 522 | 9.39 |
| 1961 | 57 | 9 | 6 | 3 | 532 | 523 | 9.33 |
| 1962 | 58 | 15 | 13 | 2 | 545 | 530 | 9.39 |
| 1963 | 59 | 10 | 9 | 1 | 554 | 544 | 9.39 |
| 1964 | 60 | 10 | 8 | 2 | 562 | 552 | 9.37 |
| 1965 | 61 | 7 | 4 | 3 | 566 | 559 | 9.28 |
| 1966 | 62 | 25 | 22 | 3 | 588 | 563 | 9.48 |
| 1967 | 63 | 14 | 10 | 4 | 598 | 584 | 9.49 |
| 1968 | 64 | 16 | 11 | 5 | 609 | 593 | 9.51 |
| 1969 | 65 | 25 | 19 | 6 | 628 | 603 | 9.66 |
| 1970 | 66 | 46 | 35 | 11 | 663 | 617 | 10.04 |
| 1971 | 67 | 17 | 10 | 7 | 673 | 656 | 10.04 |
| 1972 | 68 | 30 | 18 | 12 | 691 | 661 | 10.16 |
| 1973 | 69 | 14 | 11 | 3 | 702 | 688 | 10.17 |
| 1974 | 70 | 32 | 22 | 10 | 724 | 692 | 10.34 |
| 1975 | 71 | 38 | 20 | 18 | 744 | 706 | 10.47 |
| 1976 | 72 | 25 | 11 | 14 | 755 | 730 | 10.49 |
| 1977 | 73 | 33 | 23 | 10 | 778 | 745 | 10.66 |
| 1978 | 74 | 18 | 7 | 11 | 785 | 767 | 10.61 |
| 1979 | 75 | 17 | 3 | 14 | 788 | 771 | 10.51 |
| 1980 | 76 | 30 | 15 | 15 | 803 | 773 | 10.56 |
| 1981 | 77 | 43 | 30 | 13 | 833 | 790 | 10.82 |
| 1982 | 78 | 35 | 23 | 12 | 856 | 821 | 10.97 |
| 1983 | 79 | 18 | 7 | 11 | 863 | 845 | 10.92 |
| 1984 | 80 | 47 | 28 | 19 | 891 | 844 | 11.14 |
| 1985 | 81 | 18 | 8 | 10 | 899 | 881 | 11.10 |
| 1986 | 82 | 38 | 23 | 15 | 922 | 884 | 11.24 |

Table 8

**Growth Rate of Authors in Different Decades**

| Decade | Authors 'A' | Σ A | Growth Rate per Year |
|---|---|---|---|
| 1905-1909* | 63 | 63 | 10.5 |
| 1910-1919 | 92 | 155 | 9.2 |
| 1920-1929 | 77 | 232 | 7.7 |
| 1930-1939 | 105 | 337 | 10.5 |
| 1940-1949 | 100 | 437 | 10.0 |
| 1950-1959 | 89 | 526 | 8.9 |
| 1960-1969 | 137 | 663 | 13.7 |
| 1970-1979 | 140 | 803 | 14.0 |
| 1980-1986** | 119 | 922 | 19.83 |

* Block of 5 Years
** Block of 6 Years

# Figure 6
# Growth Of The Authors

### Figure 6.1 - Decadewise No. of Authors

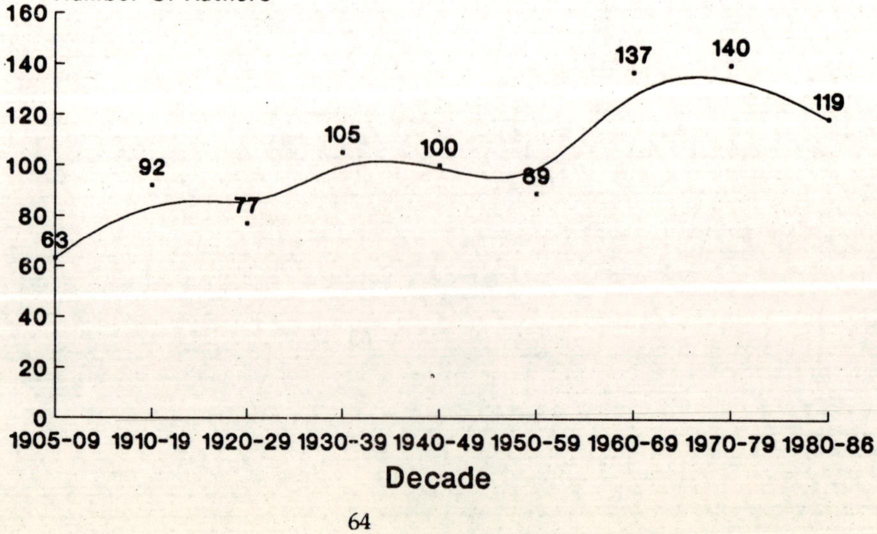

Figure 6.2 - Emerging Trend of Authors

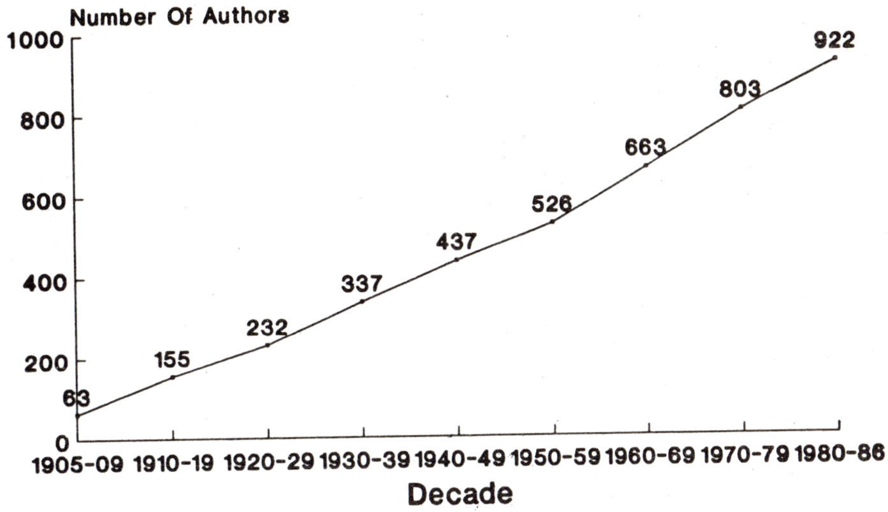


Figure 6.3 - Growth Rate of Authors

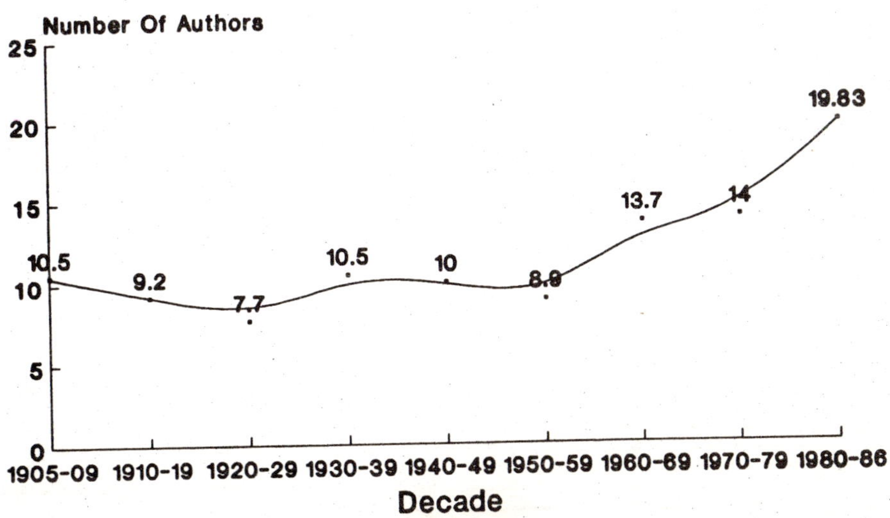

## REFERENCES

1. G.N. Gilbert, "Measuring the Growth of Science : A Review of Indicators of Scientific Growth," *Scientometrics*, 1, (1978), 9-34.

2. Daryl E. Chubin and K.E. Studer, "Knowledge and Structures of Scientific Growth," *Scientometrics*, 1, (1979), 185.

3. *Proceedings of the Seminar on Learned Periodical Publication in India Past, Present and Future* (New Delhi: India International Centre, 1989).

4. Dale Baker, "Recent Trends in the Growth of the Chemical Literature," *Chemical and Engineering*, 54, (1976), 23-27.

5. Kenneth O. May, "Growth and Quality of the Mathematical Literature," *ISIS*, 59, (1968), 363-71.

# 5. AUTHORSHIP PATTERN

Productivity of authors in terms of scientific papers was determined by Alfred J Lotka in 1926 as a quantitative measure of the frequency distribution of scientific productivity (1). According to the general formula for the Lotka's Law, the frequency of authors making i contribution(Ni) is inversely proportional to the number of contributions(i). Mathematically, it can be expressed as:-

$Ni \alpha 1/i$

$Ni=C.i^{-r}$ Where C is proportionality constant and *r* is an exponent constant.

Lotka's equation is determined by three parameters which are stated below:

(i) the number of scientists with minimal productivity (authors with single paper each)

(ii) the maximal productivity of scientist (i max.)

(iii) the characteristic exponent

## 5.1 APPLICATION OF LOTKA'S LAW IN HISTORY OF SCIENCE

To apply the Lotka's Law the author index and listing of co-authorship were generated from the computerised data base on History of Science in India and the frequency of authors making i number of contributions was plotted against the number of contributions as presented in Table 9. The exponent constant and the proportionality constant were calculated to determine the estimated values of authors' frequency and their contributions.

Table 9 reveals that if the frequency of the authors increases, the number of contributions decreases, whereas if the frequency of the authors decreases, the number of contributions increases.

To prove the applicability of the Lotka's Law in History of Science, various constants of Lotka's Law equation were calculated and the Kolmogorov-Smirnov test (2-5) was performed as given below:

$Ni=1/i.C$

$=C.i^{-r}$ Where C is a proportionality constant and *r* is an index constant

for i=1, then i=1
N(1)=C.1
667=C

for i=2
$N(2)=667x2^{-r}=136$
136 =667/2
log 136=log 667 - log2
*r*=log 667 - log 136/log 2
=2.82 - 2.131/0.3 =2.3

for i = 3
$N(3) = 50 = 667 \times 3^{-r}$
*r*=log 667 - log 50/log 3
= 2.82 - 1.69/0.48 = 2.3

for i = 4

N(4) = 28 = 667 x $4^{-r}$

$r$ = log 667 - log 28/log 4

=2.82 - 1.45/0.06=2.28

for i = 5

N(5) = 5 = 667x$5^{-r}$

$r$= log 667 - log 28/log 4

= 2.82 - 1.44/0.6=2.3

N(6) = 12=667x$12^{-r}$

$r$= log 667-log 12/log6

= 2.82-1.07/0.78=2.24

N(7) = 7=667x$7^{-r}$

$r$=log 667-log 7/log 7

= 2.82-0.84/0.84=2.36

N(8) = 3=667x$3^{-r}$

$r$= Log 667 - log 3/log 8

= 2.82-48/0.90=2.6

N(9) = 3=667x$3^{-r}$

$r$=log 667-log 3/log 9

= 2.82=0.48/0.95=2.46

N(10)=1=667x$1^{-r}$

$r$ = log 667-log 1/log 10

2.82-0/1=2.82

N(11)=3=667x$3^{-r}$

$r$=log 667-log 3/log 11

= 2.82-0.48/1.04=2.25

N(12)= 3=667x$3^{-r}$

$r$=log 667-log 3/log 12

=2.82-0.48/1.08=2.17

N(13)= 3=667x$3^{-r}$

$r$ = log 667-log 3/log 13

= 2.82-0.48/1.11=2.11

N(14)=1=667x$1^{-r}$

$r$ = log 667-log 1/log 14

=2.82-0/1.15=2.45

N(15)=1=667x$1^{-r}$

$r$=log 667-log 1/log 15

= 2.82-0/1.18=2.39

N(17)=1=667x$1^{-r}$

$r$ = log 667-log 1/log 17

= 2.82-0/1.23=2.29

N(18)=2=667x$2^{-r}$

$N(18)=2=667x2^{-r}$

$r$=log 667-log 2/log 18

=2.82-0.30/1.25=2.01

$N(20)=1=667x1^{-r}$

$r$=log 667 - log 1/log 20

=2.282-0/1.3=2.17

$N(27)=1=667x1^{-r}$

$r$ = log 667-log 1/log 27

=2.82-0/1.43=1.97

$N(47)=1=667x1^{-r}$

$r$= log 667-log 1/log 47

=2.82-0/1.67=1.69

$N(50)=1=667x1^{-r}$

$r$= log 667-log 1/log 50

=2.82-0/1.69=1.67

The average value of $r$ was found to be 2.27

$\therefore \Sigma r$-2.27

On the basis of $r$=2.27, the estimated frequencies of authors were calculated as given below and presented in Table 10.

Estimated $Ni=C/i^r$ Whereas C=667 and $r$ = 2.27

for i=1

$Ni=667/1^{2.27}=667$

Ni=667/1 =667

Estimated Ni, for i=2

$Ni=677/2^{2.27}$ =667/6.49 = 102

**Table 9**
**Contribution of Authors***

| No. of Contributions/ | No. of Authors | % of Authors | Total Contributions | % of Contributions |
|---|---|---|---|---|
| 1 | 667 | 71.720 | 667 | 38.289 |
| 2 | 136 | 14.624 | 272 | 15.614 |
| 3 | 50 | 5.376 | 150 | 8.611 |
| 4 | 28 | 3.010 | 112 | 6.429 |
| 5 | 5 | 0.538 | 25 | 1.435 |
| 6 | 12 | 1.290 | 72 | 4.133 |
| 7 | 7 | 0.753 | 49 | 2.813 |
| 8 | 3 | 0.322 | 24 | 1.378 |
| 9 | 3 | 0.322 | 27 | 1.550 |
| 10 | 1 | 0.107 | 10 | 0.574 |
| 11 | 3 | 0.322 | 33 | 1.894 |
| 12 | 3 | 0.322 | 36 | 2.066 |
| 13 | 3 | 0.322 | 39 | 2.239 |
| 14 | 1 | 0.107 | 14 | 0.804 |
| 15 | 1 | 0.107 | 15 | 0.861 |
| 17 | 1 | 0.107 | 17 | 0.976 |
| 18 | 2 | 0.215 | 36 | 2.066 |
| 20 | 1 | 0.107 | 20 | 1.148 |
| 27 | 1 | 0.107 | 27 | 1.550 |

| | | | | |
|---|---|---|---|---|
| 47 | 1 | 0.107 | 47 | 2.698 |
| 50 | 1 | 0.107 | 50 | 2.870 |
| Total | 930 | 100 | 1742 | 100 |

* Author May Occur As First Author Or Co-Author

**Table 10**

**Frequency of Estimated Authors**

| No. of Contributions/ Author | Value of *r* | Estimated Frequency of Author | % Authors |
|---|---|---|---|
| 1 | ... | 667 | 75.19 |
| 2 | 2.3 | 102 | 11.49 |
| 3 | 2.3 | 34 | 3.84 |
| 4 | 2.28 | 28 | 3.16 |
| 5 | 2.3 | 17 | 1.93 |
| 6 | 2.24 | 11 | 1.25 |
| 7 | 2.36 | 8 | 0.90 |
| 8 | 2.6 | 5 | 0.56 |
| 9 | 2.46 | 4 | 0.45 |
| 10 | 2.82 | 3 | 0.35 |
| 11 | 2.25 | 2 | 0.22 |
| 12 | 2.17 | 2 | 0.22 |
| 13 | 2.11 | 1 | 0.11 |
| 14 | 2.45 | 1 | 0.11 |
| 15 | 2.39 | 1 | 0.11 |
| 17 | 2.29 | 1 | 0.11 |
| 18 | 2.01 | 0 | 0 |
| 20 | 2.17 | 0 | 0 |
| 27 | 1.97 | 0 | 0 |
| 47 | 1.69 | 0 | 0 |
| 50 | 2.27 | 0 | 0 |
| | Σ 2.27 | 887 | 100 |

The Kolmogorov-Smirnov test used to test where the maximum deviation, D, is

D=max/Fo(X)-Sn(X)/

where Fo(X) is the estimated relative frequency and Sn(X) is the observed relative frequency of a sample of N observations. Since the Kolmogorov-Smirnov D max of 0.035 given in Table 11 is smaller than the level of significance (p=0.01) of 0.534, Lotka's Law was found to be applicable to this sample of authors of History of Science.

In the graphical form, the Lotka's Law was plotted taking both observed and estimated values in Figure 7 as number of articles against the number of authors contributing the articles on a logarithmic scale. The curves followed the Lotka's pattern with slight deviation.

## 5.2 COLLABORATIVE PATTERN

It is the study of scientific co-authorship, research productivity and the professionalisation to form collaborative group(6-11). The collaborative pattern in History of Science was drawn from the author index generated on the author field from the data base on History of Science. The authorship as a whole and amongst highly productive authors in History of Science was presented in Tables 12-13 and Figure 8.

**Table 11**

**Frequency of Observed and Estimated Number of Authors (Kolmogorov-Smirnov test)**

| No. of Contri-but'ons | Observed Cum. Freq. | Observed Relative Freq.Sn(x) | Estimated Cum. Freq. | Estimated Relative Freq.Fo(x) | Deviation D= Fo(x)-Sn(x) |
|---|---|---|---|---|---|
| 1 | 667 | 0.717 | 667 | 0.752 | 0.035 |
| 2 | 803 | 0.863 | 769 | 0.866 | 0.003 |
| 3 | 853 | 0.917 | 803 | 0.905 | 0.012 |
| 4 | 881 | 0.947 | 831 | 0.936 | 0.011 |
| 5 | 886 | 0.952 | 848 | 0.956 | 0.004 |
| 6 | 898 | 0.965 | 859 | 0.968 | 0.003 |
| 7 | 905 | 0.973 | 867 | 0.977 | 0.004 |
| 8 | 908 | 0.976 | 872 | 0.983 | 0.007 |
| 9 | 911 | 0.979 | 876 | 0.987 | 0.008 |
| 10 | 912 | 0.980 | 879 | 0.990 | 0.010 |
| 11 | 915 | 0.983 | 881 | 0.993 | 0.010 |
| 12 | 918 | 0.987 | 883 | 0.995 | 0.008 |
| 13 | 921 | 0.990 | 884 | 0.996 | 0.006 |
| 14 | 922 | 0.991 | 885 | 0.997 | 0.006 |
| 15 | 923 | 0.992 | 886 | 0.998 | 0.006 |
| 17 | 924 | 0.993 | 887 | 1.000 | 0.007 |
| 18 | 926 | 0.995 | | | |
| 20 | 927 | 0.996 | | | |
| 27 | 928 | 0.997 | | | |
| 47 | 929 | 0.998 | | | |
| 50 | 930 | 1.000 | | | |

D=Max | Fo(x)-Sn(x) | 0.035

At the 0.01 level of Significance,

K-S Statistics $= 1.63/\sqrt{N}$ N=930

$=1.63/\sqrt{930}$

$=1.63/30.5$

$=0.0534$ D<0.0534

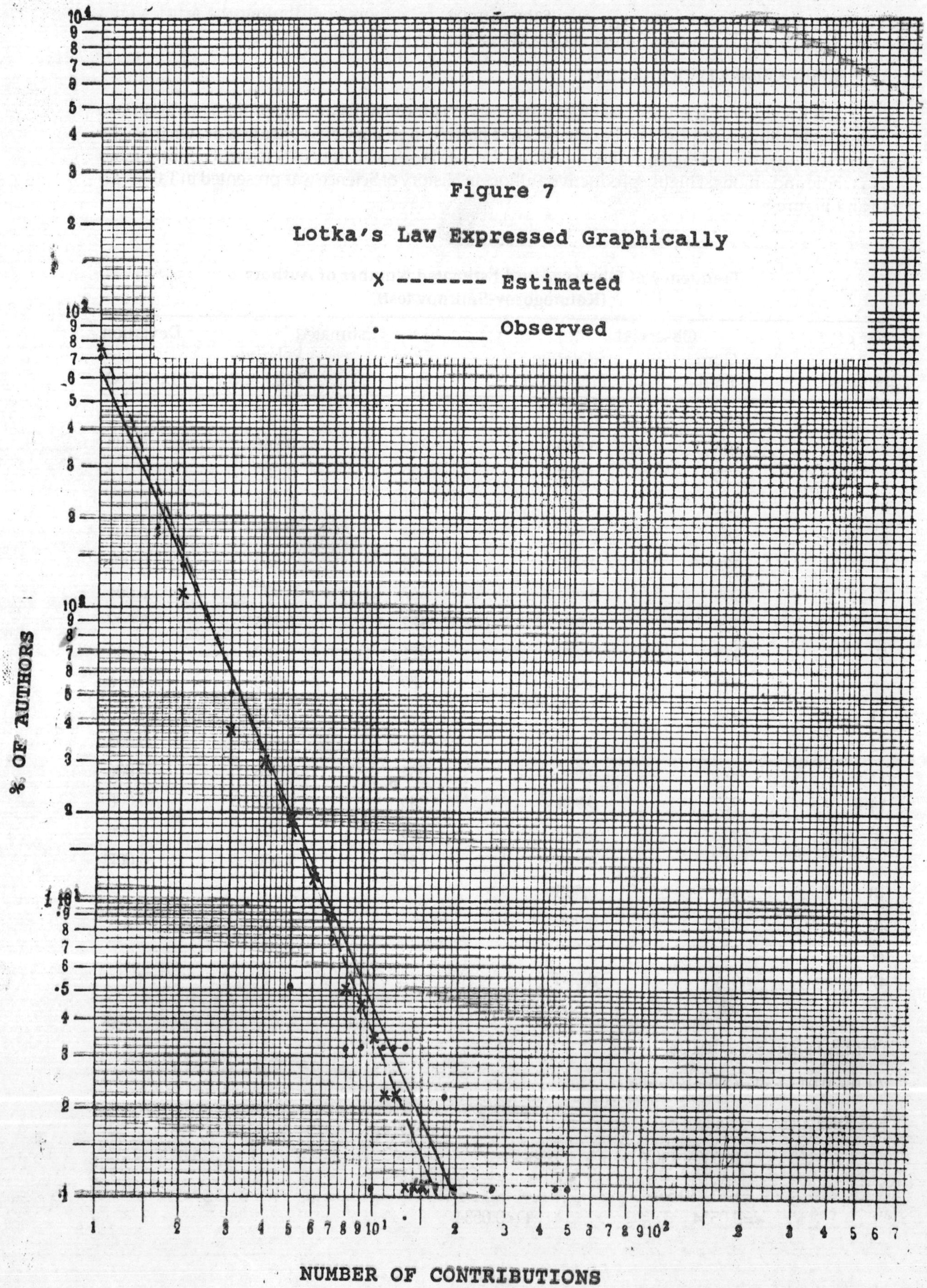

Figure 7
Lotka's Law Expressed Graphically
X ------- Estimated
• _______ Observed
% OF AUTHORS
NUMBER OF CONTRIBUTIONS

It is evident from the collaborative pattern as shown in Table 12 and Figure 8 that contributors to the literature of History of Science in India, worked mostly individually. It was found that 86.871 percent of all the contributions were made by single authors, whereas authors in collaboration whether by two (10.45%), three (1.27%) or four (0.19%) contributed only 11.91%. In total, 1569 contributions were made by 930 contributors, the average contribution per author being 1.65. Nineteen items (1.21%) turned out to be of corporate authorship and it was found that all these items came from the journal 'ISIS'. It was also found that the authors who contributed in collaboration, have contributed singly also.

**Table 12**

**Collaborative Authorship**

| Authorship | # of Contributions | % of Contributions | Cumulative % |
|---|---|---|---|
| Single authors | 1363 | 86.871 | 86.871 |
| Two authors | 164 | 10.452 | 97.323 |
| Three authors | 20 | 1.275 | 98.598 |
| Four authors | 3 | 0.191 | 98.789 |
| Corporate authors | 19 | 1.211 | 100.000 |
| Total | 1569 | | |

**Figure 8**

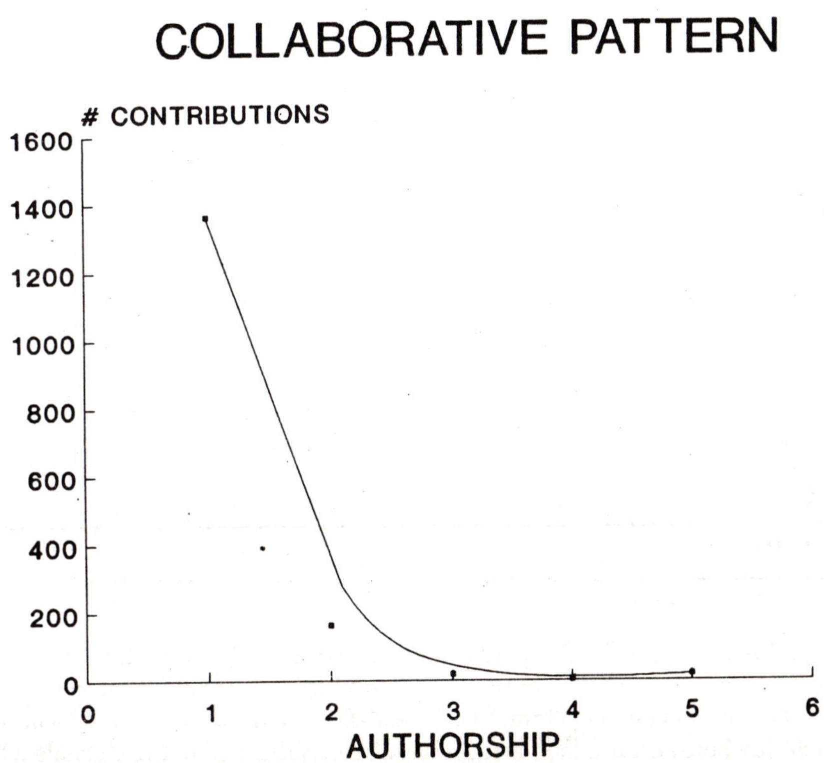


Structure of collaboration in elite is shown in Table 13. In general collaboration occurred in response to professionalisation. Scientists who collaborate exhibit higher productivity. N. Annandale who contributed maximum number of papers in collaboration, was the most productive of all the authors studied, with his total contributions of 50 papers. Similarly, S.L.Hora who ranked second with 6 papers in collaboration was also found to be ranking second among the highly productive authors, with his total of 47 papers. However, there are a few exceptions as well, for example S.D.

Sharma who contributed 5 papers in collaboration, ranked second last in the productivity, with a total of only 11 papers. The next author B. Datta who collaborated in 3 papers, came third last in the productivity. Three authors: A.K.Bag, D.C.Phillott, S.N. Sen collaborated in one paper each.

**Table 13**

**Co-Authorship of Highly Productive Authors**

| Author | No. of Papers as co-author | Total Papers |
|---|---|---|
| Annandale N | 8 | 50 |
| Hora S L | 6 | 47 |
| Sharma SD | 5 | 11 |
| Burkill I H | 4 | 27 |
| Datta B | 3 | 7 |
| Biswas K | 2 | 10 |
| Sen S N | 1 | 11 |
| Phillott D C | 1 | 31 |
| Bag A K | 1 | 13 |
| Chaudhuri S | 0 | 14 |
| Das S R | 0 | 18 |
| Dasgupta HC | 0 | 12 |
| Gershenowitz H | 0 | 17 |
| Gupta R C | 0 | 18 |
| Hooper D | 0 | 13 |
| Mahdihassan S | 0 | 11 |
| Prashad B | 0 | 12 |
| Sarma K V | 0 | 20 |
| Sarton G | 0 | 15 |
| Sengupta P C | 0 | 11 |

But the higher productivity was shown by some high productive authors, who collaborated more; it was due to the nature of the works produced by them. Though the top high productive authors - N. Annandale, S.L. Hora, I.H. Burkill, S.D. Sharma, worked in basic sciences, these have been included here because their articles have been published in the journals which are pioneers to trace the History of Science in India and moreover, their works today got historicity to look into the scientific developments in the beginning of the twentieth century. On the other hand, there are some authors namely S. Chaudhuri, S.R.Das, H.C.Dasgupta who have not at all contributed in collaboration, but have worked in the History of Science, have also made significant contributions, with only 14, 18 and 12 papers respectively.

## REFERENCES


1. A.J.Lotka, "The Frequency Distribution of Scientific Productivity," *Journal of Washington Academy of Sciences*, 16, No.12 (1926), 317-23.

2. Larry J. Murphy, "Lotka's Law in the Humanities?," *Journal of the American Society for Information Science*, 24 (1973), 461-62.

3. Alan E. Schorr, "Lotka's Law and Map Librarianship," *Journal of the American Society for Information Science*, 26 (1975), 189-90.

4. Henery Voos, "Lotka and Information Science," *Journal of the American Society for Information Science*, 25, No.4 (1974), 270-72.

5. Paul D. Allison, and others, "Lotka's Law: A Problem in the Interpretation and Applications," *Social Studies of Science*, 6, No.2 (1976), 269-76.

6. D. B. Beaver and R. Rosen, "Studies in Scientific Collaboration: Part I. The Professional Origins of Scientific Co-authorship," *Scientometrics*, 1, No. 1(1978), 65-84.

7. D.B. Beaver and R. Rosen, "Studies in Scientific Collaboration: Part II. Scientific Co-authorship, Research Productivity and Visibility in the French Scientific Elite 1799-1830," *Scientometrics*, 1, No. 2 (1979), 133-49.

8. D.B. Beaver and R. Rosen, "Studies in Scientific Collaboration: Part III. Professionalisation and the Natural History of Modern Scientific Co-authorship," *Scientometrics*, 1, No. 3 (1979), 231-45.

9. M. Smith, "The Trend Toward Multiple Authorship in Psychology," *American Psychologist*, 13 (1958), 596-99.

10. G. Daniels, "The Process of Professionalisation in American Science: The Emergent Period 1820-1869," *ISIS*, 58 (1967), 151-66.

11. B. L. Clarke, "Multiple Authorship Trends in Scientific Papers," *Science*, 143, No.3608, (1964), 822-24.

# 6. HIGHLY PRODUCTIVE AUTHORS

The contributions of the authors in terms of number of papers published on the subject is an indicator of the productivity of the author. The higher the contribution, more the productivity of the author. The category of highly productive authors depends upon a number of factors such as subject development, number of institutes, workers in the field, time frame, etc. In keeping with the line of 'CSIR Research Output: A Bibliometric Analysis', an author with number of contributions equal to or greater than (= or > 10) was considered to be highly productive.

Out of a total of 922 authors studied on History of Science, 20 authors were found to be highly productive as shown in Table 14 in a decreasing order of productivity, along with the area of specialisation of the author concerned. N. Annandale who acquired first rank contributed 50 papers in the subject; most of his papers dealt with the observations on freshwater fauna. S.L.Hora second in the rank contributed 47 papers; his papers also dealt with the observational aspects of fish. K.V.Sarma fourth in rank basically dealt with the History of Science in the area of astronomy. Besides the Indian authors five foreign authors also contributed to History of Science in India namely, I.H. Burkill, H. Gershenowitz, G.Sarton, D.Hooper, D.C.Phillott with 27, 17, 15, 13 and 13 number of contributions respectively. S.Chaudhuri and G.Sarton contributed on Library and Information Science also. K.Biswas, with 10 papers, mainly on various aspects of Botanical Sciences, was found to be lowest in the rank tally.

**Table 14**

**Ranking of Highly Productive Authors**

| Rank | Name of Author | Number of Papers | Subject of Specialisation |
|---|---|---|---|
| 1 | N.Annandale | 50 | Freshwater Fauna |
| 2 | S.L.Hora | 47 | Fish |
| 3 | I.H.Burkill | 27 | Pollination |
| 4 | K.V.Sarma | 20 | Astronomy - Sources |
| 5 | S.R.Das | 18 | Astronomy - Calendar |
| 5 | R.C.Gupta | 18 | Sine, Trignometry |
| 6 | H.Gershenowitz | 17 | Lamarck |
| 7 | G.Sarton | 15 | Hist. Sci. - Decimal Inf. |
| 8 | S.Chaudhuri | 14 | Lib. Sci. - Bibliography |
| 9 | A.K.Bag | 13 | Sine, Trignometry |
| 9 | D.Hooper | 13 | Medicine |
| 9 | D.C.Phillott | 13 | Falcon |
| 10 | H.C.Dasgupta | 12 | Paleontology |
| 10 | B.Datta | 12 | Algebra |
| 10 | B.Prashad | 12 | Mollusca |
| 10 | S.N.Sen | 12 | Astronomy |

| | | | |
|---|---|---|---|
| 11 | S.Mahdihassan | 11 | Medicine |
| 11 | P.C.Sengupta | 11 | Dating |
| 11 | S.D.Sharma | 11 | Time |
| 12 | K.Biswas | 10 | Botany |

## 6.1 HIGHLY PRODUCTIVE AUTHORS AT A GLANCE

Table 15 presents a birds' eyeview of highly productive authors in terms of number of years an author continued to contribute, the total number of papers contributed during that period, along with total number of pages contributed, total number of references used, total number of journals in which papers were contributed and number of main subject areas studied. Besides this, this table also shows the number of contributions made by an author per year, along with the length of paper and the references used on the average. It was found that only three authors D.C.Phillott, I.H.Burkill, and H. Gershenowitz, contributed more than 2 papers per year on an average, where as only four authors - D.Hooper, S.L.Hora, S.Mahdihassan, and N.Annandale contributed between 2 to 1 paper per year on an average, average length of the paper was found to be ranging from the minimum of 3.7 pages to the maximum of 43.35 pages; whereas number of references used per paper on an average ranged from 1.5 to 69.7. None of the authors contributed their papers in more than 4 journals. H.Gershenowitz with exception contributed each of his 7 papers in only one main subject area, all others contributed in 2 to 5 main subject areas.

## 6.2 DISTRIBUTION OF CONTRIBUTIONS OF INDIVIDUAL AUTHORS

Details of distribution of contributions of the highly productive authors with respect to number of contributions, number of pages, number of references in one year, along with the subject on which the author contributed and the name of journals in which contributions were made are shown individually in twenty Tables (i.e.,16-35) one for each author. Each of these tables reveals gradual growth in number of contributions, number of pages contributed and the number of references used, with the increase in number of years.

Length of contributions in terms of years is shown in Figures 9 to 27. D.C. Phillott has minimum 2 years of contribution but his average contribution per year is the highest (i.e., 6.5) B.Datta contributed for 59 years and S.N. Sen contributed for 53 year with 0.2 average contribution per year. It was found that if the length of contributions in terms of years increases, but either with decreasing number of contributions or with gaps in the contributions, the value of average contribution per year gets decreased as in the case of N.Annandale, S.L.Hora, S.R.Das, G.Sarton, K.Biswas. On the other hand if length of contributions in terms of years gets decreased, but the papers are being published on the peak formation of contribution, then the value of average contribution per year gets increased, as in the case of H.Gershenowitz, D.Hooper, I.H.Burkill, S.Mahdihassan. The growth of publications increased during 1940 - 1950 and 1980 onwards as there were large number of highly productive authors then.

### 6.2.1 Patterns of Contributions

Patterns of contributions yearwise as well as growth in year, along with the emerging trends of contributions are depicted in the form of graphs in Figures (i.e.,from Fig.9 to Fig. 27), one figure for one author, with the exception of D.C.Phillott, who contributed only for 2 years, and did not justify the curve formation. K.V.Sarma shows the distinct normal curve pattern in his contributions - 11th to 15th year, being the peak years of his contributions. Reduced height of the peak for each of S.N.Sen, H.C.Dasgupta, P.C.Sengupta indicates that for each of the three, peak years of contributions were small - just one year in this case. A.K. Bag shows a sudden half peak formation and then a long steady line of contributions, indicating that the peak years of his contributions were the initial ones followed by contributions steadily later on.

Contribution in the mixed curve pattern shows the high productive behaviour of contributions by the authors. N. Annandale, S.L.Hora, G.Sarton, I.H.Burkill, H.Gershenowitz, B.Datta,

R.C.Gupta, and B.Prashad have some similarity in their contribution curves. There are sharp peaks in the beginning of the curve and then afterwards their contributions do not maintain that height, but form small peaks; these peaks also add to the productivity account. H.Gershenowitz shows two peaks in his contributions and a slight downward slope, which indicates his consistency in high productivity through out his creative period.

Contributions of each of the four authors - D.Hooper, S.Mahdihassan, S.D.Sharma, and K.Biswas follow the stray curve patterns indicating that each of these remained steady in the beginning for about 5 to 10 years, but none could achieve peak year of contributions.

Contribution trend in case on K.V. Sarma shows a gradual increase. S.N.Sen and H.C.Dasgupta show a slight downward trend in their contributions. P.C.Sengupta, S.R.Das, S.Chaudhuri show a sharp downward trend in their contributions. A.K. Bag shows a slow downward trend in his contributions.

N.Annandale, S.L.Hora, and I.H.Burkill show a steep downward trend of the contributions, whereas B.Datta and G.Sarton show a slight downward trend in their contributions. H.Gershenowitz is the only exception in whose case the trend remains almost a straight line. All others show an upward trend in their contributions, with sharp upward trend in case of S.D.Sharma.

Except where there is a single peak in the contribution curve, the cumulative graphs do not show the plateau region formation. For K.V.Sarma cumulative contribution graph shows distinct pattern with plateau region at the end. A.K.Bag, S.Mahdihassan, and K.Biswas show a sharp upward straight line at the end of the curve, which indicates the absence of distinct peak years of contributions.

### 6.2.2 Subject Scattering

Subjects mapped in articles of highly productive authors individually are shown in the Figures 28.1 to 28.20. These have been described in terms of three categories - core subjects, related subjects, and alien subjects. Among the highly productive authors, ranking of subject specialists in different subjects is shown in Table 36. As is evident from this table, K.V.Sarma, H.Gershenowitz, I.H.Burkill, G.Sarton, S.Chaudhuri, R.C.Gupta, D.Hooper, and N.Annandale turned out to be ranking first as Astronomer, Biologist, Botanist, Historian of Science, Librarian, Mathematician, Physician, and Zoologist respectively.

It was found that although the authors concentrated mainly in three or four areas in their core subjects, they wrote in related subjects too. Most of them have also glanced into alien subjects, the subjects which do not fall within the ambit of their core subjects. For instance, N. Annandale for whom core subject was Zoology, worte on Bismer Weighing Beam, and Loom Weaving Apparatus also; S.L.Hora fish expert wrote on Scape goat, and Birds; A.K.Bag mathematician also scanned Technology area; S.R.Das astronomer wrote on Plant Fossils; K.V.Sarma astronomer also wrote on Food-Dietary, and Medicinal Properties; B.Prashad zoologist wrote on Artesian, etc.

### 6.2.3 Journal Preference

Journals preferred by highly productive authors for contributing their articles are presented in Table 37 and shown in Figure 29. Journal preference by authors who preferred two or more than two journals for contributing their publications have been shown individually by means of Figures 30.1 to 30.11 also. The overall journal preference for publications adopted by authors turned out to be as JPASB > IJHS > JASBSC > IHQ > VIJ > ISIS > JASBL > JAS with 162, 78, 27, 25, 24, 19, 14, and 5 papers respectively. The JPASB appears to have been preferred the most, because of its being the sole available journal for a long time in the concerned subject, and for the same reason, out of the journals studied, IJHS - the exclusively available journal in the History of Science, ranked second in the order. Out of a total 20 authors found to be highly productive, each of eight authors - A.K. Bag, I.H.Burkill, H.C. Dasgupta, H.Gershenowitz, D.Hooper, D.C.Phillott, B.Prashad, and G.Sarton preferred only one journal for contributing his total publications; the journal preferred was IJHS, JPASB, JPASB, IJHS, JPASB, JPASB, JPASB, ISIS respectively. Each of six authors - K.Biswas, R.C.Gupta, S.Mahdihassan, S.D.Sharma, P.C.Sengupta, K.V.Sarma chose only two journals for

## Table 15

## Highly Productive Authors at a Glance

| AUTHOR | DURATION OF CONTRIBUTION IN YEARS | NO. OF CONTRIBUTIONS | AV. CONTRIBUTION | TOTAL NO. OF PAGES | AV. LENGTH OF PAPER | TOTAL NO. OF REFERENCES USED | AV. REFERENCES PER PAPER | JOURNAL DISTRIBUTION | NO. OF MAIN SUBJECT AREAS STUDIED |
|---|---|---|---|---|---|---|---|---|---|
| ANNANDALE N | 46 | 50 | 1.087 | 515 | 10.300 | 284 | 5.680 | 1 | 4 |
| BAG AK | 21 | 13 | 0.619 | 110 | 8.500 | 249 | 19.200 | 1 | 4 |
| BISWAS K | 25 | 10 | 0.400 | 162 | 16.200 | 188 | 18.800 | 2 | 4 |
| BURKILL IH | 11 | 27 | 2.454 | 272 | 10.074 | 51 | 1.890 | 1 | 3 |
| CHAUDHURI S | 22 | 14 | 0.636 | 173 | 12.360 | 976 | 69.714 | 4 | 4 |
| DAS SR | 45 | 18 | 0.400 | 229 | 12.722 | 415 | 23.050 | 4 | 4 |
| DASGUPTA HC | 23 | 12 | 0.521 | 44 | 3.700 | 153 | 12.750 | 1 | 2 |
| DATTA B | 59 | 11 | 0.203 | 259 | 23.545 | 655 | 59.545 | 4 | 2 |
| GERSHENOWITZ H | 7 | 17 | 2.428 | 141 | 8.300 | 371 | 21.820 | 1 | 1 |
| GUPTA RC | 20 | 18 | 0.900 | 186 | 10.333 | 359 | 19.940 | 2 | 5 |
| HOOPER D | 7 | 13 | 1.857 | 67 | 5.154 | 19 | 1.500 | 1 | 2 |
| HORA SL | 47 | 47 | 1.388 | 400 | 8.511 | 397 | 8.450 | 3 | 4 |
| MAHDIHASSAN S | 9 | 11 | 1.222 | 60 | 5.454 | 129 | 11.730 | 2 | 3 |
| PHILLOTT DC | 2 | 31 | 6.500 | 80 | 6.154 | 99 | 7.615 | 1 | 2 |
| PRASHAD B | 15 | 12 | 0.800 | 54 | 4.500 | 68 | 5.700 | 1 | 2 |
| SARMA KV | 23 | 20 | 0.869 | 867 | 43.350 | 304 | 15.200 | 2 | 3 |
| SARTON G | 38 | 15 | 0.394 | 174 | 11.600 | 187 | 12.470 | 1 | 2 |
| SEN SN | 53 | 11 | 0.207 | 225 | 20.454 | 540 | 49.100 | 3 | 2 |
| SENGUPTA PC | 20 | 11 | 0.550 | 148 | 13.454 | 238 | 21.640 | 2 | 4 |
| SHARMA SD | 14 | 11 | 0.786 | 129 | 11.730 | 349 | 31.730 | 2 | 3 |

## Table 16

### Distribution of Contributions of N. Annandale

| Year of Contribution (Y) | ΣY | No. of Contribution (C) | ΣC | No. of Pages (P) | ΣP | No. of Reference (R) | ΣR | Journal | Subject |
|---|---|---|---|---|---|---|---|---|---|
| 1905 | 1 | 5 | 5 | 4 | 4 | 2 | 2 | JPASB | Worm |
| | | | | 60 | 64 | 2 | 4 | JPASB | Snakes |
| | | | | 7 | 71 | 3 | 7 | JPASB | snakes |
| | | | | 6 | 77 | 4 | 11 | JPASB | Dugong |
| | | | | 3 | 80 | | 11 | JPASB | Birds |
| 1906 | 2 | 14 | 19 | 10 | 90 | | 11 | JPASB | Freshwater fauna |
| | | | | 4 | 94 | | 11 | JPASB | Freshwater fauna |
| | | | | 2 | 96 | | 11 | JPASB | Freshwater fauna |
| | | | | 4 | 100 | 2 | 13 | JPASB | Tortoise, Herpetology |
| | | | | 2 | 102 | | 13 | JPASB | Indo-pacific water animals |
| | | | | 2 | 104 | | 13 | JPASB | Herpetology |
| | | | | 4 | 108 | 2 | 15 | JPASB | Freshwater fauna |
| | | | | 14 | 122 | | 15 | JPASB | Insects |
| | | | | 1 | 123 | | 15 | JPASB | Earwigs |
| | | | | 3 | 126 | 5 | 20 | JPASB | Freshwater fauna |
| | | | | 8 | 134 | 9 | 29 | JPASB | Freshwater fauna |
| | | | | 4 | 138 | 9 | 38 | JPASB | Freshwater fauna |
| | | | | 5 | 143 | 4 | 42 | JPASB | Freshwater fauna |
| | | | | 2 | 145 | | 42 | JPASB | Zoology new species |
| 1907 | 3 | 5 | 24 | 12 | 157 | | 42 | JPASB | Sponges-freshwater |
| | | | | 2 | 159 | 1 | 43 | JPASB | Freshwater fauna |
| | | | | 3 | 162 | 1 | 44 | JPASB | Zoology |
| | | | | 3 | 165 | 2 | 46 | JPASB | Freshwater fauna |
| | | | | 11 | 176 | 12 | 58 | JPASB | freshwater fauna |
| 1908 | 4 | 1 | 25 | 2 | 178 | | 58 | JPASB | Diptera |
| 1912 | 8 | 1 | 26 | 15 | 193 | 28 | 86 | JPASB | Freshwater fauna of India - Re |
| 1913 | 9 | 8 | 34 | 1 | 194 | 9 | 95 | JPASB | Polyzoa |
| | | | | 7 | 201 | 2 | 97 | JPASB | Lake Tiberias-Biology |
| | | | | 4 | 205 | 4 | 101 | JPASB | Leeches |
| | | | | 2 | 207 | | 101 | JPASB | Sponge-larva |
| | | | | 18 | 225 | 11 | 112 | JPASB | Crabs |
| | | | | 11 | 236 | | 112 | JPASB | Tiberias lake |
| | | | | 34 | 270 | 13 | 125 | JPASB | Geology and fauna of limestone |
| | | | | 32 | 302 | 36 | 161 | JPASB | Sponges of Tiberias |
| 1914 | 10 | 1 | 35 | 8 | 310 | 4 | 165 | JPASB | Mud-Turttes |
| 1915 | 11 | 1 | 36 | 40 | 350 | 72 | 237 | JPASB | Lake Tiberias-Biology |
| 1918 | 14 | 2 | 38 | 2 | 352 | 1 | 238 | JPASB | Bismer weighing beam-Darjeelin |
| | | | | 6 | 358 | 14 | 252 | JPASB | Mollusca |
| 1919 | 15 | 1 | 39 | 31 | 389 | 1 | 253 | JPASB | Vegetation of Seistan |
| 1920 | 16 | 3 | 42 | 3 | 392 | | 253 | JPASB | Loom weaving apparatus |
| | | | | 7 | 399 | 3 | 256 | JPASB | Mollusca |
| | | | | 16 | 415 | 7 | 263 | JPASB | Anthropometric |
| 1922 | 18 | 2 | 44 | 28 | 443 | | 263 | JPASB | Freshwater fauna-Advances in |
| | | | | 3 | 447 | | 263 | JPASB | Mollusca |
| 1923 | 19 | 3 | 47 | 3 | 450 | | 263 | JPASB | Fish-Review of Americans works |
| | | | | 13 | 463 | | 263 | JPASB | Mollusca |
| | | | | 24 | 487 | 11 | 274 | JPASB | Mollusca |
| 1950 | 46 | 3 | 50 | 13 | 500 | 7 | 281 | JPASB | Herpetology-oriental |
| | | | | 2 | 502 | 3 | 284 | JPASB | Hydra |
| | | | | 13 | 515 | | 284 | JPASB | Herpetology-oriental |
| Total | 46 | | 50 | | 515 | | 284 | | |
| Average | — | — | 1.087 | 10.3 | | 5.68 | | | |

## Table 17

## Distribution of Contributions of A.K. Bag

| Year of Contribution (Y) | $\Sigma$Y | No. of Contributions (C) | $\Sigma$C | No. of Pages (P) | $\Sigma$P | No. of References (R) | $\Sigma$R | Journal | Subject |
|---|---|---|---|---|---|---|---|---|---|
| 1966 | 1 | 2 | 2 | 7 | 7 | 13 | 13 | IJHS | Binomial theorem |
| | | | | 9 | 16 | 17 | 30 | IJHS | Trignometrical series |
| 1969 | 4 | 2 | 4 | 4 | 20 | 7 | 37 | IJHS | Sources in Astron.& math |
| | | | | 7 | 27 | 19 | 56 | IJHS | Sine table |
| 1972 | 7 | 1 | 5 | 4 | 31 | 3 | 59 | IJHS | Sine table |
| 1975 | 10 | 1 | 6 | 11 | 42 | 29 | 88 | IJHS | Maths-Al-Biruni |
| 1976 | 11 | 1 | 7 | 4 | 46 | 7 | 95 | IJHS | Sin, cosine series |
| 1977 | 12 | 1 | 8 | 16 | 62 | 32 | 127 | IJHS | Indeterminate equations |
| 1980 | 15 | 1 | 9 | 15 | 77 | 8 | 135 | IJHS | Biblio.-Indian liter.1400-1800 |
| 1982 | 17 | 1 | 10 | 9 | 86 | 28 | 163 | IJHS | Technology |
| 1984 | 19 | 1 | 11 | 9 | 95 | 11 | 174 | IJHS | Indeterminate equations |
| 1985 | 20 | 1 | 12 | 9 | 104 | 58 | 232 | IJHS | Astronomy |
| 1986 | 21 | 1 | 13 | 6 | 110 | 17 | 249 | IJHS | Ibn Sina |
| Total — — — | 21 | | 13 | | 110 | | 249 | | |
| Average — — — — — | | 0.619 | | 8.5 | | 19.2 | | | |

## Table 18

## Distribution of Contributions of K. Biswas

| Year of Contribution (Y) | $\Sigma$Y | No. of Contributions (C) | $\Sigma$C | No. of Pages (P) | $\Sigma$P | No. of References (R) | $\Sigma$R | Journal | Subject |
|---|---|---|---|---|---|---|---|---|---|
| 1924 | 1 | 1 | 1 | 8 | 8 | 0 | 0 | JPASB | Algae |
| 1930 | 7 | 1 | 2 | 8 | 16 | 41 | 41 | JPASB | Filtered water - organisms |
| 1932 | 9 | 1 | 3 | 19 | 35 | 0 | 41 | JPASB | Conifers |
| 1936 | 12 | 1 | 4 | 5 | 40 | 0 | 41 | JPASB | Diatoms |
| 1942 | 18 | 1 | 5 | 4 | 44 | 0 | 41 | JASBSC | Algae |
| 1943 | 19 | 1 | 6 | 5 | 49 | 6 | 47 | JASBSC | Fern |
| 1946 | 21 | 1 | 7 | 38 | 87 | 45 | 92 | JASBSC | Oil yielding tree |
| 1947 | 22 | 1 | 8 | 40 | 127 | 52 | 144 | JASBSC | Lichen |
| 1950 | 25 | 2 | 10 | 19 | 146 | 24 | 168 | JASBSC | Seaweed |
| | | | | 16 | 162 | 20 | 188 | JASBSC | Dioscoreas - dietary |
| Total — — — | 25 | | 10 | | 162 | | 188 | | |
| Average — — — — — | | 0.4 | | 16.2 | | 18.8 | | | |

## Table 19

### Distribution of Contributions of I.H. Burkill

| Year of Contribution (Y) | ΣY | No. of Contributions (C) | ΣC | No. of Pages (P) | ΣP | No. of References (R) | ΣR | Journal | Subject |
|---|---|---|---|---|---|---|---|---|---|
| 1906 | 1 | 6 | 6 | 3 | 3 | 9 | 9 | JPASB | Parasite plant |
| | | | | 19 | 22 | | 9 | JPASB | Plant |
| | | | | 19 | 41 | | 9 | JPASB | Swertia-plant |
| | | | | 4 | 45 | 1 | 10 | JPASB | Pollination |
| | | | | 6 | 51 | | 10 | JPASB | Pollination |
| | | | | 5 | 56 | | 10 | JPASB | Pollination |
| 1907 | 2 | 9 | 15 | 20 | 76 | 4 | 14 | JPASB | Inflorescence |
| | | | | 4 | 80 | | 14 | JPASB | Swertia |
| | | | | 6 | 86 | | 14 | JPASB | Mango |
| | | | | 2 | 88 | | 14 | JPASB | Bamboo-fish trap |
| | | | | 10 | 98 | 5 | 19 | JPASB | Pollination-cotton |
| | | | | 3 | 101 | 1 | 20 | JPASB | Inflorescence |
| | | | | 2 | 103 | | 20 | JPASB | Fruit |
| | | | | 2 | 105 | 1 | 21 | JPASB | Dye plant |
| | | | | 6 | 111 | | 22 | JPASB | Leaf structure |
| 1908 | 3 | 2 | 17 | 17 | 128 | 6 | 28 | JPASB | Pollination |
| | | | | 35 | 163 | 3 | 31 | JPASB | Pollination |
| 1909 | 4 | 4 | 21 | 5 | 168 | 4 | 35 | JPASB | Drug L. dentata |
| | | | | 15 | 183 | | 35 | JPASB | Drug-coptis |
| | | | | 5 | 188 | | 35 | JPASB | Oil plant |
| | | | | 7 | 195 | | 35 | JPASB | Pollination cotton |
| 1911 | 6 | 3 | 24 | 2 | 197 | 2 | 37 | JPASB | Jute plant |
| | | | | 3 | 200 | | 37 | JPASB | Bulbils |
| | | | | 2 | 202 | | 37 | JPASB | Swertias |
| 1914 | 9 | 1 | 25 | 37 | 239 | 5 | 42 | JPASB | Dioscoreas plant |
| 1916 | 11 | 2 | 27 | 27 | 266 | 9 | 51 | JPASB | Pollination |
| | | | | 6 | 272 | | 51 | JPASB | Terai forests |
| Total — — — | 11 | | 27 | | 272 | | 51 | | |
| Average — — — — | | 2.454 | | 10.074 | | 1.89 | | | |

## Table 20

### Distribution of Contributions of S. Chaudhuri

| Year of Contribution (Y) | ΣY | No. of Contributions (C) | ΣC | No. of Pages (P) | ΣP | No. of References (R) | ΣR | Journal | Subject |
|---|---|---|---|---|---|---|---|---|---|
| 1950 | 1 | 1 | 1 | 26 | 26 | | 0 | JASBL | Bibliography |
| 1952 | 3 | 3 | 4 | 7 | 33 | 22 | 22 | IHQ | Fauna in Ramayana |
| | | | | 8 | 41 | 56 | 78 | IHQ | Fauna in Ramayana |
| | | | | 10 | 51 | 102 | 180 | IHQ | Fauna in Ramayana |
| 1953 | 4 | 4 | 8 | 8 | 59 | 162 | 342 | IHQ | Fauna in Ramayana |
| | | | | 10 | 69 | 189 | 531 | IHQ | Fauna in Ramayana |
| | | | | 9 | 78 | 230 | 761 | IHQ | Fauna in Ramayana |
| | | | | 8 | 86 | 128 | 889 | IHQ | Fauna in Ramayana |
| 1954 | 5 | 1 | 9 | 6 | 92 | 37 | 926 | IHQ | Fauna in Ramayana |
| 1956 | 7 | 2 | 11 | 43 | 135 | | 926 | JASBSC | Bibliography |
| | | | | * | | * | | JAS | Index to AS 1788-1953 |
| 1961 | 12 | 1 | 12 | 25 | 160 | 5 | 931 | IHQ | Place-name |
| 1970 | 21 | 1 | 13 | * | 160 | | 931 | JAS | Index to AS 1954-1968 |
| 1971 | 22 | 1 | 14 | 13 | 173 | 45 | 976 | JAS | Raw silk - trade |
| Total — — — | 22 | | 14 | | 173 | | 976 | | |
| Average — — — — — | | 0.636 | | 12.36 | | 69.714 | | | |

* Index

## Table 21

### Distribution of Contributions of S.R. Das

| Year of Contribution (Y) | ΣY | No. of Contributions (C) | ΣC | No. of Pages (P) | ΣP | No. of References (R) | ΣR | Journal | Subject |
|---|---|---|---|---|---|---|---|---|---|
| 1925 | 1 | 1 | 1 | 20 | 20 | | 0 | IHQ | Astronomy - History |
| 1927 | 3 | 4 | 5 | 11 | 31 | 24 | 24 | JPASB | Equinoxes |
| | | | | 20 | 51 | 36 | 60 | IHQ | Numerals |
| | | | | 5 | 56 | 9 | 69 | IHQ | Astronomy |
| | | | | 23 | 79 | 48 | 117 | IHQ | Numerals |
| 1928 | 4 | 5 | 10 | 18 | 97 | 23 | 140 | JPASB | Eclipses |
| | | | | 14 | 111 | 53 | 193 | IHQ | Astronomy - Instruments |
| | | | | 29 | 140 | 103 | 296 | IHQ | Calendar |
| | | | | 14 | 154 | 42 | 338 | IHQ | Seasons |
| | | | | 10 | 164 | | 338 | IHQ | Astronomy |
| 1930 | 6 | 1 | 11 | 15 | 179 | | 338 | ISIS | Astronomy |
| 1931 | 7 | 1 | 12 | 13 | 192 | 19 | 357 | IHQ | Astronomy - Chronology |
| 1932 | 8 | 2 | 14 | 13 | 205 | 32 | 389 | IHQ | Astronomy - Jaina |
| | | | | 6 | 211 | 10 | 399 | IHQ | Astronomy - Jaina |
| 1933 | 9 | 2 | 16 | 6 | 217 | 3 | 402 | ISIS | Calendar |
| | | | | 5 | 222 | 11 | 413 | IHQ | Time |
| 1934 | 10 | 1 | 17 | 5 | 227 | 2 | 415 | IHQ | Calendar |
| 1969 | 45 | 1 | 18 | 2 | 229 | | 415 | JAS | Plant fossil |
| Total — — — | 45 | | 18 | | 229 | | 415 | | |
| Average — — — — | | 0.4 | | 12.722 | | 23.05 | | | |

## Table 22

### Distribution of Contributions of H.C. Dasgupta

| Year of Contribution (Y) | ΣY | No. of Contributions (C) | ΣC | No. of Pages (P) | ΣP | No. of References (R) | ΣR | Journal | Subject |
|---|---|---|---|---|---|---|---|---|---|
| 1908 | 1 | 1 | 1 | 3 | 3 | 3 | 3 | JPASB | Geology/Hill |
| 1909 | 2 | 1 | 2 | 1 | 4 | 3 | 6 | JPASB | Cretaceous rocks |
| 1910 | 3 | 2 | 4 | 2 | 6 | 7 | 13 | JPASB | Palaeontological-river |
| | | | | 3 | 9 | | 13 | JPASB | - |
| 1912 | 5 | 1 | 5 | 2 | 11 | 5 | 18 | JPASB | Crystallized slag |
| 1915 | 8 | 1 | 6 | 5 | 16 | 17 | 35 | JPASB | Palaeontological notes |
| 1917 | 10 | 1 | 7 | 6 | 22 | 30 | 65 | JPASB | Palaeontology |
| 1923 | 16 | 1 | 8 | 6 | 28 | 23 | 88 | JPASB | Fossil Pectinidae |
| 1924 | 17 | 1 | 9 | 3 | 31 | 11 | 99 | JPASB | Fossil |
| 1926 | 19 | 1 | 10 | 3 | 34 | 6 | 105 | JPASB | Palaeontology |
| 1928 | 21 | 1 | 11 | 7 | 41 | 35 | 140 | JPASB | Fossil in river |
| 1930 | 23 | 1 | 12 | 3 | 44 | 13 | 153 | JPASB | Fossil-Dinosaur |
| Total — — — | 23 | | 12 | | 44 | | 153 | | |
| Average — — — — | | 0.521 | | | 3.7 | 12.75 | | | |

## Table 23

## Distribution of Contributions of B. Datta

| Year of Contribution (Y) | ΣY | No. of Contributions (C) | ΣC | No. of Pages (P) | ΣP | No. of References (R) | ΣR | Journal | Subject |
|---|---|---|---|---|---|---|---|---|---|
| 1926 | 1 | 1 | 1 | 18 | 18 | 84 | 84 | JPASB | Pie |
| 1927 | 2 | 2 | 3 | 7 | 25 | 18 | 102 | JPASB | Arithmetical operations |
| 1929 | 4 | 1 | 4 | 11 | 36 | 11 | 113 | IHQ | Numbers |
| 1930 | 5 | 2 | 6 | 34 | 70 | 105 | 218 | IHQ | Hindu Ganita - Development |
| 1932 | 7 | 1 | 7 | 8 | 78 | 56 | 274 | JPASB | Geometrical figures - rect. |
| 1933 | 8 | 1 | 8 | 10 | 88 | 36 | 310 | IHQ | Bhaskar |
| 1980 | 55 | 1 | 9 | 9 | 97 | 32 | 342 | ISIS | Mahavira to Sridhara |
| 1983 | 58 | 1 | 10 | 14 | 111 | 12 | 354 | ISIS | Algebra - Narayana |
| 1984 | 59 | 1 | 11 | 68 | 179 | 278 | 632 | IJHS | Geometry - Hindu |
| | | | | 70 | 249 | 0 | 632 | IJHS | Hindu trignometry |
| | | | | 10 | 259 | 23 | 655 | IJHS | Caiculus |
| Total — — — | 59 | | 11 | | 259 | | 655 | | |
| Average — — — — | | 0.203 | | 23.545 | | 59.545 | | | |

## Table 24

## Distribution of Contributions of H. Gershenowitz

| Year of Contribution (Y) | ΣY | No. of Contributions (C) | ΣC | No. of Pages (P) | ΣP | No. of References (R) | ΣR | Journal | Subject |
|---|---|---|---|---|---|---|---|---|---|
| 1978 | 1 | 3 | 3 | 7 | 7 | 41 | 41 | IJHS | Lamarckism |
| | | | | 5 | 12 | 11 | 52 | IJHS | Neo-lamarckians |
| | | | | 6 | 18 | 19 | 71 | IJHS | Lamarck |
| 1979 | 2 | 4 | 7 | 9 | 27 | 25 | 96 | IJHS | Lamarckian-psychoanal.theory |
| | | | | 8 | 35 | 20 | 116 | IJHS | Neo-lamarckian |
| | | | | 9 | 44 | 24 | 140 | IJHS | Lamarck |
| | | | | 6 | 50 | 20 | 160 | IJHS | Darwinist |
| 1980 | 3 | 2 | 9 | 6 | 56 | 14 | 174 | IJHS | Lamarck |
| | | | | 6 | 62 | 12 | 186 | IJHS | Lamarck and Napoleon |
| 1981 | 4 | 1 | 10 | 9 | 71 | 23 | 209 | IJHS | Neo-lamarckian |
| 1983 | 6 | 4 | 14 | 8 | 79 | 16 | 225 | IJHS | Lamarckian |
| | | | | 17 | 96 | 43 | 268 | IJHS | Lamarckian |
| | | | | 8 | 104 | 26 | 294 | IJHS | Lamarckian |
| | | | | 12 | 116 | 24 | 318 | IJHS | Lamarck and Encyclopedias |
| 1984 | 7 | 3 | 17 | 4 | 120 | 13 | 331 | IJHS | Neo-lamarckian |
| | | | | 11 | 131 | 13 | 344 | IJHS | Lamarck |
| | | | | 10 | 141 | 27 | 371 | IJHS | Lamarck |
| Total — — — | 7 | | 17 | | 141 | | 371 | | |
| Average — — — — | | 2.428 | | 8.3 | | 21.82 | | | |

## Table 25

## Distribution of Contributions of R.C. Gupta

| Year of Contribution (Y) | $\Sigma Y$ | No. of Contributions (C) | $\Sigma C$ | No. of Pages (P) | $\Sigma P$ | No. of References (R) | $\Sigma R$ | Journal | Subject |
|---|---|---|---|---|---|---|---|---|---|
| 1967 | 1 | 1 | 1 | 16 | 16 | 23 | 23 | IJHS | Sine |
| 1969 | 3 | 1 | 2 | 13 | 29 | 33 | 56 | IJHS | Interpolation II order |
| 1971 | 5 | 1 | 3 | 9 | 38 | 9 | 65 | IJHS | Sine |
| 1972 | 6 | 1 | 4 | 6 | 44 | 10 | 75 | IJHS | Sine |
| 1974 | 8 | 3 | 7 | 8 | 52 | 6 | 81 | IJHS | Sine, Cosine, Arcs |
| | | | | 14 | 66 | 18 | 99 | IJHS | Sine,cosine-Add & sub theories |
| | | | | 14 | 80 | 7 | 106 | IJHS | Triangle |
| 1975 | 9 | 1 | 8 | 9 | 89 | 20 | 126 | IJHS | Circumference |
| 1976 | 10 | 1 | 9 | 10 | 99 | 15 | 141 | IJHS | Sine |
| 1977 | 11 | 1 | 10 | 7 | 106 | 22 | 163 | IJHS | Mathematical rules |
| 1978 | 12 | 1 | 11 | 17 | 123 | 64 | 227 | IJHS | Sinus totus |
| 1979 | 13 | 2 | 13 | 7 | 130 | 4 | 231 | IJHS | Interpolation II order |
| | | | | 8 | 138 | 13 | 244 | VIJ | Time |
| 1980 | 14 | 2 | 15 | 6 | 144 | 16 | 260 | IJHS | Jyotipatti |
| | | | | 6 | 150 | 10 | 270 | VIJ | Citrabhanu |
| 1983 | 17 | 1 | 16 | 16 | 166 | 59 | 329 | IJHS | Numerals |
| 1986 | 20 | 2 | 18 | 9 | 175 | 15 | 344 | IJHS | Pie |
| | | | | 11 | 186 | 15 | 359 | IJHS | Equalization |
| Total — — — | 20 | | 18 | | 186 | | 359 | | |
| Average — — — — — | | 0.9 | | 10.333 | | 19.94 | | | |

## Table 26

## Distribution of Contributions of D. Hooper

| Year of Contribution (Y) | $\Sigma Y$ | No. of Contributions (C) | $\Sigma C$ | No. of Pages (P) | $\Sigma P$ | No. of References (R) | $\Sigma R$ | Journal | Subject |
|---|---|---|---|---|---|---|---|---|---|
| 1906 | 1 | 1 | 1 | 3 | 3 | 1 | 1 | JPASB | Pottery |
| 1907 | 2 | 2 | 3 | 4 | 7 | | | JPASB | Well-water |
| | | | | 3 | 10 | | | JPASB | Fats (Carbohydrates) |
| 1908 | 3 | 2 | 5 | 2 | 12 | | | JPASB | Fats (Carbohydrates) |
| | | | | 1 | 13 | | | JPASB | Oil |
| 1909 | 4 | 2 | 7 | 6 | 19 | 12 | 13 | JPASB | |
| | | | | 4 | 23 | | | JPASB | |
| 1910 | 5 | 2 | 9 | 3 | 26 | | | JPASB | Medicinal lizards |
| | | | | 16 | 42 | | | JPASB | Materia medica |
| 1911 | 6 | 3 | 12 | 10 | 52 | | | JPASB | Phosphorus |
| | | | | 6 | 58 | | | JPASB | Yams |
| | | | | 5 | 63 | 5 | 18 | JPASB | Milk-products |
| 1912 | 7 | 1 | 13 | 4 | 67 | 1 | 19 | JPASB | Ash |
| Total — — — | 7 | | 13 | | 67 | | 19 | | |
| Average — — — — — | | 1.857 | | 5.154 | | 1.5 | | | |

## Table 27

## Distribution of Contributions of S.L. Hora

| Year of Contribution (Y) | Σ Y | No. of Contributions (C) | Σ C | No. of Pages (P) | Σ P | No. of References (R) | Σ R | Journal | Subject |
|---|---|---|---|---|---|---|---|---|---|
| 1922 | 1 | 3 | 3 | 4 | 4 | 3 | 3 | JPASB | Fish |
| | | | | 3 | 7 | | 3 | JPASB | Fish |
| | | | | 7 | 14 | 21 | 24 | JPASB | Fish-Tadpoles |
| 1923 | 2 | 3 | 6 | 3 | 17 | | 25 | JPASB | Fish & fossil - Rev. |
| | | | | 5 | 22 | 5 | 30 | JPASB | Garra |
| | | | | 9 | 31 | 10 | 40 | JPASB | Geckos & Tree-frogs |
| 1926 | 5 | 9 | 15 | 9 | 40 | | 40 | JPASB | Fish drawings |
| | | | | 2 | 42 | 2 | 42 | JPASB | |
| | | | | 1 | 43 | | 42 | JPASB | Scapegoat |
| | | | | 6 | 49 | 6 | 48 | JPASB | Fish |
| | | | | 4 | 53 | 6 | 54 | JPASB | Loach-Hermaphrodite |
| | | | | 4 | 57 | 3 | 57 | JPASB | Implements-fishing |
| | | | | 2 | 59 | 1 | 58 | JPASB | Polyzoon |
| | | | | 6 | 65 | 7 | 65 | JPASB | Fish drawing |
| | | | | 17 | 82 | 11 | 76 | JPASB | |
| 1927 | 6 | 3 | 18 | 3 | 85 | 2 | 78 | JPASB | Reprod.-Lunar period (Insects) |
| | | | | 2 | 87 | | 78 | JPASB | Birds |
| | | | | 3 | 90 | | 78 | JPASB | Fish |
| 1928 | 7 | 1 | 19 | 4 | 94 | | 78 | JPASB | Loach |
| 1931 | 10 | 2 | 21 | 13 | 107 | 4 | 82 | JPASB | Fish Manuscript |
| | | | | ? | | ? | | JPASB | |
| 1932 | 11 | 1 | 22 | 9 | 116 | | 82 | JPASB | Fish-trap |
| 1933 | 12 | 3 | 25 | 4 | 120 | 7 | 89 | JPASB | Chemical action of mucus-fish |
| | | | | 7 | 127 | 12 | 101 | JPASB | Wild animals-worship |
| | | | | 16 | 143 | 11 | 112 | JPASB | Rains of fishes in India |
| 1934 | 13 | 1 | 26 | 15 | 158 | | 112 | JPASB | Jiol Machh-Trade(Live fish) |
| 1935 | 14 | 1 | 27 | 7 | 165 | 15 | 127 | JASBSC | Fish-locomotion |
| 1940 | 19 | 2 | 29 | 7 | 172 | 6 | 133 | JASBSC | Fish taranomy |
| | | | | 20 | 192 | 41 | 174 | JASBSC | Hilsa fish |
| 1942 | 21 | 1 | 30 | 8 | 200 | 10 | 184 | JASBSC | Fish |
| 1943 | 22 | 1 | 31 | 15 | 215 | 15 | 199 | JASBSC | Fish from Iraq |
| 1945 | 24 | 1 | 32 | 19 | 234 | 11 | 210 | JASBSC | Fish-farms |
| 1948 | 27 | 2 | 34 | 6 | 240 | 5 | 215 | JASBSC | Sanskrit names of fish |
| | | | | 4 | 244 | 3 | 218 | JASBSC | Fish & fisheries-Anci.Hindu |
| 1950 | 29 | 1 | 35 | 14 | 258 | 17 | 235 | JASBSC | Fish & fisheries-Anci Hindu |
| 1951 | 30 | 2 | 37 | 25 | 283 | 38 | 273 | JASBSC | Fish & fisheries-Anci.Hindu |
| | | | | 10 | 293 | 9 | 282 | JASBSC | Fishery in irrigation tanks |
| 1952 | 31 | 4 | 41 | 6 | 299 | 7 | 289 | JASBSC | Fish-evolution |
| | | | | 3 | 302 | | 289 | JASBSC | Lac-insect |
| | | | | 7 | 309 | 15 | 304 | JASBSC | Fish in Ramayana |
| | | | | 19 | 328 | | 304 | JASBSC | Fisheries |
| 1953 | 32 | 1 | 42 | 15 | 343 | 37 | 341 | JASBSC | Fish & fisheries-Anci.Hindu |
| 1954 | 33 | 3 | 45 | 15 | 358 | | 341 | JASBSC | Hilsa Fish |
| | | | | 4 | 362 | 4 | 345 | JASBSC | Hilsa fish |
| | | | | 9 | 371 | | 345 | JASBSC | Hilsa fish |
| 1955 | 34 | 2 | 47 | 13 | 384 | 8 | 353 | JASBSC | Fish in Jatka Sculptures |
| | | | | 16 | 400 | 44 | 397 | JASBSC | Fish in Jatka Tales |
| Total — — — | 34 | | 47 | | 400 | | 397 | | |
| Average — — — — | | 1.388 | | 8.511 | | 8.45 | | | |

## Table 28

### Distribution of Contributions of S. Mahdihassan

| Year of Contribution (Y) | ΣY | No. of Contributions (C) | ΣC | No. of Pages (P) | ΣP | No. of References (R) | ΣR | Journal | Subject |
|---|---|---|---|---|---|---|---|---|---|
| 1978 | 1 | 2 | 2 | 6 | 6 | 18 | 18 | IJHS | Drug - medicinal plant |
| | | | | 5 | 11 | 3 | 21 | VIJ | Time |
| 1979 | 2 | 1 | 3 | 3 | 14 | 8 | 29 | VIJ | Pottery |
| 1980 | 3 | 1 | 4 | 7 | 21 | 11 | 40 | IJHS | Humors - Trodosa |
| 1981 | 4 | 2 | 6 | 7 | 28 | 14 | 54 | IJHS | Liquor |
| | | | | 13 | 41 | 27 | 81 | IJHS | Alchemy |
| 1983 | 6 | 1 | 7 | 1 | 42 | 3 | 84 | IJHS | Kim - Purusa - Etymology |
| 1984 | 7 | 2 | 9 | 3 | 45 | 6 | 90 | IJHS | Wheat - word |
| | | | | 3 | 48 | 4 | 94 | IJHS | Tridosa |
| 1986 | 9 | 2 | 11 | 6 | 54 | 19 | 113 | IJHS | Soma |
| | | | | 6 | 60 | 16 | 129 | IJHS | Lac |
| Total — — — | 9 | | 11 | | 60 | | 129 | | |
| Average — — — — | | 1.222 | | 5.454 | | 11.73 | | | |

## Table 29

### Distribution of Contributions of D.C. Phillott

| Year of Contribution (Y) | ΣY | No. of Contributions (C) | ΣC | No. of Pages (P) | ΣP | No. of References (R) | ΣR | Journal | Subject |
|---|---|---|---|---|---|---|---|---|---|
| 1907 | 1 | 9 | 9 | 2 | 2 | 1 | 1 | JPASB | Falcon |
| | | | | 4 | 6 | 11 | 12 | JPASB | Raven |
| | | | | 3 | 9 | 29 | 41 | JPASB | Falcon |
| | | | | 5 | 14 | 8 | 49 | JPASB | Falcon |
| | | | | 7 | 21 | 21 | 70 | JPASB | Falcon |
| | | | | 5 | 26 | 13 | 83 | JPASB | Merlin |
| | | | | 7 | 33 | 1 | 84 | JPASB | Birds/Heron |
| | | | | 3 | 36 | 4 | 88 | JPASB | Heron |
| | | | | 4 | 40 | 2 | 90 | JPASB | Secret language |
| 1908 | 2 | 4 | 13 | 6 | 46 | | 90 | JPASB | Falcon |
| | | | | 7 | 53 | 9 | 99 | JPASB | Falcon |
| | | | | 25 | 78 | | 99 | JPASB | Indian birds-vacabulary |
| | | | | 2 | 80 | | 99 | JPASB | Hawks |
| Total — — — | 2 | | 13 | | 80 | | 99 | | |
| Average — — — — | | 6.5 | | 6.154 | | 7.615 | | | |

## Table 30

## Distribution of Contributions of B. Prashad

| Year of Contribution (Y) | ΣY | No. of Contributions (C) | ΣC | No. of Pages (P) | ΣP | No. of References (R) | ΣR | Journal | Subject |
|---|---|---|---|---|---|---|---|---|---|
| 1916 | 1 | 1 | 1 | 5 | 5 | | 0 | JPASB | Pond life-seasonal conditions |
| 1918 | 3 | 1 | 2 | 6 | 11 | 14 | 14 | JPASB | Mollusca |
| 1919 | 4 | 1 | 3 | 4 | 15 | 1 | 15 | JPASB | Fish |
| 1920 | 5 | 2 | 5 | 7 | 22 | 3 | 18 | JPASB | Mollusca |
| | | | | 4 | 26 | | 18 | JPASB | Artisian |
| 1922 | 7 | 2 | 7 | 5 | 31 | 5 | 23 | JPASB | Genetics |
| | | | | 1 | 32 | 5 | 28 | JPASB | Nomenclature |
| 1923 | 8 | 1 | 8 | 6 | 38 | 8 | 36 | JPASB | Mollusca |
| 1926 | 11 | 2 | 10 | 1 | 39 | 2 | 38 | JPASB | Mollusca |
| | | | | 2 | 41 | | 38 | JPASB | Mollusca |
| 1927 | 12 | 1 | 11 | 3 | 44 | 4 | 42 | JPASB | Indian Fishes-Publication Date |
| 1930 | 15 | 1 | 12 | 10 | 54 | 26 | 68 | JPASB | Zoological Res. - in India |
| Total — — — | 15 | | 12 | | 54 | | 68 | | |
| Average — — — — — | | 0.8 | | 4.5 | | 5.7 | | | |

## Table 31

## Distribution of Contributions of K.V. Sarma

| Year of Contribution (Y) | ΣY | No. of Contributions (C) | ΣC | No. of Pages (P) | ΣP | No. of References (R) | ΣR | Journal | Subject |
|---|---|---|---|---|---|---|---|---|---|
| 1963 | 1 | 1 | 1 | 33 | 33 | | 0 | VIJ | Astron. Comput. - Kerala |
| 1965 | 3 | 1 | 2 | 44 | 77 | | 0 | VIJ | Paramesvara |
| 1966 | 4 | 1 | 3 | 40 | 117 | | 0 | VIJ | Paramesvara |
| 1969 | 7 | 1 | 4 | 14 | 131 | 24 | 24 | VIJ | Food-diet. & Medicinal Prop. |
| 1970 | 8 | 1 | 5 | 28 | 159 | | 24 | VIJ | Nilakantha Somayaji |
| 1972 | 10 | 1 | 6 | 96 | 255 | | 24 | VIJ | Astron.& Math - Kerala |
| 1973 | 11 | 3 | 9 | 112 | 367 | | 24 | VIJ | Bibliography & subject index |
| | | | | 32 | 399 | | 24 | VIJ | Index to VIJ, 1962-72 |
| | | | | 56 | 455 | | 24 | VIJ | True moon |
| 1976 | 14 | 4 | 13 | 20 | 475 | 12 | 36 | VIJ | Astronomy-vedic |
| | | | | 4 | 479 | 1 | 37 | VIJ | Astrology-scientific basis |
| | | | | 31 | 510 | | 37 | VIJ | Nilakantha Somayaji |
| | | | | 54 | 564 | | 37 | VIJ | Nilakantha Somayaji |
| 1977 | 15 | 4 | 17 | 6 | 570 | 19 | 56 | VIJ | Aryabhata's Astron. - Kerala |
| | | | | 41 | 611 | | 56 | VIJ | Acyuta Pisarati |
| | | | | 15 | 626 | 27 | 83 | VIJ | Aryabhata's Astron. - sphere |
| | | | | 90 | 716 | 170 | 253 | VIJ | Nilakantha Somayaji |
| 1978 | 16 | 1 | 18 | 7 | 723 | 6 | 259 | VIJ | Astron.& Math. - Kerala Texts |
| 1979 | 17 | 1 | 19 | 124 | 847 | | 259 | VIJ | Hindu Astronomy |
| 1985 | 23 | 1 | 20 | 20 | 867 | 45 | 304 | IJHS | Astronomy - Source Materials |
| Total — — — | 23 | | 20 | | 867 | | 304 | | |
| Average — — — — — | | 0.869 | | 43.35 | | 15.2 | | | |

## Table 32

### Distribution of Contributions of G. Sarton

| Year of Contribution (Y) | ΣY | No. of Contributions (C) | ΣC | No. of Pages (P) | ΣP | No. of References (R) | ΣR | Journal | Subject |
|---|---|---|---|---|---|---|---|---|---|
| 1920 | 1 | 1 | 1 | 12 | 12 | 6 | 6 | ISIS | Synthetic Bibliography-HS |
| 1935 | 16 | 2 | 3 | 18 | 30 | | 6 | ISIS | Catlin G |
| | | | | 92 | 122 | 119 | 125 | ISIS | Decimal |
| 1936 | 17 | 1 | 4 | 4 | 126 | 10 | 135 | ISIS | Decimal |
| 1937 | 18 | 1 | 5 | 2 | 128 | | 135 | ISIS | Decimal |
| 1938 | 19 | 1 | 6 | 2 | 130 | | 135 | ISIS | Catlin G |
| 1944 | 25 | 1 | 7 | 5 | 135 | 9 | 144 | ISIS | Grassmann, Historian of Sc |
| 1947 | 28 | 2 | 9 | 3 | 138 | | 144 | ISIS | Qualifi. of Teachers in HS |
| | | | | 2 | 140 | | 144 | ISIS | HS in London Univ |
| 1949 | 30 | 1 | 10 | 3 | 143 | | 144 | ISIS | Qualifi. of Teachers in HS |
| 1950 | 31 | 2 | 12 | 10 | 153 | 14 | 158 | ISIS | Reviewing of books |
| | | | | 8 | 161 | | 158 | ISIS | Critical Bibliography of ISIS |
| 1951 | 32 | 1 | 13 | 1 | 162 | | 158 | ISIS | HS in India |
| 1953 | 34 | 1 | 14 | 11 | 173 | 29 | 187 | ISIS | Why ISIS? |
| 1957 | 38 | 1 | 15 | 1 | 174 | | 187 | ISIS | Publications of ISIS |
| Total— — — — | 38 | | 15 | | 174 | | 187 | | |
| Average— — — — — | | 0.394 | | 11.6 | | 12.47 | | | |

## Table 33

### Distribution of Contributions of S.N. Sen

| Year of Contribution (Y) | ΣY | No. of Contributions (C) | ΣC | No. of Pages (P) | ΣP | No. of References (R) | ΣR | Journal | Subject |
|---|---|---|---|---|---|---|---|---|---|
| 1933 | 1 | 1 | 1 | 6 | 6 | 1 | 1 | JPASB | 'Rains of fishes'-meterol. |
| 1962 | 30 | 1 | 2 | 25 | 31 | 181 | 182 | JAS | Geography |
| 1966 | 34 | 2 | 4 | 11 | 42 | 22 | 204 | IJHS | Wastern Science in India |
| | | | | 12 | 54 | 30 | 234 | IJHS | Impetus theory |
| 1969 | 37 | 1 | 5 | 12 | 66 | 10 | 244 | IJHS | Astronomy |
| 1970 | 38 | 1 | 6 | 15 | 81 | 21 | 265 | IJHS | Indian Science - Influence |
| 1972 | 40 | 1 | 7 | 27 | 108 | 20 | 285 | IJHS | Scientific works in |
| 1974 | 42 | 1 | 8 | 15 | 123 | 24 | 309 | IJHS | Epicyclic theories (Astronomy) |
| 1975 | 43 | 1 | 9 | 13 | 136 | 46 | 355 | IJHS | Astron.-lat.& longitudes |
| 1982 | 50 | 1 | 10 | 17 | 153 | 45 | 400 | IJHS | Geographical coordinates |
| 1985 | 53 | 1 | 11 | 72 | 225 | 140 | 540 | IJHS | European languages - Studies |
| Total — — — | 53 | | 11 | | 225 | | 540 | | |
| Average — — — — | | 0.207 | | 20.454 | | 49.1 | | | |

## Table 34

## Distribution of Contributions of P.C. Sengupta

| Year of Contribution (Y) | ΣY | No. of Contributions (C) | ΣC | No. of Pages (P) | ΣP | No. of References (R) | ΣR | Journal | Subject |
|---|---|---|---|---|---|---|---|---|---|
| 1937 | 1 | 1 | 1 | 19 | 19 | 40 | 40 | JASBL | Astronomy |
| 1938 | 2 | 4 | 5 | 21 | 40 | 60 | 100 | JASBSC | Date Calculation |
| | | | | 20 | 60 | 24 | 124 | JASBSC | Days |
| | | | | 9 | 69 | 11 | 135 | JASBSC | Science of spring |
| | | | | 9 | 78 | 8 | 143 | JASBSC | Indra became Maghavan |
| 1941 | 5 | 2 | 7 | 8 | 86 | 13 | 156 | JASBL | Time |
| | | | | 23 | 109 | 20 | 176 | JASBL | Eclipse |
| 1942 | 6 | 1 | 8 | 16 | 125 | 19 | 195 | JASBL | Gupta Era |
| 1950 | 14 | 1 | 9 | 12 | 137 | 31 | 226 | JASBL | Chronology - Res. |
| 1952 | 16 | 1 | 10 | 1 | 138 | | | JASBL | Chronology - Res.Note |
| 1956 | 20 | 1 | 11 | 10 | 148 | 12 | 238 | JASBL | Mahabharta - Astro.Data |
| Total — — — — — | 20 | | 11 | | 148 | | 249 | | |
| Average — — — — | | 0.55 | | 13.454 | | 21.64 | | | |

## Table 35

## Distribution of Contributions of S.D. Sharma

| Year of Contribution (Y) | ΣY | No. of Contributions (C) | ΣC | No. of Pages (P) | ΣP | No. of References (R) | ΣR | Journal | Subject |
|---|---|---|---|---|---|---|---|---|---|
| 1972 | 1 | 1 | 1 | 4 | 4 | 6 | 6 | IJHS | Date(Tithis) |
| 1975 | 4 | 1 | 2 | 4 | 8 | 4 | 10 | VIJ | Transition - Multipole |
| 1977 | 6 | 2 | 4 | 8 | 16 | 34 | 44 | IJHS | Time |
| | | | | 12 | 28 | 30 | 74 | IJHS | Science of Sciatherics (season determ. |
| 1979 | 8 | 1 | 5 | 5 | 33 | 26 | 100 | IJHS | Circumference |
| 1980 | 9 | 1 | 6 | 11 | 44 | 29 | 129 | IJHS | Time |
| 1981 | 10 | 1 | 7 | 12 | 56 | 42 | 171 | IJHS | Saptagrahi(astronomy) |
| 1984 | 13 | 1 | 8 | 9 | 65 | 23 | 194 | IJHS | Pluto |
| 1985 | 14 | 3 | 11 | 14 | 79 | 27 | 221 | IJHS | Astronomy |
| | | | | 24 | 103 | 73 | 294 | IJHS | Eclipse |
| | | | | 26 | 129 | 55 | 349 | IJHS | Astronomical - observatories |
| Total — — — — | 14 | | 11 | | 129 | | 349 | | |
| Average — — — — — | | 0.786 | | 11.73 | | 31.73 | | | |

**FIGURE 9**
YEARWISE CONTRIBUTIONS, CUMULATED CONTRIBUTIONS AND EMERGING TREND OF CONTRIBUTIONS OF ANNANDALE N

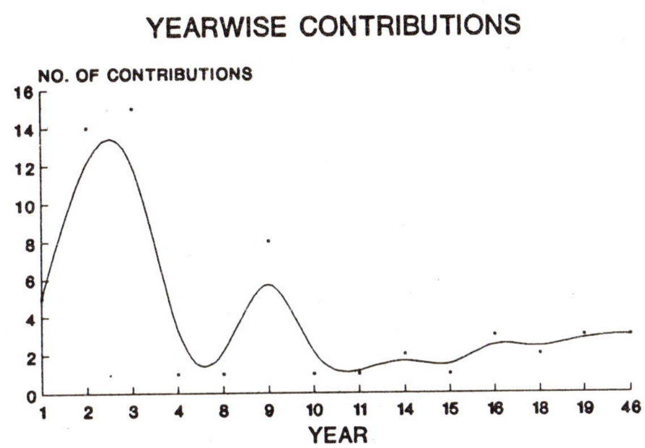


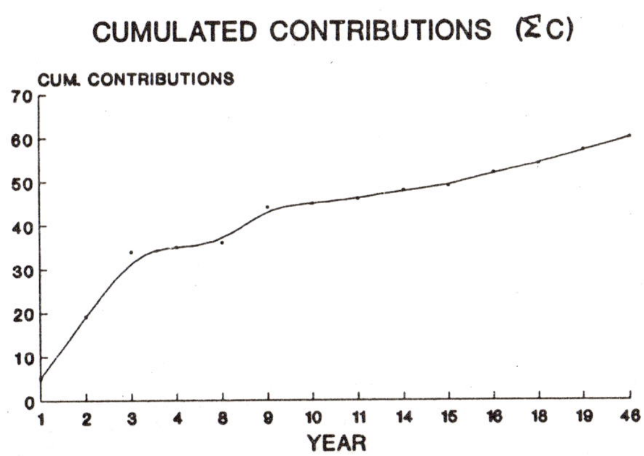


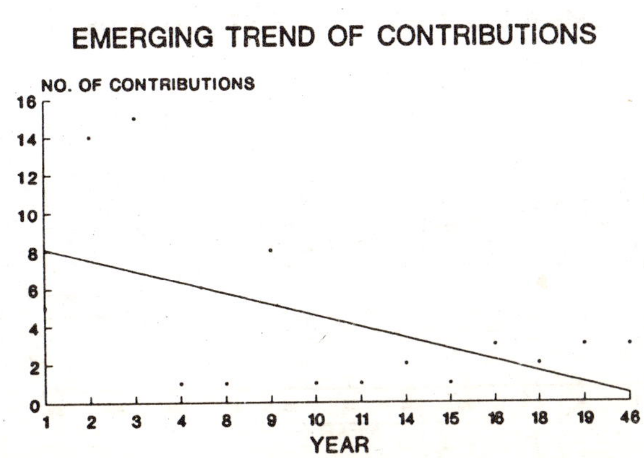

**Figure 10**

YEARWISE CONTRIBUTIONS, CUMULATED CONTRIBUTIONS AND EMERGING TREND OF CONTRIBUTIONS OF BAG AK

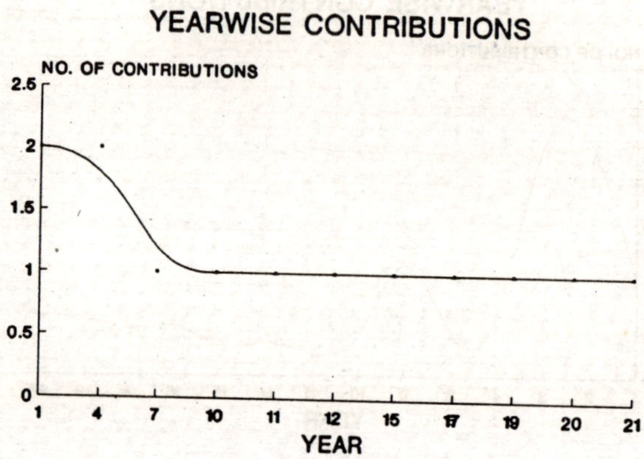


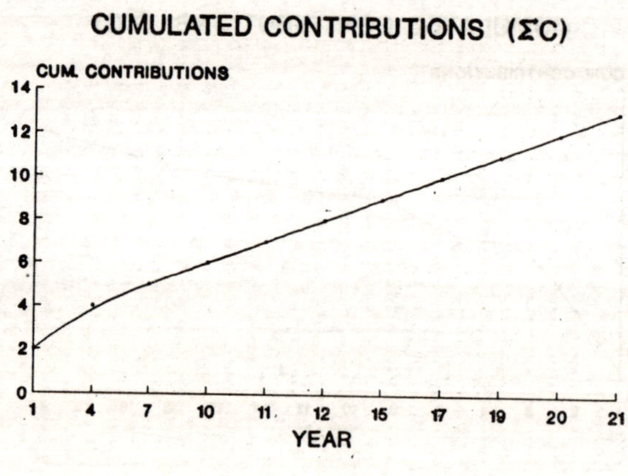


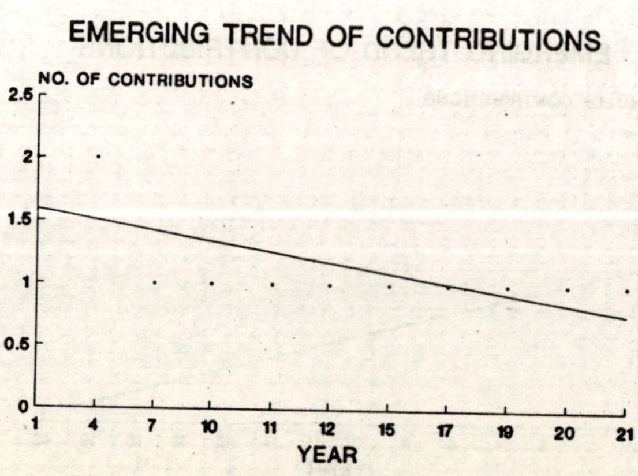

**Figure 11**

YEARWISE CONTRIBUTIONS, CUMULATED CONTRIBUTIONS AND EMERGING TREND OF CONTRIBUTIONS OF BISWAS K

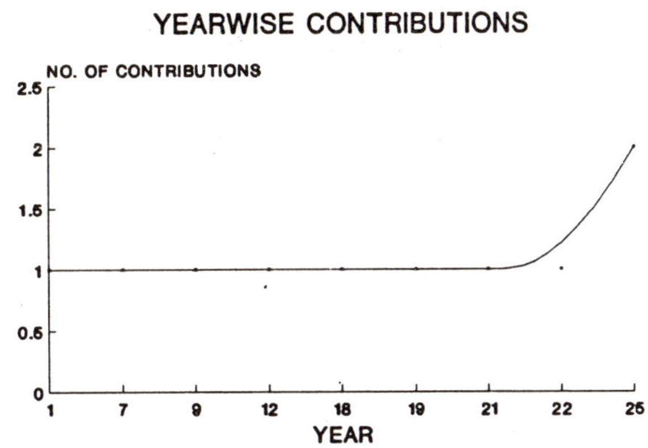


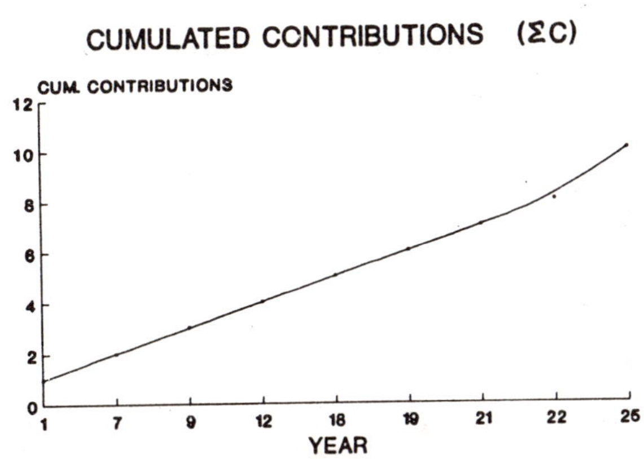


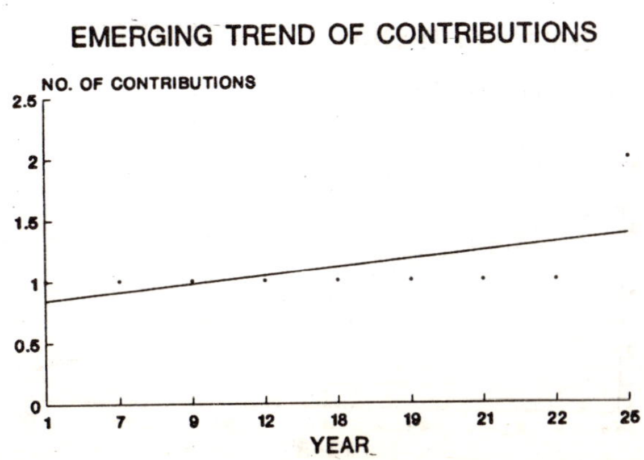

**Figure 12**

YEARWISE CONTRIBUTIONS, CUMULATED CONTRIBUTIONS AND EMERGING TREND OF CONTRIBUTIONS OF BURKILL IH

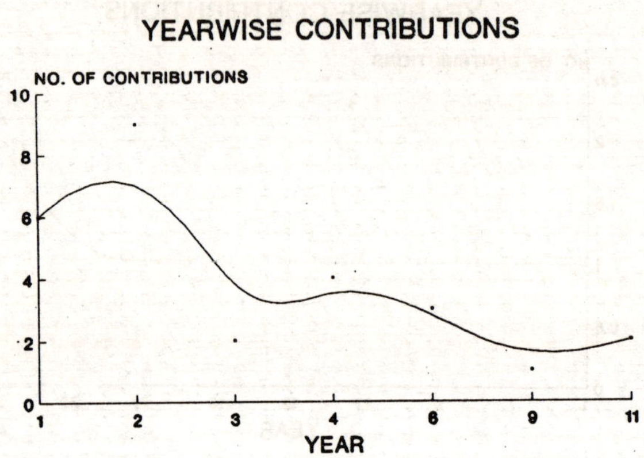


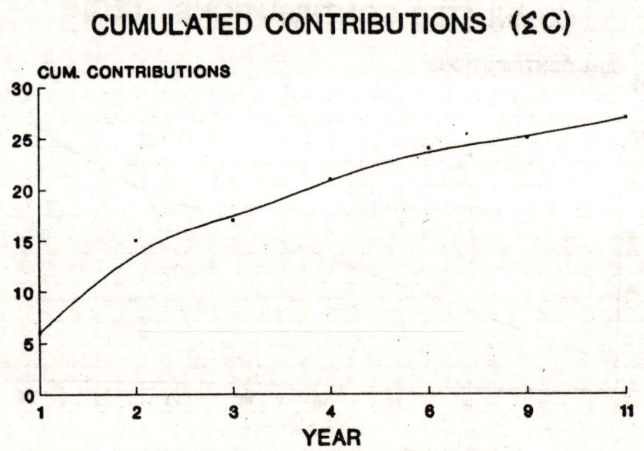


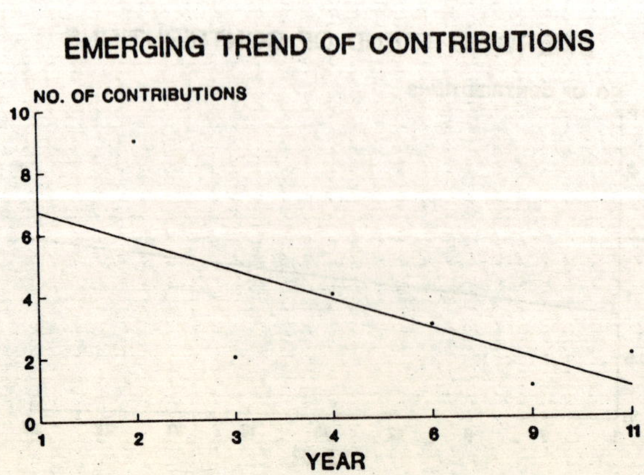

**Figure 13**

YEARWISE CONTRIBUTIONS, CUMULATED CONTRIBUTIONS AND EMERGING TREND OF CONTRIBUTIONS OF CHAUDHURI S

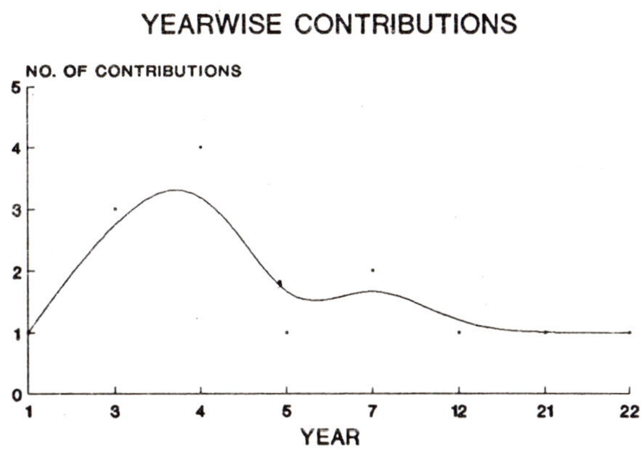


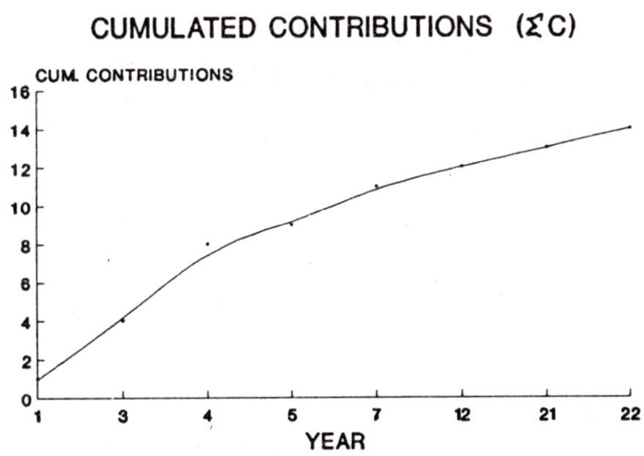


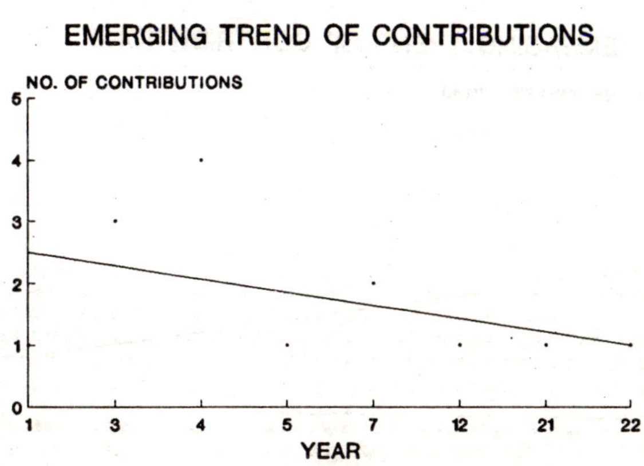

**Figure 14**

YEARWISE CONTRIBUTIONS, CUMULATED CONTRIBUTIONS AND EMERGING TREND OF CONTRIBUTIONS OF DAS SR

YEARWISE CONTRIBUTIONS

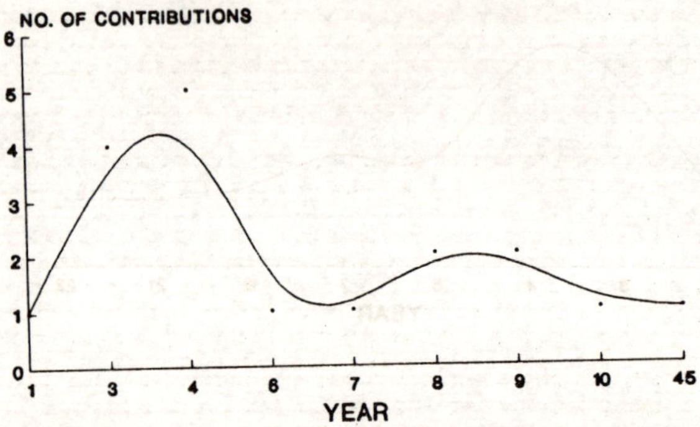


CUMULATED CONTRIBUTIONS (ΣC)

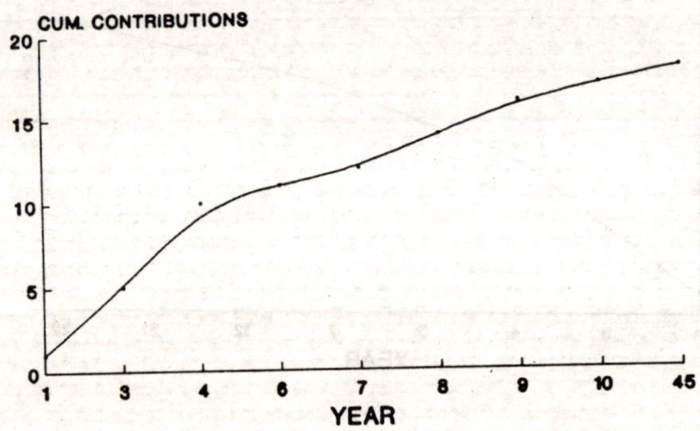


EMERGING TREND OF CONTRIBUTIONS

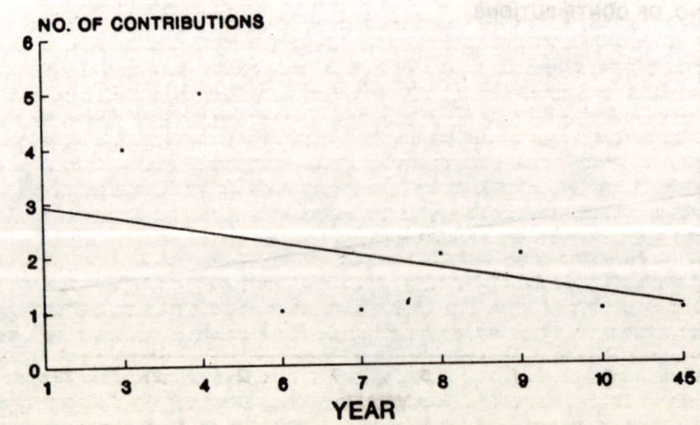

Figure 15

YEARWISE CONTRIBUTIONS, CUMULATED CONTRIBUTIONS AND EMERGING TREND OF CONTRIBUTIONS OF DASGUPTA HC

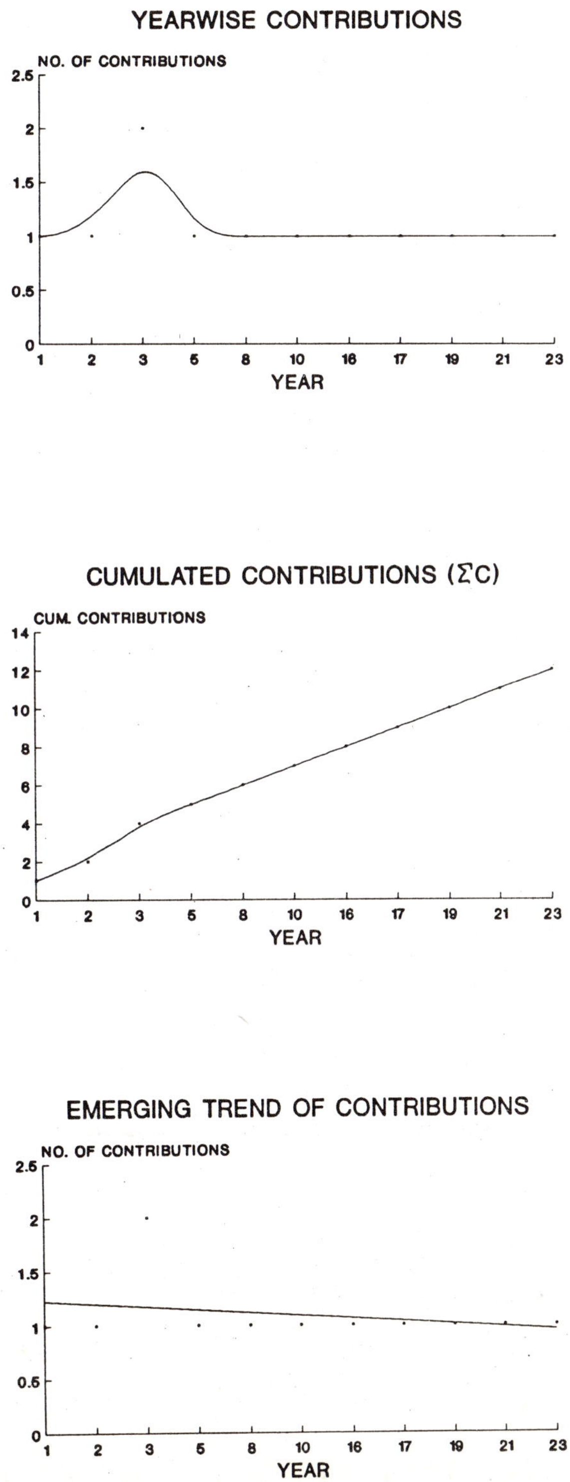

**Figure 16**

YEARWISE CONTRIBUTIONS, CUMULATED CONTRIBUTIONS AND EMERGING TREND OF CONTRIBUTIONS OF DATTA B

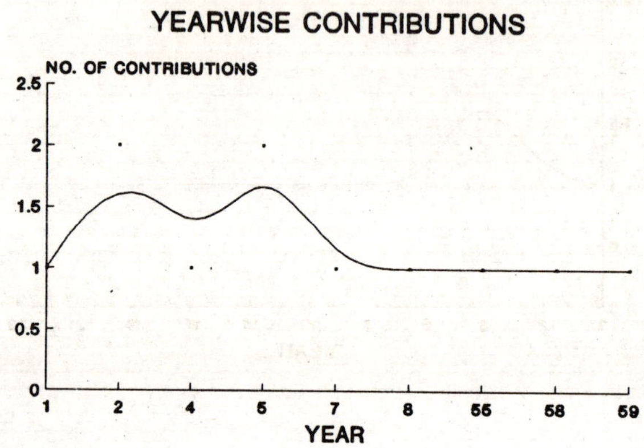


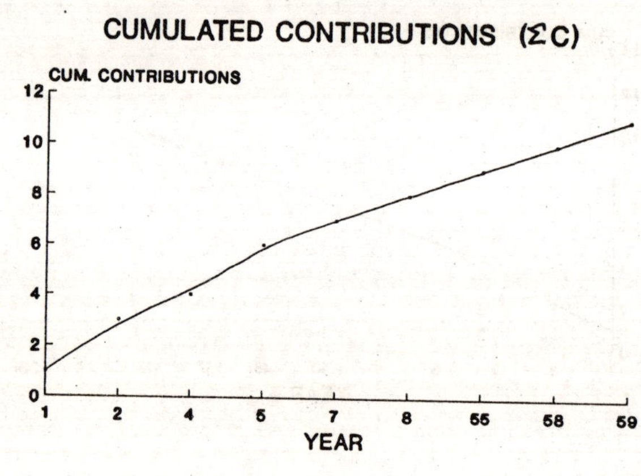


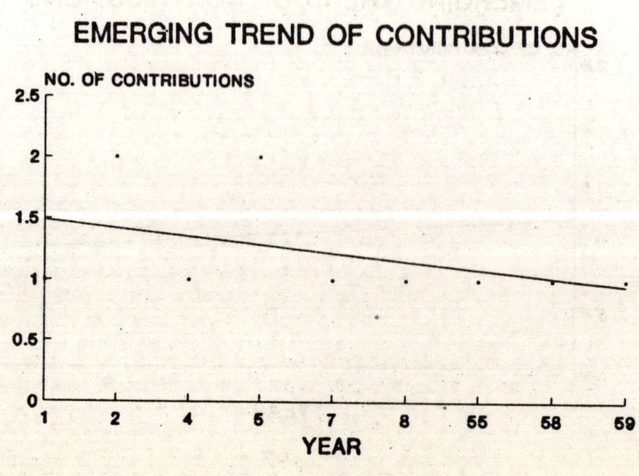

**Figure 17**

YEARWISE CONTRIBUTIONS, CUMULATED CONTRIBUTIONS AND EMERGING TREND OF CONTRIBUTIONS OF GERSHENOWITZ H

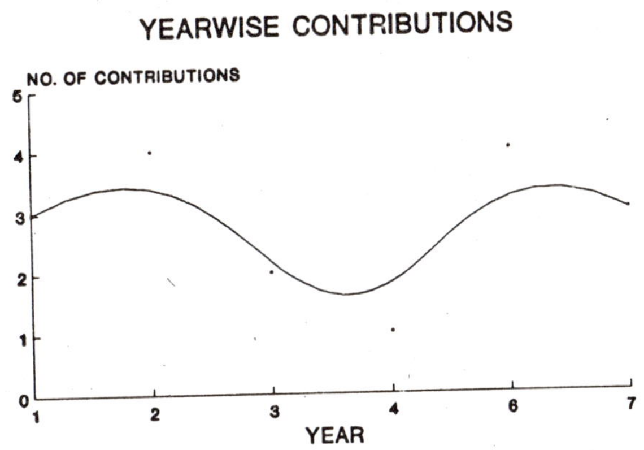


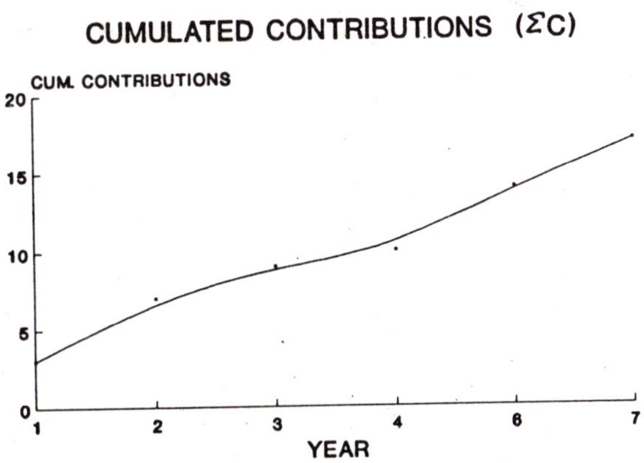


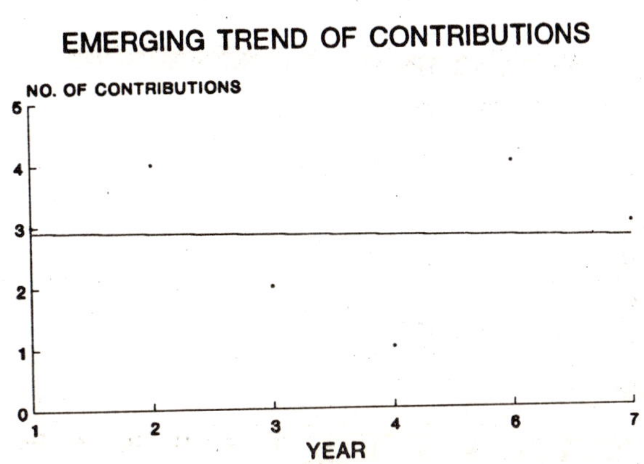

**Figure 18**

YEARWISE CONTRIBUTIONS, CUMULATED CONTRIBUTIONS AND EMERGING TREND OF CONTRIBUTIONS OF GUPTA RC

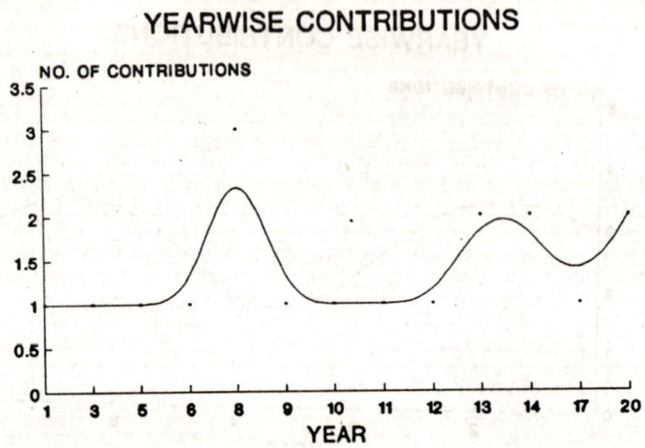


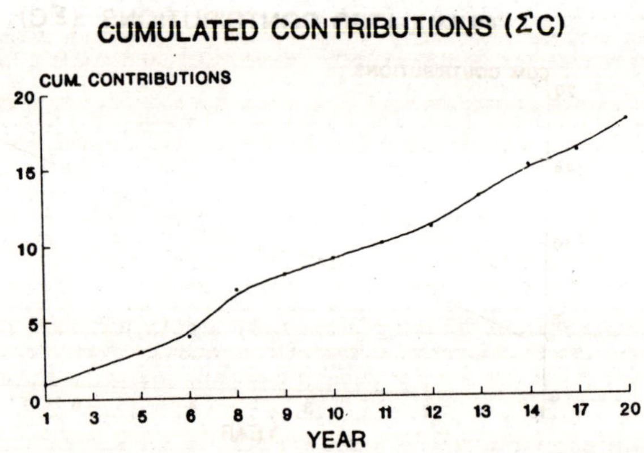


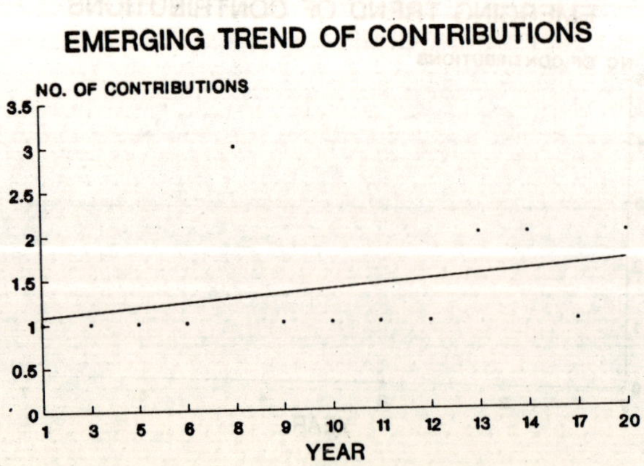

Figure 19

YEARWISE CONTRIBUTIONS, CUMULATED CONTRIBUTIONS AND EMERGING TREND OF CONTRIBUTIONS OF HOOPER D

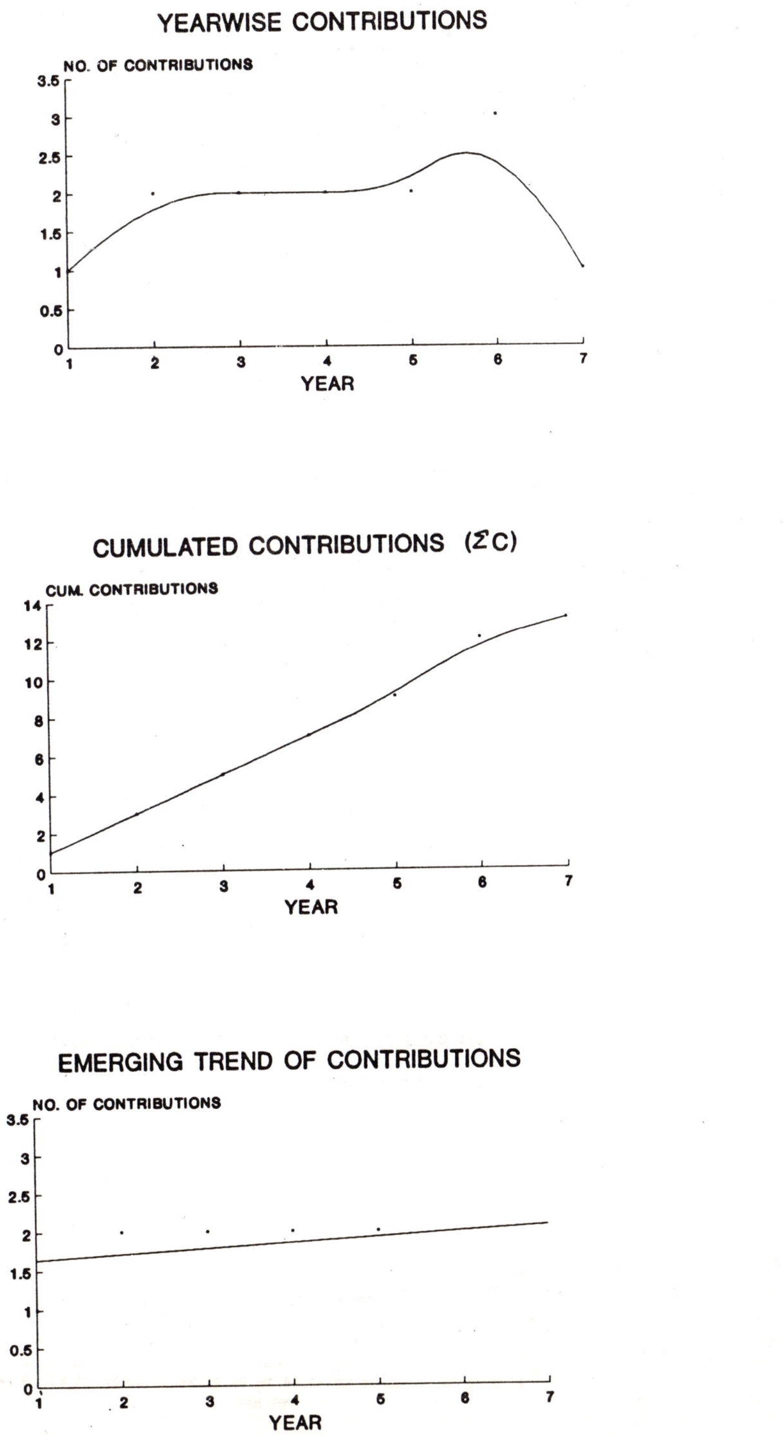

**Figure 20**

YEARWISE CONTRIBUTIONS, CUMULATED CONTRIBUTIONS AND EMERGING TREND OF CONTRIBUTIONS OF HORA SL

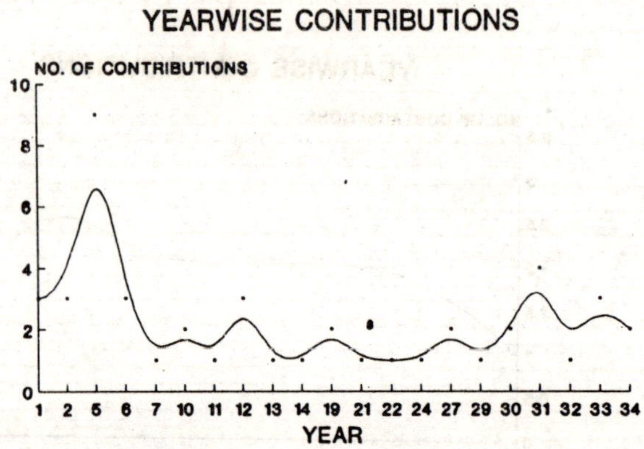


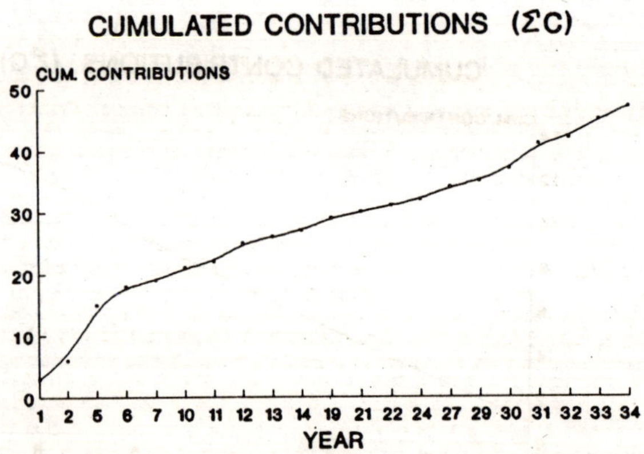


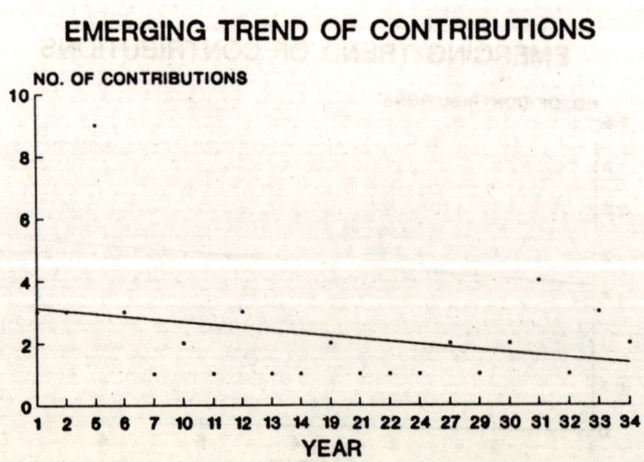

## Figure 21

YEARWISE CONTRIBUTIONS, CUMULATED CONTRIBUTIONS AND EMERGING TREND OF CONTRIBUTIONS OF MAHDIHASSAN S

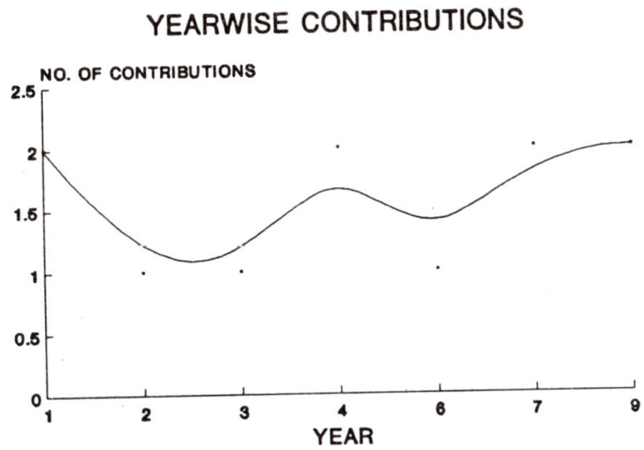


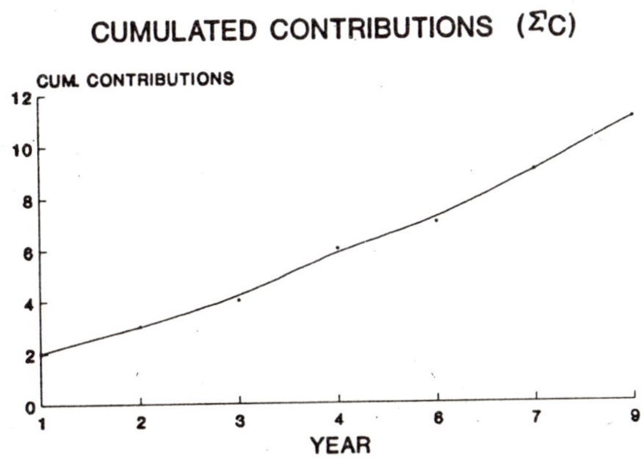


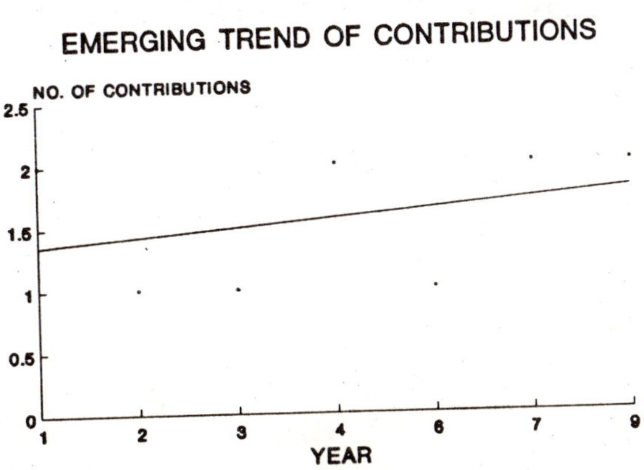

**Figure 22**

YEARWISE CONTRIBUTIONS, CUMULATED CONTRIBUTIONS AND EMERGING TREND OF CONTRIBUTIONS OF PRASHAD B

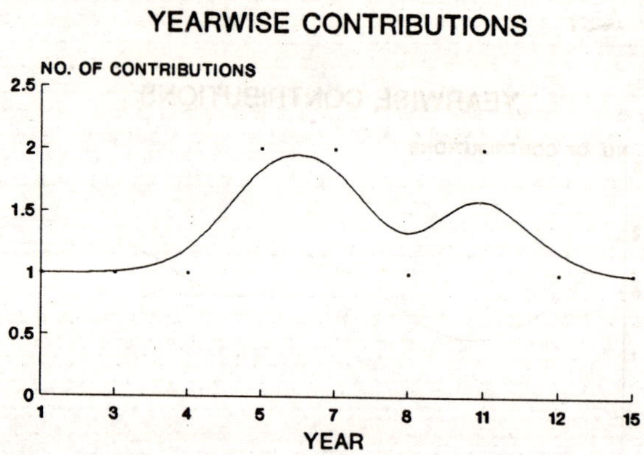


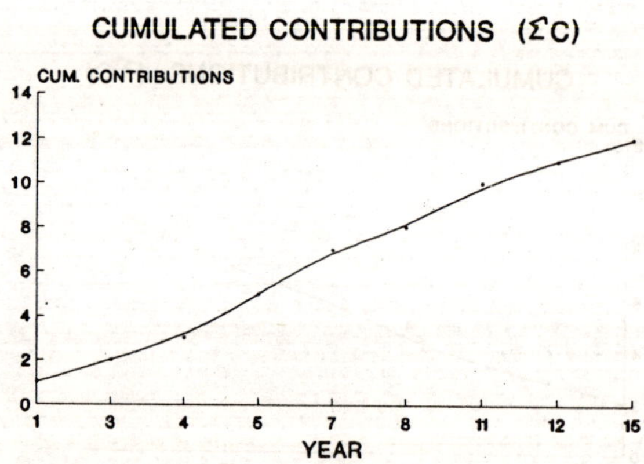


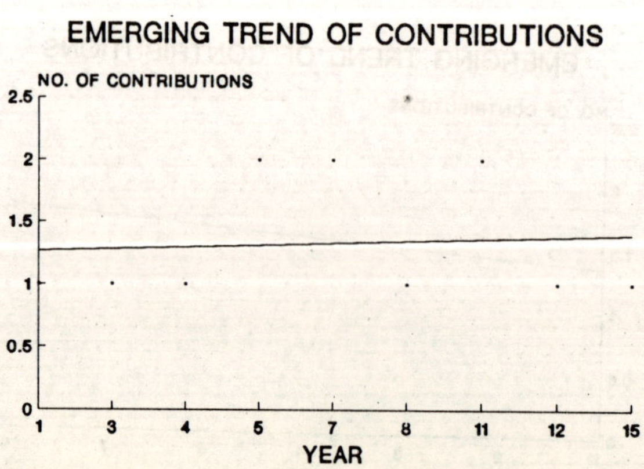

**Figure 23**

YEARWISE CONTRIBUTIONS, CUMULATED CONTRIBUTIONS AND EMERGING TREND OF CONTRIBUTIONS OF SARMA KV

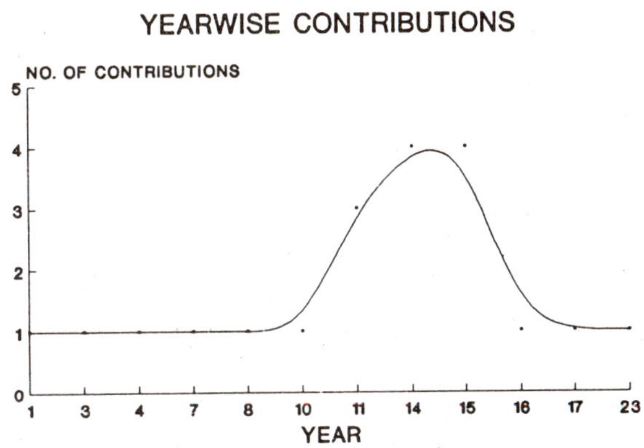


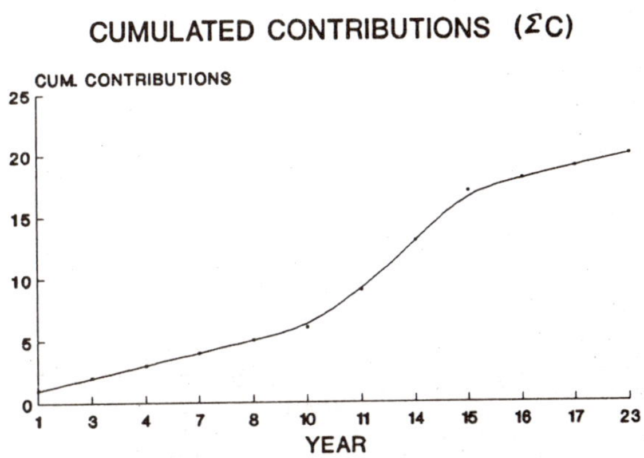


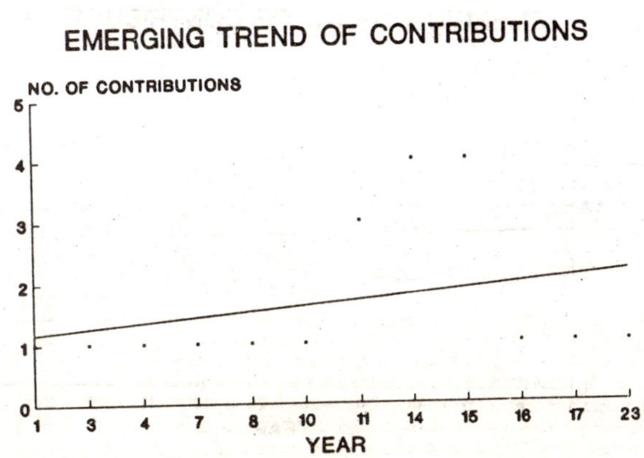

**Figure 24**

YEARWISE CONTRIBUTIONS, CUMULATED CONTRIBUTIONS AND EMERGING TREND OF CONTRIBUTIONS OF SARTON G

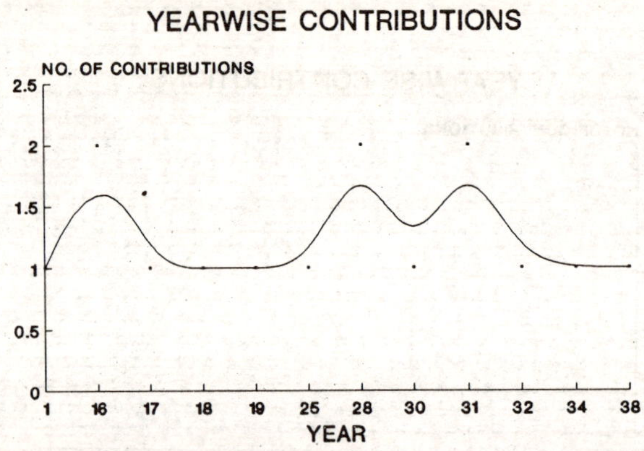


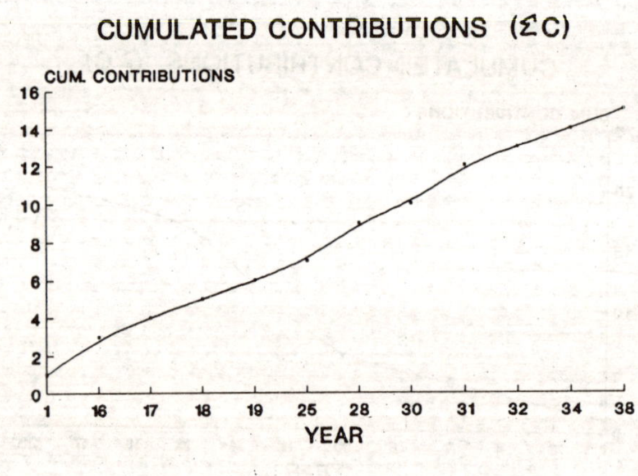


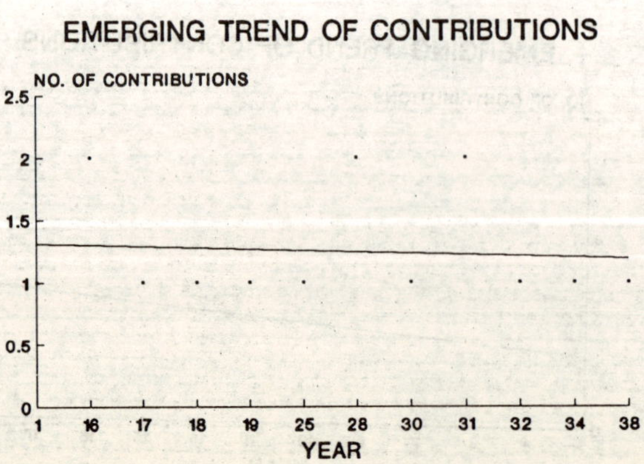

**Figure 25**

YEARWISE CONTRIBUTIONS, CUMULATED CONTRIBUTIONS AND EMERGING TREND OF CONTRIBUTIONS OF SEN SN

YEARWISE CONTRIBUTIONS

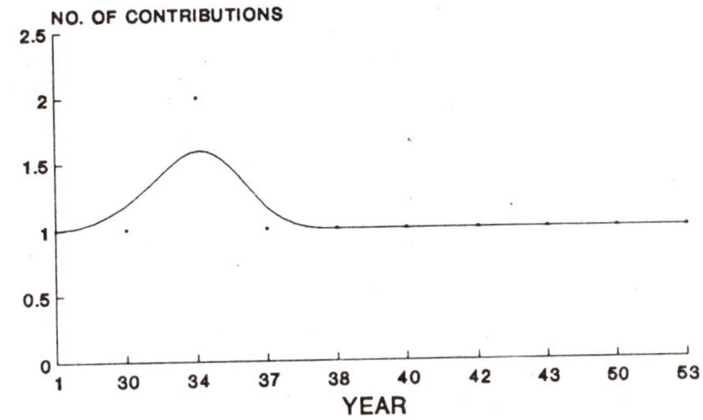


CUMULATED CONTRIBUTIONS (ΣC)

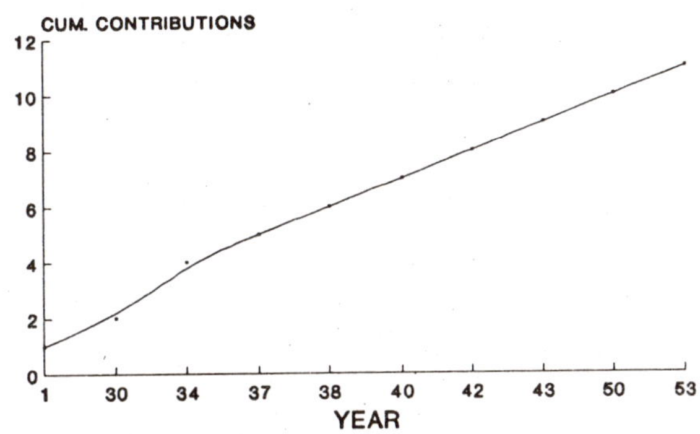


EMERGING TREND OF CONTRIBUTIONS

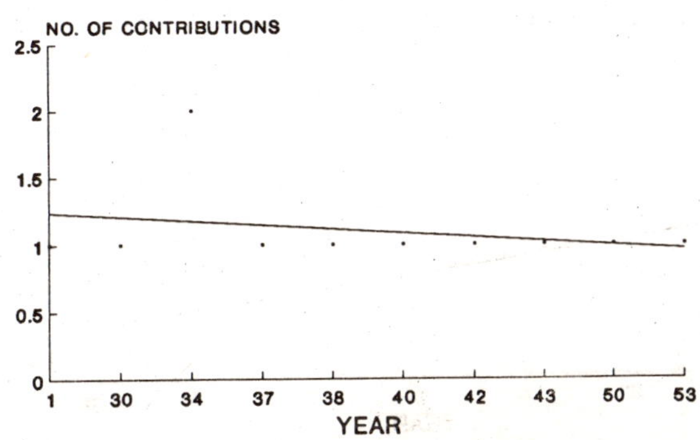

**Figure 26**

YEARWISE CONTRIBUTIONS, CUMULATED CONTRIBUTIONS AND EMERGING TREND OF CONTRIBUTIONS OF SENGUPTA PC

YEARWISE CONTRIBUTIONS

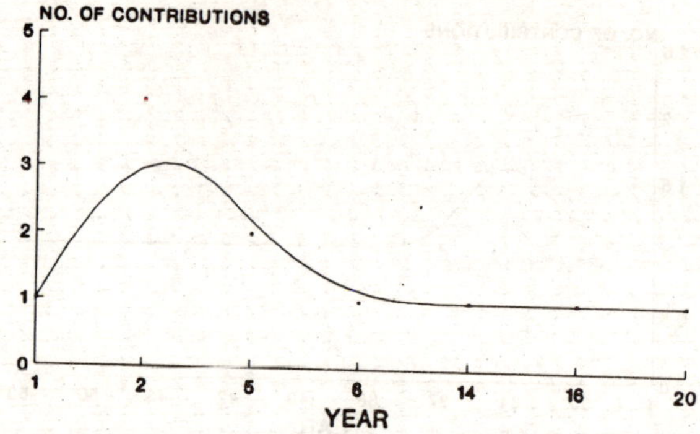


CUMULATED CONTRIBUTIONS (ΣC)

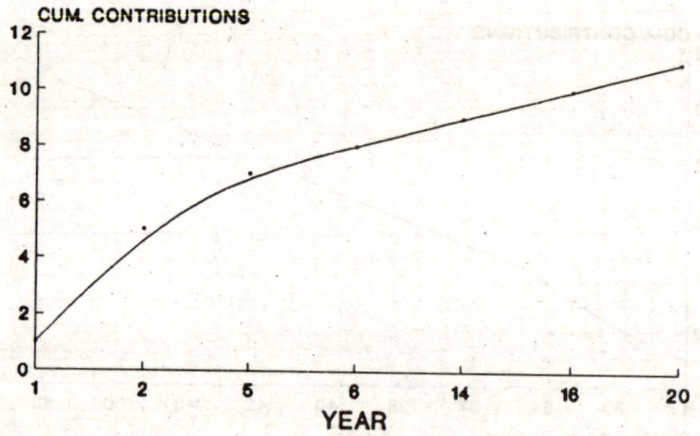


EMERGING TREND OF CONTRIBUTIONS

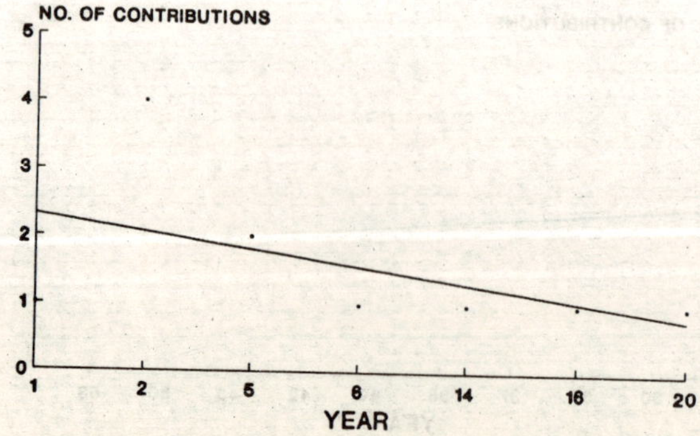

Figure 27

YEARWISE CONTRIBUTIONS, CUMULATED CONTRIBUTIONS AND EMERGING TREND OF CONTRIBUTIONS OF SHARMA SD

YEARWISE CONTRIBUTIONS

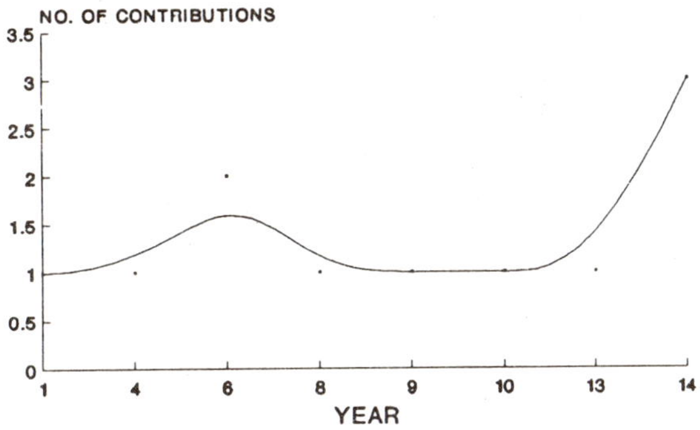


CUMULATED CONTRIBUTIONS (ΣC)

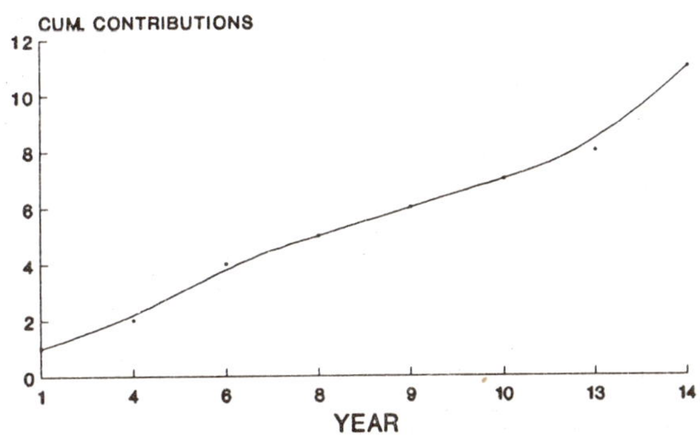


EMERGING TREND OF CONTRIBUTIONS

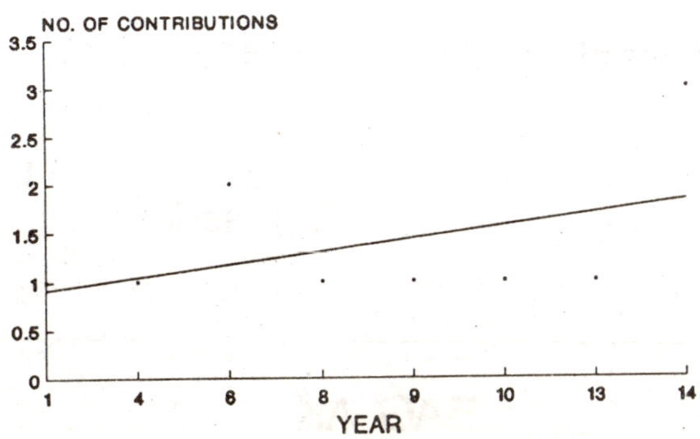

Figure 28.1

**Mapping of Subjects in Articles of Highly Productive Authors Individually has been Described in Twenty Figures Ranging from Figure 28.1 to Figure 28.20**

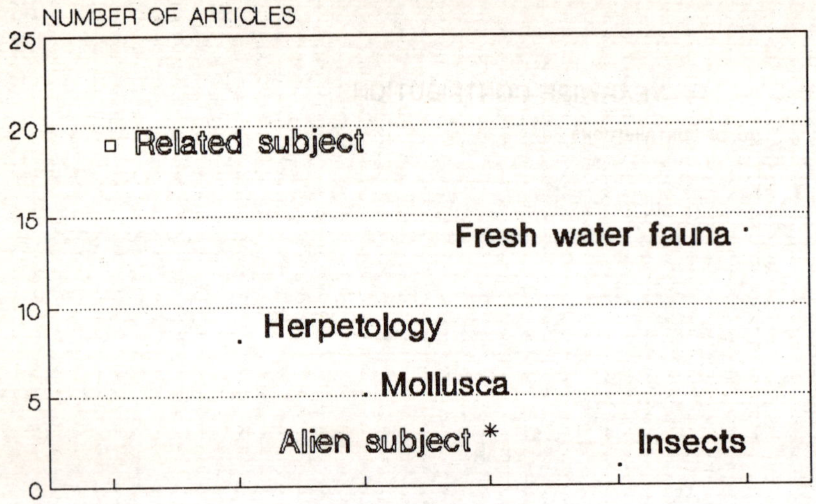


ANNANDALE N

Figure 28.2

Mapping of Subjects in Articles of A.K. Bag

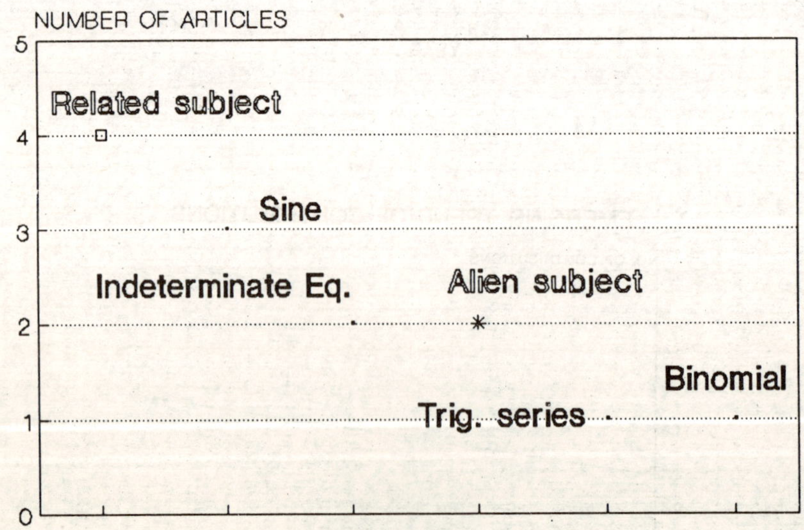


BAG AK

Figure 28.3

**Mapping of Subjects in Articles of K. Biswas**

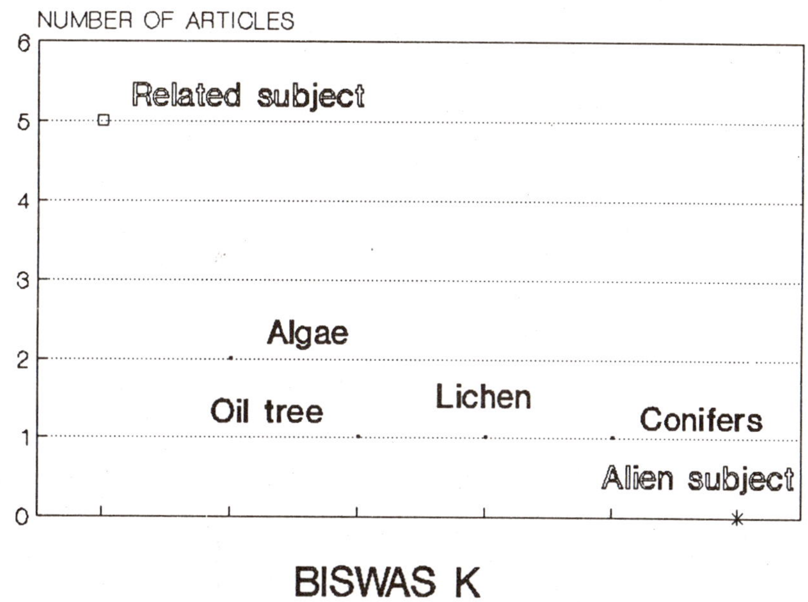


Figure 28.4

**Mapping of Subjects in Articles of I.H. Burkill**

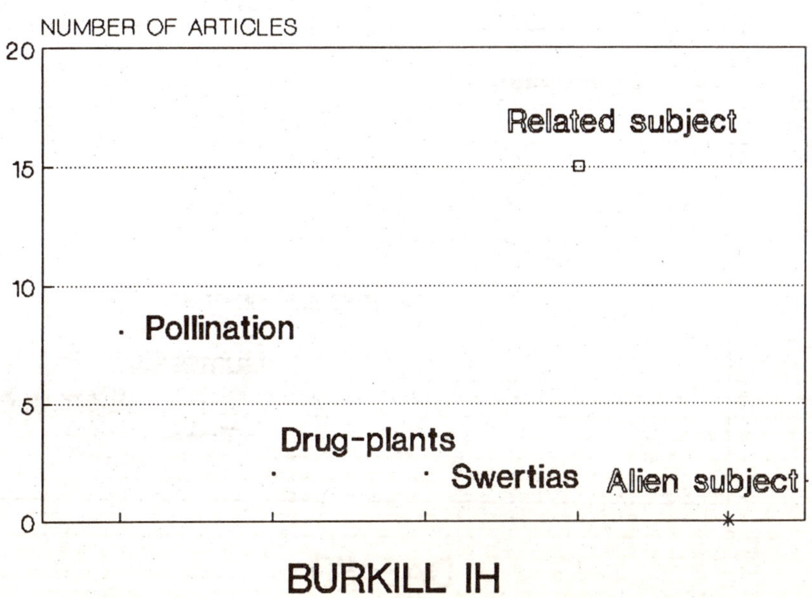

Figure 28.5

Mapping of Subjects in Articles of S. Chaudhuri

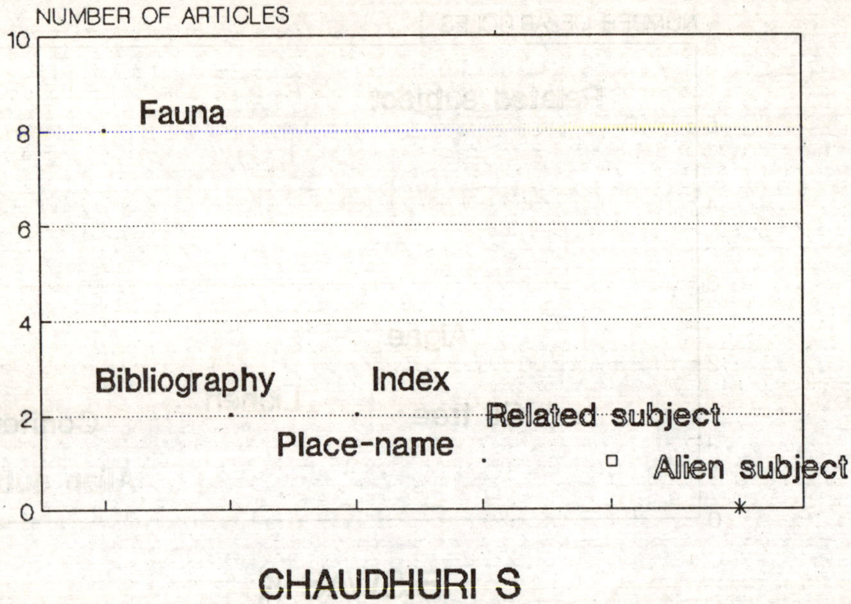


Figure 28.6

Mapping of Subjects in Articles of S.R. Das

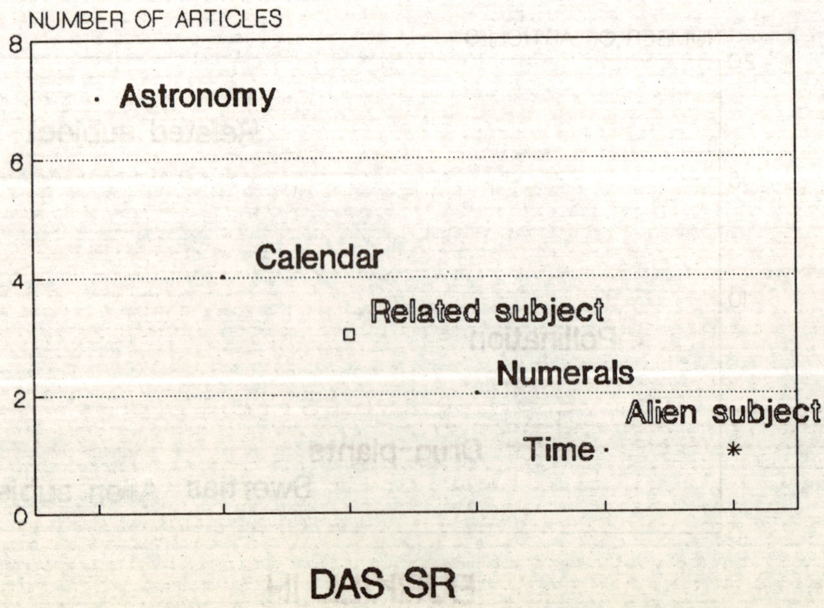

Figure 28.7

Mapping of Subjects in Articles of
H.C. Dasgupta

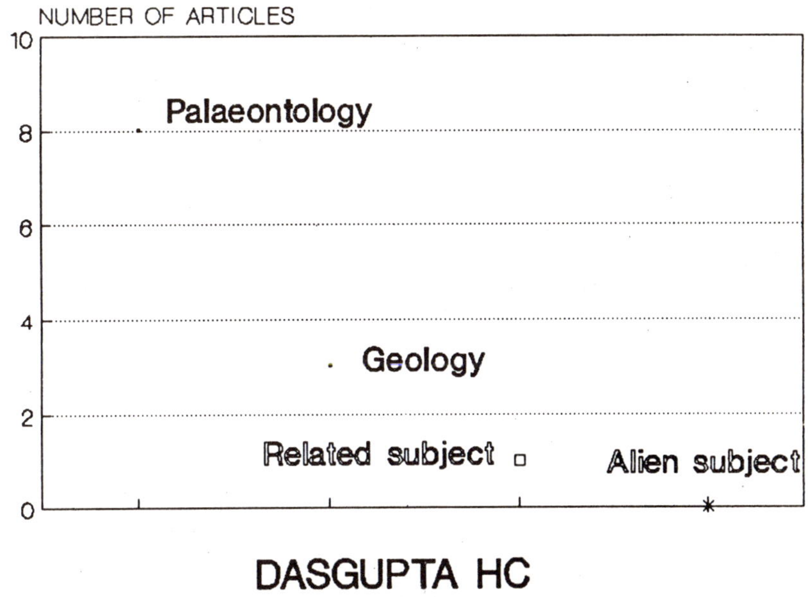


Figure 28.8

Mapping of Subjects in Articles of
B. Datta

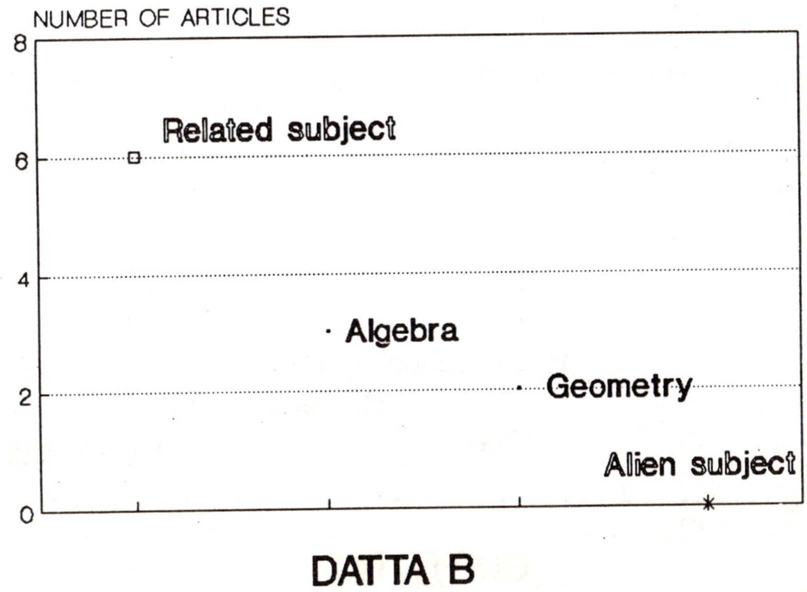

Figure 28.9

Mapping of Subjects in Articles of
H. Gershenowitz

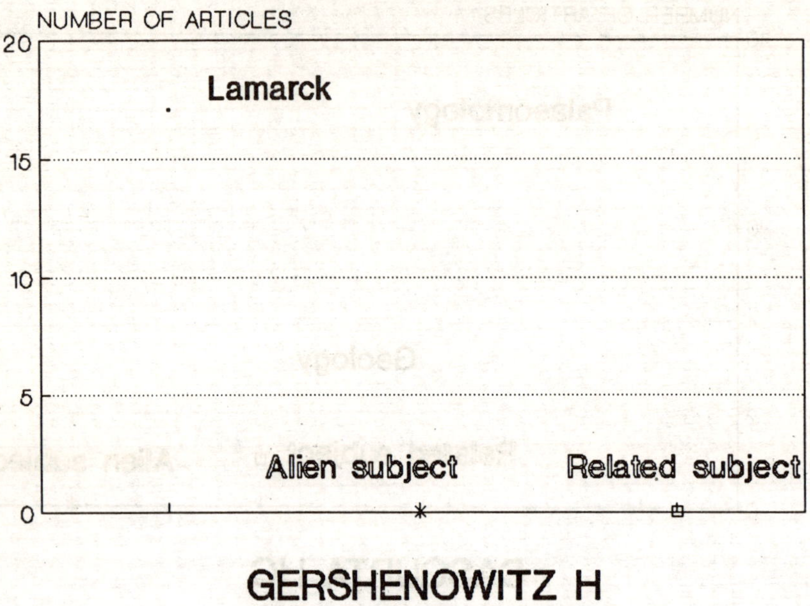


Figure 28.10

Mapping of Subjects in Articles of
R.C. Gupta

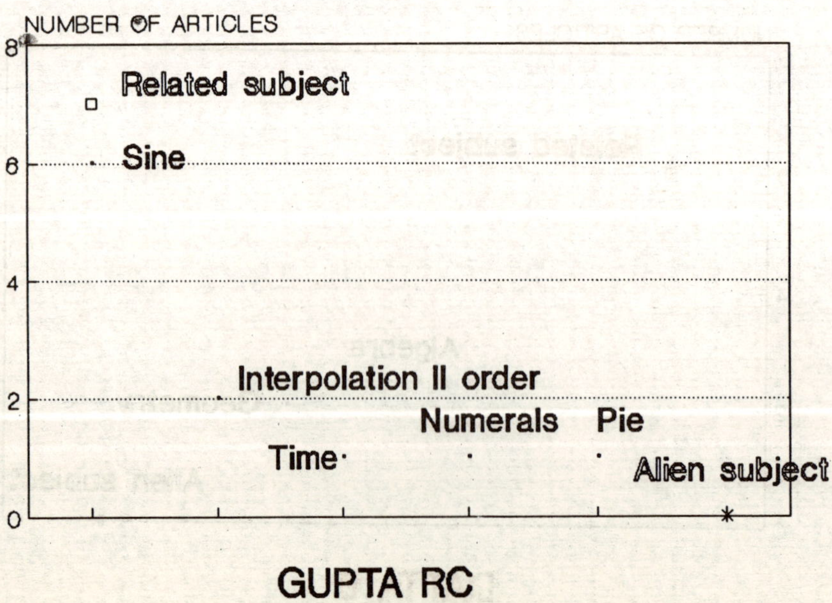

Figure 28.11

Mapping of Subjects in Articles of D. Hooper

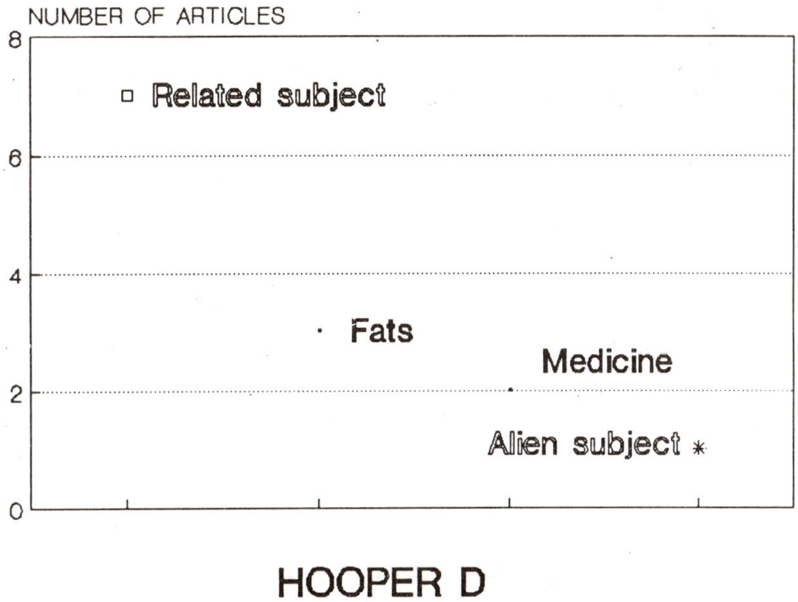


Figure 28.12

Mapping of Subjects in Articles of S.L. Hora

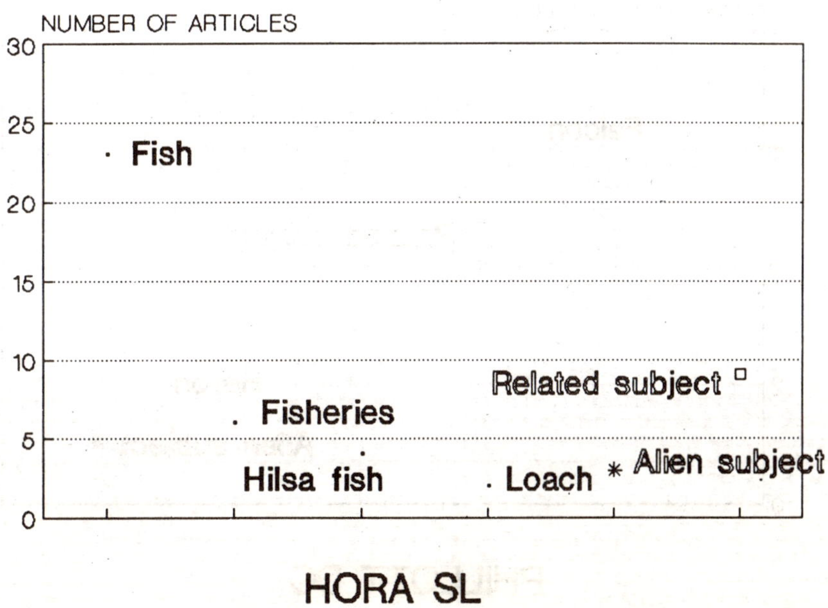

Figure 28.13

Mapping of Subjects in Articles of S. Mahdihassan

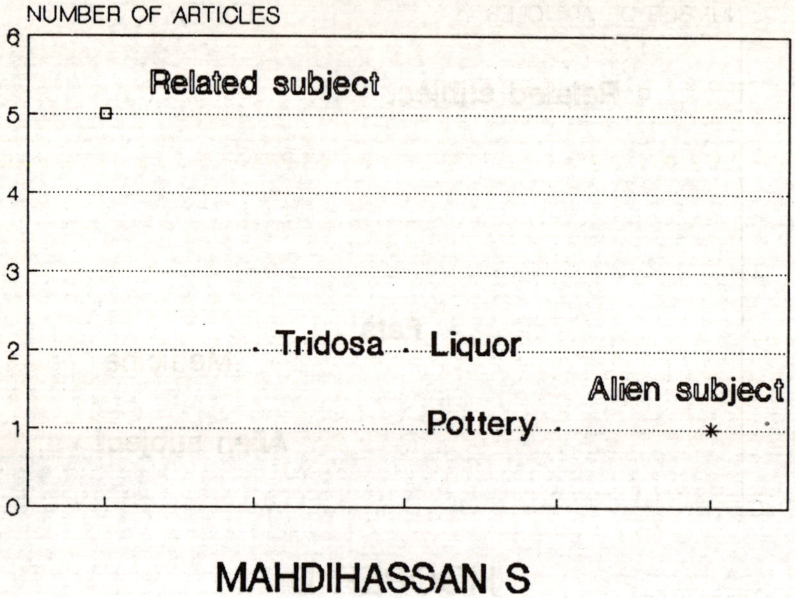


Figure 28.14

Mapping of Subjects in Articles of D.C. Phillott

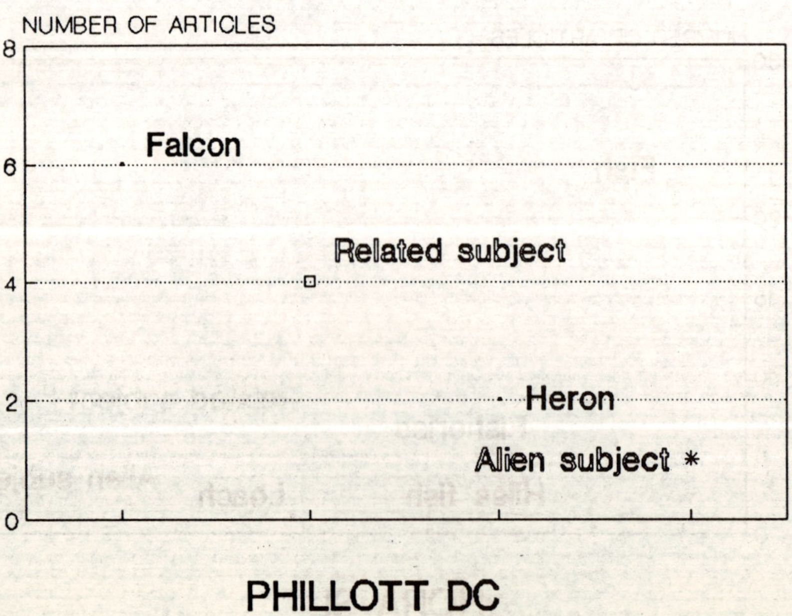

Figure 28.15

Mapping of Subjects in Articles of
B. Prashad

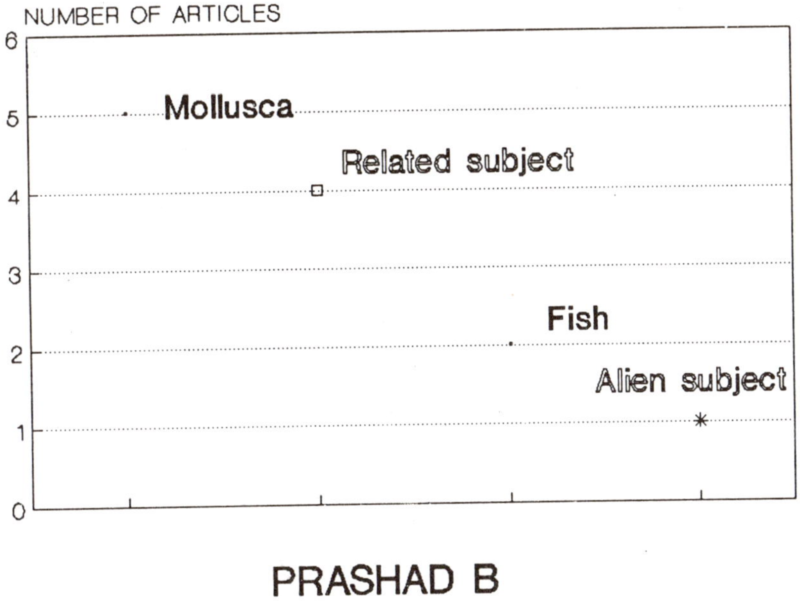


Figure 28.16

Mapping of Subjects in Articles of
K.V. Sarma

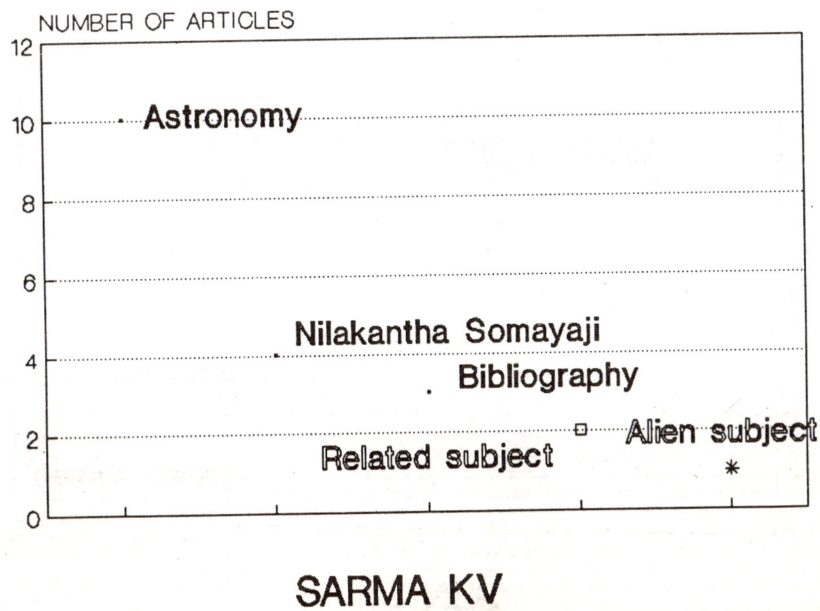

Figure 28.17

**Mapping of Subjects in Articles of G. Sarton**

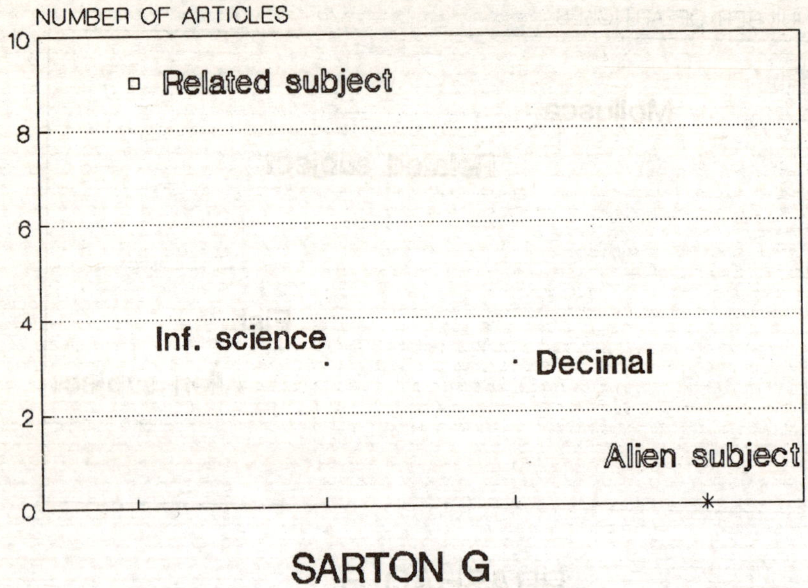


Figure 28.18

**Mapping of Subjects in Articles of S.N. Sen**

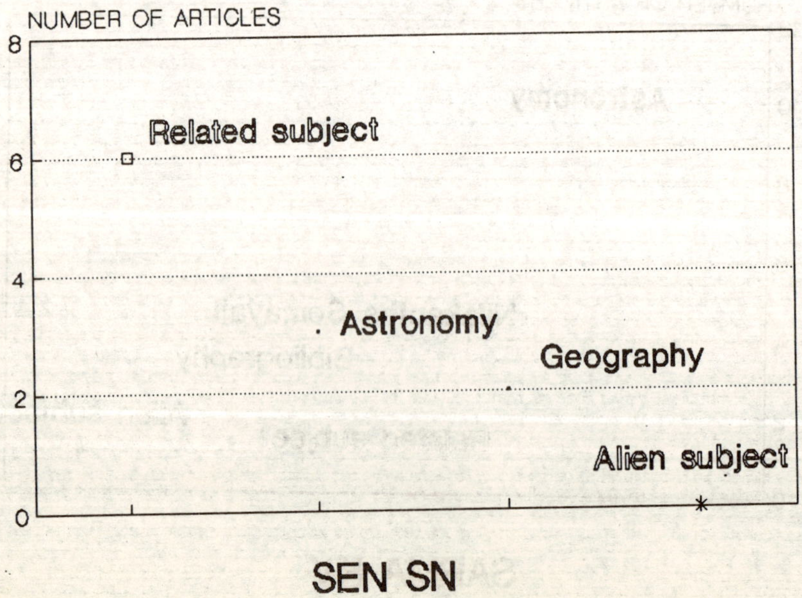

Figure 28.19

Mapping of Subjects in Articles of
P.C. Sengupta

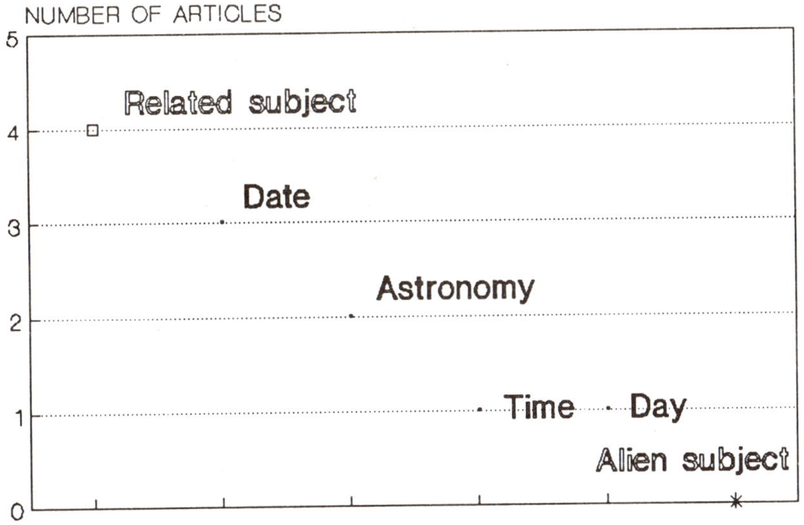


SENGUPTA PC

Figure 28.20

Mapping of Subjects in Articles of
S.D. Sharma

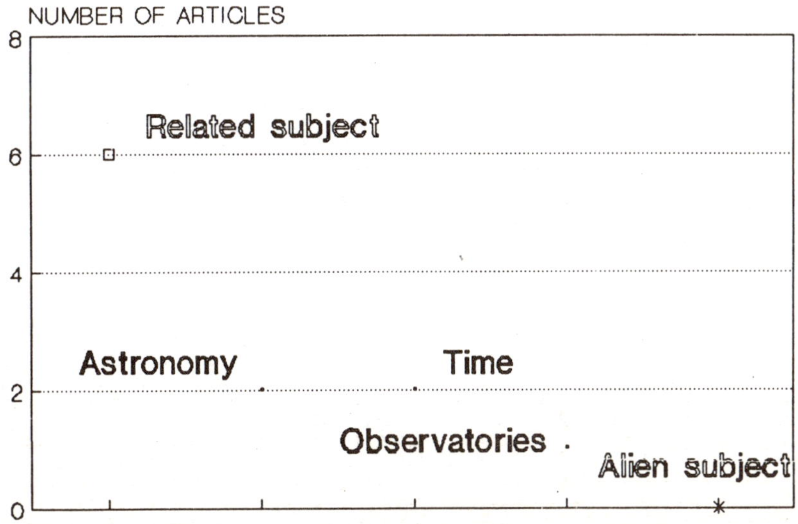


SHARMA SD

**Table 36**

**Ranking of Subject Specialists**

| | Subject Rank |
|---|---|
| **Astronomer** | |
| Sarma KV | 1 |
| Das SR | 2 |
| Sengupta PC | 3 |
| Sharma SD | 4 |
| **Biologist** | |
| Gershenowitz H | 1 |
| Dasgupta HC | 2 |
| **Botanist** | |
| Burkill IH | 1 |
| Biswas K | 2 |
| **Historian of Science** | |
| Sarton G | 1 |
| Sen SN | 2 |
| **Librarian** | |
| Chaudhuri S | 1 |
| **Mathematician** | |
| Gupta RC | 1 |
| Bag AK | 2 |
| Datta B | 3 |
| **Physician** | |
| Hooper D | 1 |
| Mahdihassan S | 2 |
| **Zoologist** | |
| Annandale N | 1 |
| Hora SL | 2 |
| Phillott DC | 3 |
| Prashad B | 4 |

## Table 37

## Preference For Journals By Highly Productive Authors Individually

| AUTHOR | NUMBER OF PAPERS in | | | | | | | | | JOURNAL PREFERENCE |
|---|---|---|---|---|---|---|---|---|---|---|
| | ASIATIC SOCIETY, BENGAL | | | | | | | | | |
| | JPASB | JASBSC | JASBL | JAS | JRAS | IHQ | ISIS | IJHS | VIJ | |
| ANNANDALE N | 50 | | | | | | | | | <JPASB> |
| BAG AK | | | | | | | | 13 | | <IJHS> |
| BISWAS K | 3 | 7 | | | | | | | | JASBSC > JPASB |
| BURKILL IH | 27 | | | | | | | | | <JPASB> |
| CHAUDHURI S | | 1 | 1 | 3 | | 9 | | | | IHQ > JAS > JASBSC = JASBL |
| DAS SR | 2 | | | 1 | | 13 | 2 | | | IHQ > JPASB = ISIS > JAS |
| DASGUPTA HC | 12 | | | | | | | | | <JPASB> |
| DATTA B | 3 | | | | | 3 | 2 | 3 | | JPASB = IHQ = IJHS > ISIS |
| GERSHENOWITZ H | | | | | | | | 17 | | <IJHS> |
| GUPTA RC | | | | | | | | 16 | 2 | IJHS > VIJ |
| HOOPER D | 13 | | | | | | | | | <JPASB> |
| HORA SL | 26 | 15 | 6 | | | | | | | JPASB > JASBSC > JASBL |
| MAHDIHASSAN S | | | | | | | | 9 | 2 | IJHS > VIJ |
| PHILLOTT DC | 13 | | | | | | | | | <JPASB> |
| PRASHAD B | 12 | | | | | | | | | <JPASB> |
| SARMA KV | | | | | | | | 1 | 19 | VIJ > IJHS |
| SARTON G | | | | | | | 15 | | | <ISIS> |
| SEN SN | 1 | | | 1 | | | | 9 | | IJHS > JPASB = JAS |
| SENGUPTA PC | | 4 | 7 | | | | | | | JASBL > JASBSC |
| SHARMA SD | | | | | | | | 10 | 1 | IJHS > VIJ |
| TOTAL | 162 | 27 | 14 | 5 | 0 | 25 | 19 | 78 | 24 | JPASB > IJHS > JASBSC > IHQ > VIJ > ISIS >JASBL > JAS |

Figure 29

## DISTRIBUTION OF CONTRIBUTIONS IN JOURNALS BY HIGHLY PRODUCTIVE AUTHORS IN TOTO

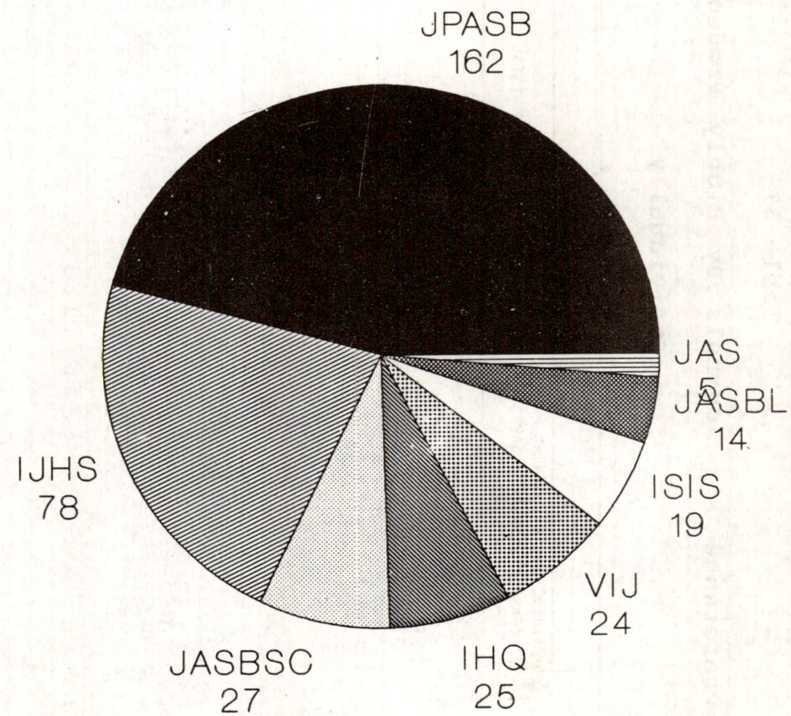

## Figure 30.1

**Distribution of contributions in Journals by Highly productive Authors Individually has been shown Diagramatically in Eleven Figures Ranging from Figure 30.1 to Figure 30.11**

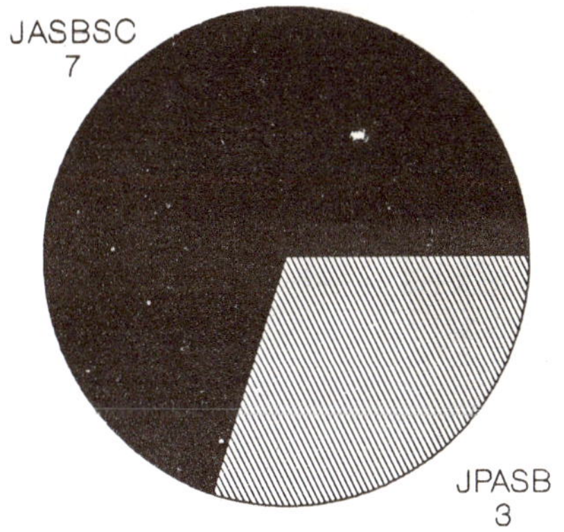


## Figure 30.2
## Distribution Of Contributions In Journals By S. CHAUDHURI

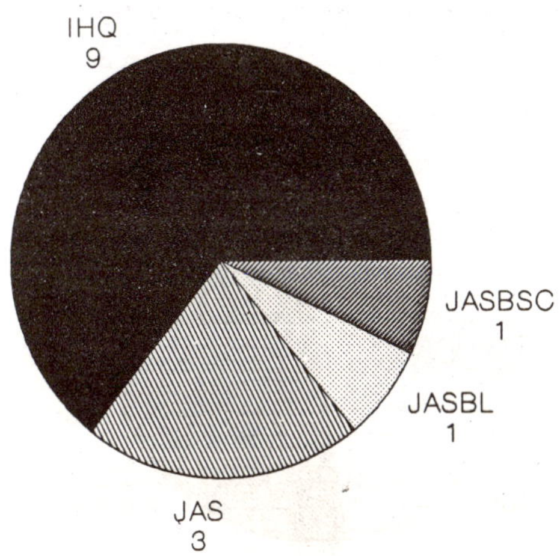

**Figure 30.3**
**Distribution Of Contributions**
**In Journals By S.R. DAS**

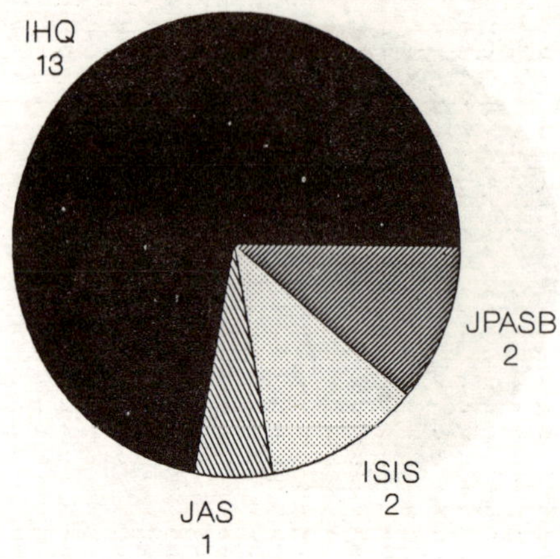


**Figure 30.4**
**Distribution Of Contributions**
**In Journals By B. DATTA**

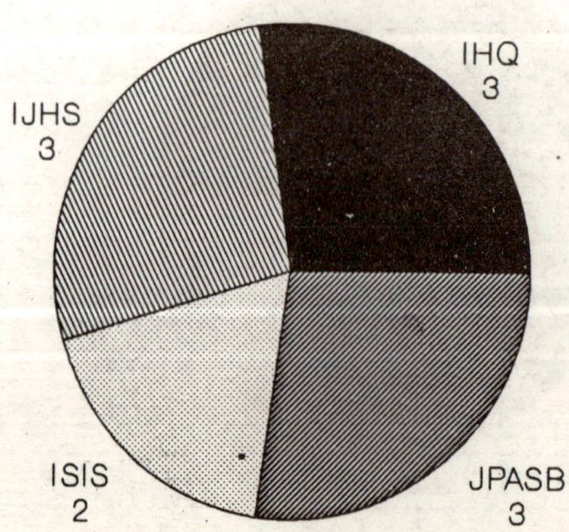

Figure 30.5
Distribution Of Contributions
In Journals By R.C. GUPTA

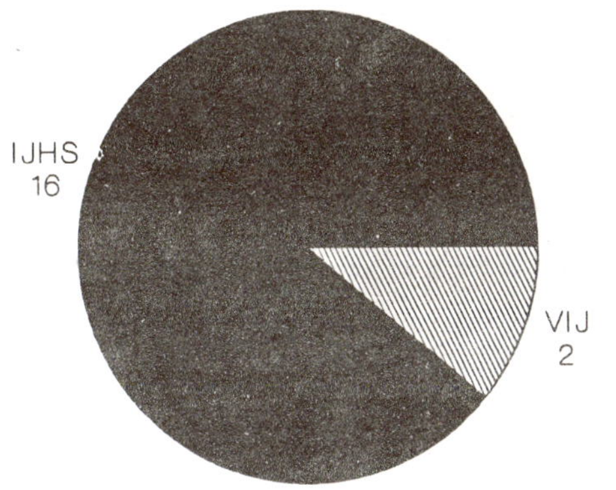


Figure 30.6
Distribution Of Contributions
In Journals By S.L. HORA

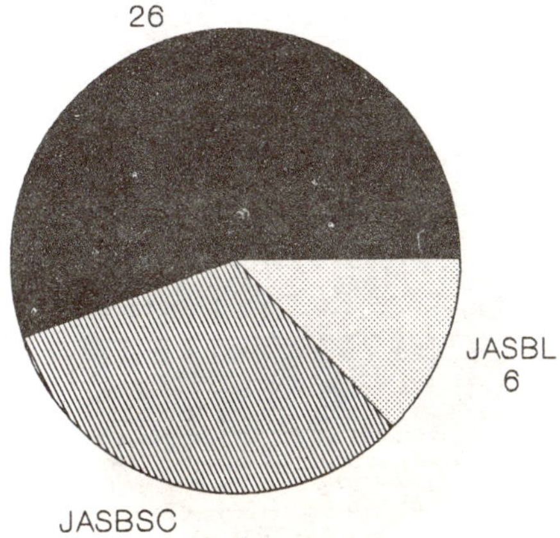

Figure 30.7
Distribution Of Contributions
In Journals By S. MAHDIHASSAN

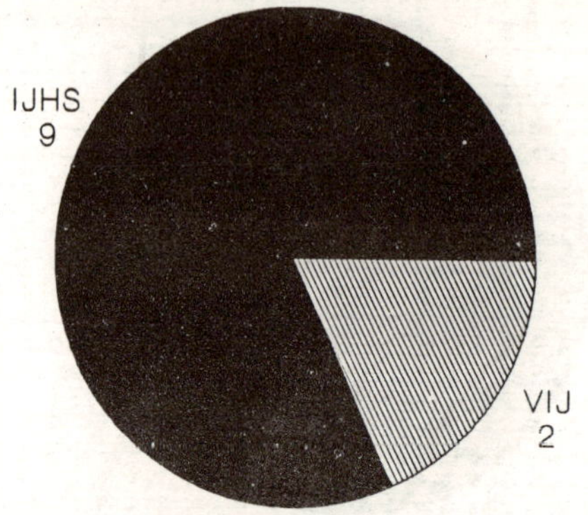


Figure 30.8
Distribution Of Contributions
In Journals By K.V. SARMA

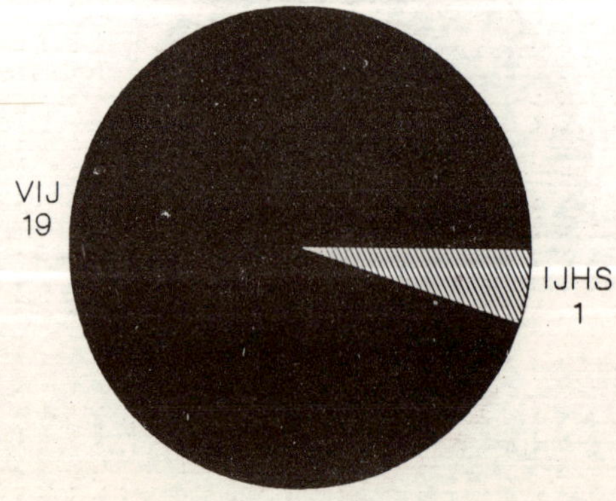

Figure 30.9
Distribution Of Contributions
In Journals By S.N. SEN

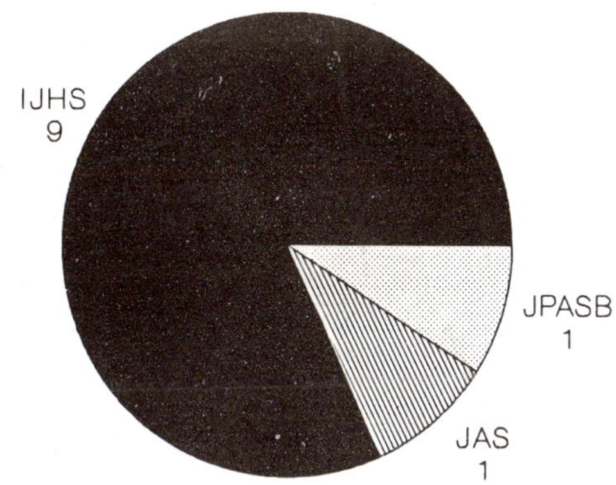


Figure 30.10
Distribution Of Contributions
In Journals By P.C. SENGUPTA

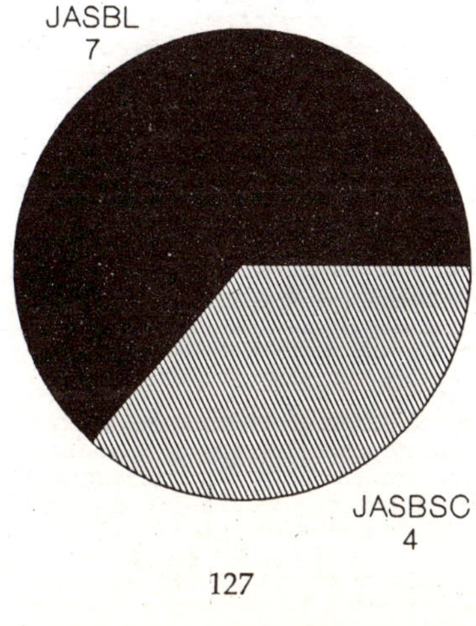

Figure 30.11
Distribution Of Contributions
In Journals By S.D. SHARMA

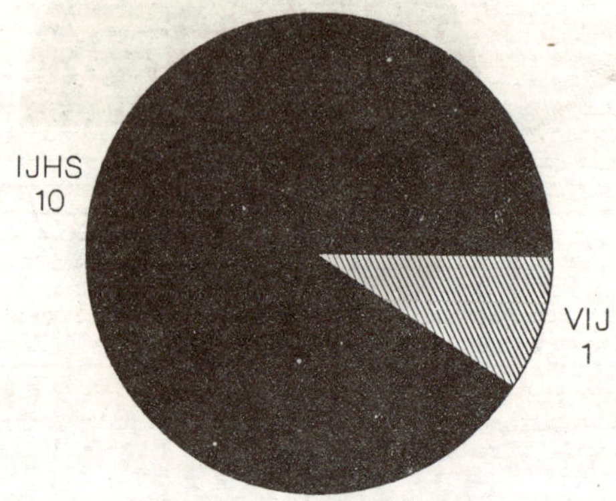

# 7. CONTENT ANALYSIS

Development of research in different scientific fields can be determined by various indicators such as number of publications, concentration of research areas, number of research institutes, etc. Number of publications is one of the most significant indicators. During the initial stages of any scientific development the number of publications in the field grows exponentially, but later saturation or even a decrease is observed. The decrease in the number of publications in a given field is caused by different causes. One of them may be the topical contents of publications which reflect processes of differentiation of scientific fields, the deepening of research in specific fields and mutual penetration of research activities. Therefore, the analysis of the topical content leads in a better way to the state and trend of research development in a particular field of science(1).

## 7.1 ANALYSIS

There are many methods of text analyses at the level of words:content-analysis, thesaurus etc. (2-3). Here, the Keyword Index from the data base on History of Science was used which made it possible to define readily the research topic in a given field. Since keywords themselves are sufficient to reflect the research topic, for the purpose of index, substantive words were picked up and all the nonsubstantive words were rejected. Nonsubstantive words are those words that convey little or no information by themselves, such as articles, prepositions, conjunctions, pronouns, and auxiliary verbs (4-6).

Substantive words were taken from the title of the articles and if it was found that title itself was not fully informative, full text of the article was perused and the words from the sub-headings were taken into consideration (7). In some cases the articles did not have even sub-headings. For this, each paragraph was perused and appropriate keywords were assigned. Sometimes the words were so specific that it was difficult to identify the field of study. In such cases, the generalised words were used. Data from the keyword index was tabulated and various indicators which reflected the development of the field were studied.

### 7.1.1 Frequency Distribution of Keywords

The frequency distribution of keywords was first given by George Kingsley Zipf (1902-1950) in 1935. The words were arranged in decreasing order of frequency so that the most frequent word got rank 1; the next most frequent was ranked 2; and so on. This set of rankordered frequency counts is shown in Table 38. The relationship of the rank-frequency distribution, namely Zipf's Law was stated as:

$$rf = c \qquad \text{...(1)}$$

where r is the rank of a word-type, f is the frequency of occurrence of the word-type, and c is a constant, dependent on the subject (8-10).

The value of the products of rank r and frequency f fall in the range 300-400 in the middle of Table 38, and it was noted that there was no prior reason to expect that the middle products rf would fall within so limited a range.

When stated algebraically, Zipf's Law is usually given in the form of an equation (1), but the law is probably most familiar in the graphic representation of a mathematically equivalent form stated below:-

$$\log rf = \log c$$

$$\log r + \log f = \log c \qquad \text{...(2)}$$

The values of log r, log f and log c were calculated as shown in Table 38. The dashed line in Figure 31 showed the display of Zipf's Law in the form of an equation(2). It was found that Zipf's Law only approximates the relationship between rank r and frequency f for the History of Science data.

### 7.1.2 Keywords with High Frequency

Table 39 presents keywords of high frequency, along with their frequencies. It was found that only 7 keywords had frequency ≥25. The word 'fish' had the highest frequency of 76. The other keywords in the decreasing order of frequency were Astronomy > Time > Bengal > Plant > Aryabhata > Botany with 60, 32, 30, 28, 26 and 26 respectively. In the first block of 1905-09, out of seven keywords only 5 occurred 18 times. All the seven keywords occurred twice, first time in 1925-29 and second time in 1970-74 in 26 and 22 number of contributions. The keyword 'fish' occurred maximum 20 times in the block year 1950-54 and this keyword was at its peak activities during 1920-1954. Plant studies were more in the beginning of the blocks years. Aryabhata, Astronomy, and Time keywords gained momentum in the blocks year 1965-1986.

### 7.1.3 Mapping of Keywords Occurring ≥ 5

The keywords occurring ≥ 5 are presented in Table 40. A total of 140 keywords occurred ≥ 5. In the frequency range of 5-9, 10-14, 15-19, and 20-24 there were 86, 25, 16, and 6 keywords respectively. The term 'Information Science' was found to be occurring 23 times in the literature studied in History of Science, whereas 'Scientometrics' occurred only 5 times in the literature. It was found that there were more terms in the field of biological sciences followed by astronomical sciences and mathematics. Technology enveloped less number of terms in mapping.

### 7.1.4 Text Value

The impact of research paper is indicated by various factors - such as the journal in which the research paper is published, the author'(s) standing and the reputation of the Institute he hails from, recency of the work, etc.(2,7,11). If the ideas expressed in the article hold originality, addition of new thoughts or approach to the origin of idea, backup the impact of the paper.

More consultations and use of references in the papers though give trends and ideas, but were supported heavily by other thoughts, these are thus far away from the creativity and originality of the contents. The views supported by fewer number of references, give an idea of originality and can be much more consulted, and reviewed. In other words, lesser is the number of references in an article of given length, more will be its text value. Text value is therefore, said to be the ratio of number of pages to the number of references.

The above discussion pertains to the text value of the research papers. There may be text value to the overall writings of the scientists. This was found out by taking ratio of the average number of pages per contribution to the average number of references per contribution. And the maximum and minimum deviations were also observed in pages and references as shown in Table 41. Out of the total of twenty highly productive authors, only five authors got text value equal to/ greater than one. I.H.Burkill received the highest text value of 10. D.Hooper and N.Annandale ranked second and third in the text value with 5 and 2 respectively. S.L.Hora ranked fourth with text value as one. G.Sarton's contributions were found to be having text value of 0.9. The text value was also determined for the author's contributions made by them during their most prolific years of contribution as shown in Table 42. As a result, as before, I.H.Burkill and D.Hooper ranked first and second, with text value of 5 and 4.2 respectively, followed by B.Prashad, N. Annandale and S.L.Hora with text value of 3.667, 2.097 and 1.417 respectively.

## Table 38

## Rank-Frequency of Occurrence of Keywords

| Word-Type | Rank r | Log r | Frequency f | Log f. | Product rf=c | Log c |
|---|---|---|---|---|---|---|
| 1 word | 1 | 0 | 76 | 1.88 | 76 | 1.88 |
| 1 word | 2 | 0.30 | 60 | 1.78 | 120 | 2.08 |
| 1 word | 3 | 0.47 | 32 | 1.50 | 96 | 1.98 |
| 1 word | 4 | 0.60 | 30 | 1.48 | 120 | 2.08 |
| 1 word | 5 | 0.69 | 28 | 1.45 | 140 | 2.15 |
| 2 words | 6.5 | 0.81 | 26 | 1.41 | 169 | 2.23 |
| 2 words | 8.5 | 0.93 | 23 | 1.36 | 195 | 2.3 |
| 1 word | 10 | 1 | 22 | 1.34 | 220 | 2.34 |
| 2 words | 11.5 | 1.06 | 21 | 1.32 | 241.5 | 2.38 |
| 1 word | 13 | 1.11 | 20 | 1.30 | 260 | 2.41 |
| 4 words | 15.5 | 1.19 | 19 | 1.28 | 294.5 | 2.47 |
| 3 words | 19 | 1.28 | 18 | 1.25 | 342 | 2.53 |
| 1 word | 21 | 1.32 | 17 | 1.23 | 357 | 2.55 |
| 2 words | 22.5 | 1.35 | 16 | 1.20 | 360 | 2.56 |
| 6 words | 26.5 | 1.42 | 15 | 1.18 | 397.5 | 2.6 |
| 2 words | 30.5 | 1.48 | 14 | 1.15 | 427 | 2.63 |
| 7 words | 35 | 1.54 | 13 | 1.11 | 455 | 2.66 |
| 5 words | 41 | 1.61 | 12 | 1.08 | 492 | 2.69 |
| 3 words | 45 | 1.65 | 11 | 1.04 | 495 | 2.69 |
| 8 words | 50.5 | 1.7 | 10 | 1 | 505 | 2.70 |
| 8 words | 58.5 | 1.77 | 9 | 0.95 | 526.5 | 2.72 |
| 8 words | 66.5 | 1.82 | 8 | 0.90 | 532 | 2.72 |
| 21 words | 81 | 1.91 | 7 | 0.84 | 567 | 2.75 |
| 24 words | 103.5 | 2.01 | 6 | 0.78 | 621 | 2.79 |
| 25 words | 127 | 2.1 | 5 | 0.69 | 635 | 2.8 |
| 48 words | 164.5 | 2.22 | 4 | 0.60 | 658 | 2.82 |
| 78 words | 227.5 | 2.35 | 3 | 0.47 | 682.5 | 2.83 |
| 162 words | 347.5 | 2.54 | 2 | 0.30 | 695 | 2.84 |
| 1134 words | 995.5 | 2.99 | 1 | 0 | 995.5 | 2.99 |

Figure 31

Observed Rank-Frequency Pairs of Keywords

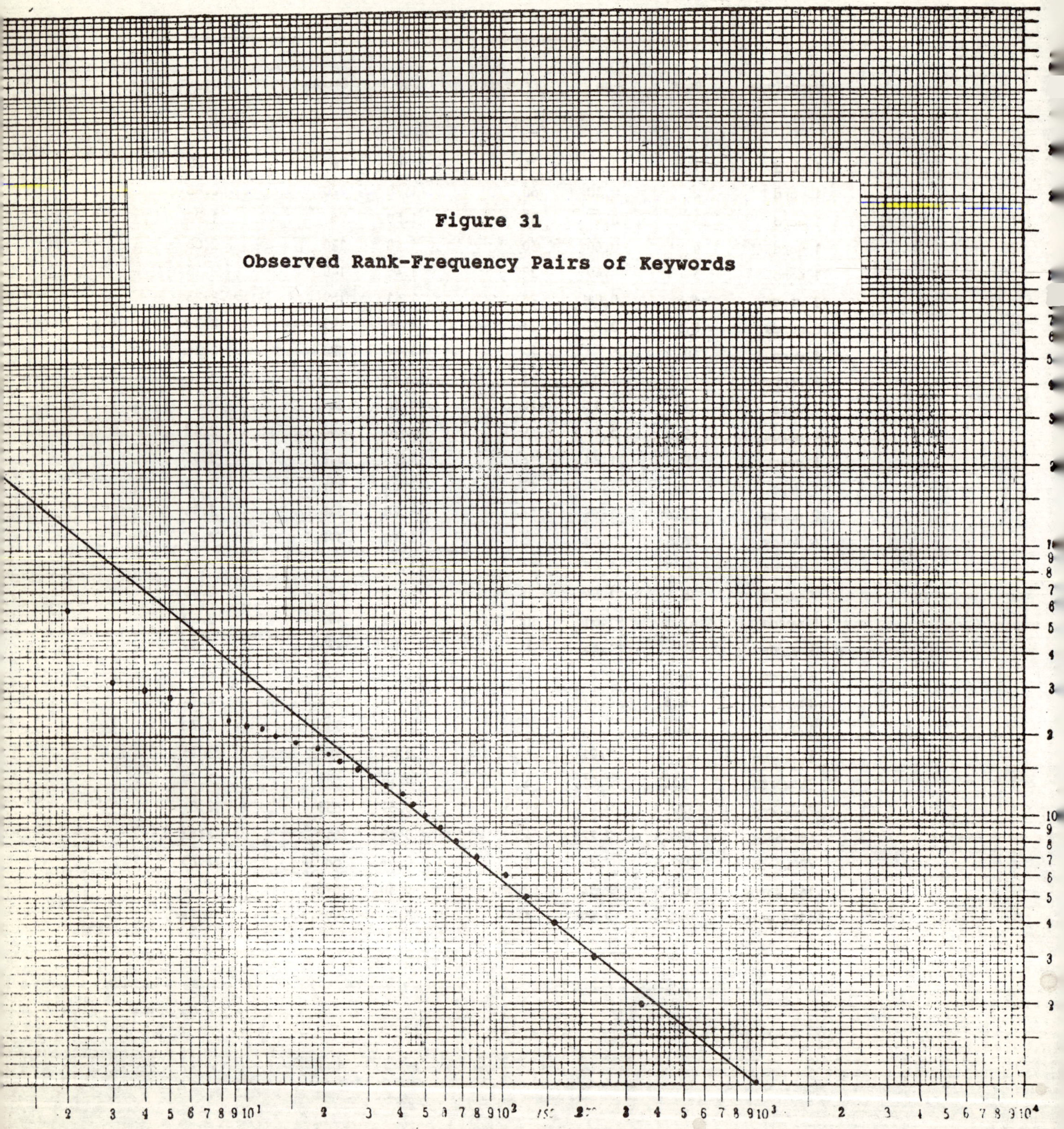

LOG RANK

## Table 39

## Keywords with High Frequencies of Occurrences Blockwise*

| Keyword | Frequency 1905-09 | 10-14 | 15-19 | 20-24 | 25-29 | 30-34 | 35-39 | 40-44 | 45-49 | 50-54 | 55-59 | 60-64 | 65-69 | 70-74 | 75-79 | 80-84 | 85-86 | Total Ocurrences |
|---|---|---|---|---|---|---|---|---|---|---|---|---|---|---|---|---|---|---|
| Aryabhata | 1 | | | | 3 | | | | | | | | 1 | 4 | 16 | | 1 | 26 |
| Botany | 2 | 3 | | 2 | 1 | 1 | | | 1 | | 1 | | 3 | 4 | 2 | 6 | | 26 |
| Plant | 10 | 1 | 3 | 1 | 1 | 3 | | 1 | | | | 1 | 1 | 3 | 1 | 2 | | 28 |
| Bengal | 3 | 1 | | 3 | 2 | 5 | 1 | 2 | | 7 | 1 | | | 1 | 3 | 1 | | 30 |
| Time | | | 1 | | 3 | 4 | 1 | 2 | 2 | 1 | 1 | 1 | 1 | 2 | 7 | 4 | 2 | 32 |
| Astronomy | | | 1 | 1 | 7 | 6 | 1 | 1 | 1 | | 1 | 3 | 6 | 7 | 13 | 7 | 5 | 60 |
| Fish | 2 | 2 | 1 | 7 | 9 | 8 | 4 | 4 | 8 | 20 | 4 | 2 | 1 | 1 | | 2 | 1 | 76 |
| Number of terms in the block year | 5 | 4 | 4 | 5 | 7 | 6 | 4 | 5 | 4 | 3 | 5 | 4 | 6 | 7 | 6 | 6 | 4 | |
| Number of contributions | 18 | 7 | 6 | 14 | 26 | 27 | 7 | 10 | 12 | 28 | 8 | 7 | 13 | 22 | 42 | 22 | 9 | 278 |

***Block Comprises of a duration of Five Years**

## Table 40
## Mapping of Keywords Occuring ≥ 5 Times

MAPPING OF KEY TERMS OCCURRING >=5

| TERMS\5 | 6 | 7 | 8 | 9 | 10 | 11 |
|---|---|---|---|---|---|---|
| BIBLIOGRAPHY | ALGEBRA | AMERICA | BIHAR | ARITHMETIC | AYURVEDIC | EDUCATION |
| CHRONOLOGY | ANATOMY | ARCHITECTURE | CYPRAEACEA | DYE | BHASKARA | AGRICULTURE |
| CLASSIFICATION | ASTROLOGY | ASSAM | MALAYAN | FOSSIL | BURMA | TRADE |
| COPPER | ATHARVAVEDA | BLOOD | MATERIA-MEDICA | MOLLUSCS | CHINA | |
| COSINE | CARAK | BUILDING | MERCURY | NUMERALS | HIMALYAS | |
| CRUSTACEAE | DARWIN | COTTON | MOHENJODARO | PIE | INDET EQUATION | |
| FICUS | EARTH | DISEASES | TRIGNOMETRY | VEDA | INDUSTRY | |
| GALILEE | GLASS | ECILPSES | VARAHAMIHIRA | ZOOLOGY | POLLINATION | |
| GANGES | HARAPPA | GALL | | | | |
| GANITA | MAHABHASYA | GREEK | | | | |
| GEOMETRY | OIL | HERMAPHROD | | | | |
| INDIA | ORISSA | IRON | | | | |
| INSECTS | PANINI | KERALA | | | | |
| IRRIGATION | PARASITE | MADRAS | | | | |
| MAHABHARATA | PHILOSOPHY | MAGIC | | | | |
| MINERAL | PAINGALA | METEOROLOGY | | | | |
| MOON | PSYCHOLOGY | PATANJALI | | | | |
| NEO-CLAMARCK | RIVER | PUNJAB | | | | |
| PAINT | SANSKRIT | SURGERY | | | | |
| PLANETS | SEX | SUSRUTA SAMHITA | | | | |
| PTOLEMY | SOMA | UNANI | | | | |
| RAIN | SOMATOLOGY | | | | | |
| SCIENTOMETRICS | WEIGHT | | | | | |
| SILK | YUNNAN | | | | | |
| SUN | | | | | | |
| COPERNICAN | EUROPE | HISTORY | CALCUTTA | BUDDHIST | DRUG | BIRDS FAUNA |
| GEOLOGY | FLORA | SCIENCE | CALENDAR | TIBERIAS | FRESHWATER | KNOWLEDGETRAN |
| METAL | FLOWER | | CHEMISTRY | | | IBN SINA LAKE |
| POTTERY | GEOGRA | | INDUS | | | MATHEMATICS |
| RAMAYANA | HILSA | | RGVEDA | | | |
| | LAMARC | | SINE | | | |
| | TANTRA | | | | | |

JAIN MEDICINE AL-BIRUNI ANTHROPOLOGY ARYABHATA PLANT BENGAL TIME ASTRONOMY FISH(ERIES)
MYSTICISM INF. SCIENCE BOTANY

## Table 41

## Authors' Contributions in terms of Pages and References

| Author (P/R) ≥0.9 | Pages per contribution | Deviation Max D page | Deviation Min D page | References per contribution | Deviation Max D reference | Deviation Min D reference |
|---|---|---|---|---|---|---|
| Annandale N (2) | 10.30 | 49.70 | 8.30 | 5.68 | 66.32 | 5.68 |
| Bag AK | 8.46 | 7.54 | 4.46 | 19.15 | 38.85 | 16.50 |
| Biswas K | 16.20 | 23.80 | 12.20 | 18.80 | 33.20 | 18.80 |
| Burkill IH (10) | 10.07 | 26.93 | 8.07 | 1.89 | 7.1[illegible] | 1.89 |
| Chaudhuri S | 12.36 | 30.64 | 6.36 | 69.71 | 160.29 | 69.71 |
| Das SR | 12.72 | 16.28 | 10.72 | 23.05 | 79.95 | 23.05 |
| Dasgupta HC | 3.70 | 3.30 | 2.70 | 12.75 | 22.25 | 12.75 |
| Datta B | 23.54 | 46.54 | 16.54 | 59.54 | 218.46 | 59.54 |
| Gershenowitz H | 8.30 | 8.70 | 4.30 | 21.82 | 21.18 | 10.82 |
| Gupta RC | 10.33 | 6.67 | 4.33 | 19.94 | 44.06 | 13.94 |
| Hooper D (5) | 5.15 | 10.85 | 4.15 | 1.50 | 10.50 | 1.50 |
| Hora SL (1) | 8.51 | 16.49 | 5.51 | 8.45 | 35.55 | 8.45 |
| Mahdihassan S | 5.45 | 7.55 | 4.45 | 11.73 | 15.27 | 8.73 |
| Phillott DC | 6.15 | 18.85 | 4.15 | 7.61 | 21.38 | 7.61 |
| Prashad B | 4.50 | 5.50 | 3.50 | 5.70 | 20.30 | 5.70 |
| Sarma KV | 33.25 | 90.65 | 29.3[illegible] | 15.20 | 154.80 | 15.20 |
| Sarton G (0.9) | 11.60 | 80.40 | 10.6 | 12.47 | 106.53 | 12.47 |
| Sen SN | 20.45 | 51.55 | 14.4[illegible] | 49.10 | 131.90 | 48.10 |
| Sengupta PC | 13.45 | 9.55 | 12.[illegible] | 21.64 | 38.36 | 21.64 |
| Sharma SD | 11.73 | 14.27 | 7.[illegible] | 31.73 | 41.27 | 27.73 |

## Table 42

## Authors' Most Prolific Years of Contributions and Ratio of Pages/References

| Author | Most Prolific Year | No. of Contributions | No. of Pages (P) | No. of references (R) | Ratio of P/R |
|---|---|---|---|---|---|
| Annandale N | 1906 | 14 | 65 | 31 | 2.097 |
| Bag AK | 1966,1969 | 2,2 | 16,11 | 30,26 | 0.533,0.423 |
| Biswas K | 1950 | 2 | 35 | 44 | 0.759 |
| Burkill IH | 1907 | 9 | 55 | 11 | 5 |
| Chaudhuri S | 1953 | 4 | 35 | 709 | 0.049 |
| Das SR | 1928 | 5 | 85 | 221 | 0.385 |
| Dasgupta HC | 1910 | 2 | 5 | 7 | 0.714 |
| Datta B | 1927,1930 | 2,2 | 18,18 | 29,92 | 0.621,0.196 |
| Gershenowitz H | 1979,1983 | 4,4 | 32,45 | 89,109 | 0.359,0.413 |
| Gupta RC | 1974 | 3 | 36 | 31 | 1.161 |
| Hooper D | 1911 | 3 | 21 | 5 | 4.2 |
| Hora SL | 1926 | 9 | 51 | 36 | 1.417 |
| Mahdihassan S | 1978,1981,1984,1986 | 2,2,2,2 | 11,20,6,12 | 21,41,10,35 | 0.524,0.489,0.6, |
| Phillott DC | 1907 | 9 | 40 | 90 | 0.444 |
| Prashad B | 1920,1922,1926 | 2,2,2 | 11,6,3 | 3,10,2 | 3.667,0.6,1.5 |
| Sarma KV | 1976,1977 | 4,4 | 109,152* | 13,216 | 8.385,0.704 |
| Sarton G | 1935,1947,1950 | 2,2,2 | 110*,5**,18 | 119,0,14 | 0.924,-,1.286 |
| Sen SN | 1966 | 2 | 23 | 52 | 0.442 |
| Sengupta PC | 1938 | 4 | 59 | 103 | 0.573 |
| Sharma SD | 1985 | 3 | 64 | 155 | 0.413 |

* Bibliographies on the subjects

** More references at the end of article

## REFERENCES

1. A.P.Trofimenko, "Scientometric Analysis of the Topical Content of Scientific Research and its Particularities," *Scientometrics*, 18, No. 5-6 (1990), 409-35.

2. P.R.Bird and M.A.Knight, "Word Count Statistics Titles of Scientific Papers." *The Information Scientist*, 9, No.2 (1975), 67-69.

3. R.Hartsough and J.Laffal, "Content Analysis of Scientific Writings," *The Journal of General Psychology*, 83 (1970), 193-206.

4. J.E.Fedorowicz, "The Theoretical Foundation of Zipf's Law and its Application to the Bibliographic Data Base Environment," *Journal of the American Society for Information Science*, 33 (1982), 285-93.

5. J.E.Fedorowicz, "A Zipfian Model of an Automatic Bibliographic System: An Approach to MEDLINE," *Journal of the American Society for Information Science*, 33 (1982), 223-32.

6. Y.Chen and F.F.Leimkuhler, "Analysis of Zipf's Law: An Index Approach," *Information Processing & Management*, 23, No. 3 (1987), 171-82.

7. J.J.Tocatlian, "Are Titles of Chemical Papers Becoming more Informative?," *Journal of the American Society for Information Science*, 21 (1970), 345-50.

8. G.K.Zipf, *Psycho-Biology of Language.* Houghton: Mifflin, 1935.

9. R.E. Wyllys, "Empirical and Theoretical Bases of Zipf's Law," *Library Trends*, 30, No.1 (1981), 53-64.

10. H.S.Sichel, "On a Distribution Law for Word Frequencies," *Journal of the American Statistical Association*, 70, No. 352 (1975), 542-47.

11. H.S.Sichel, "On a Distribution Representing Sentence-Length in Written Prose," *Journal of the Royal Statistical Society: Series A*, 137 (1974), 25-34.

# 8. CITATION ANALYSIS

Citation analysis is used as a measure of the impact of individual articles, journals, authors, etc. (1-3). It describes various types of linkages between references on the basis of citation counts, co-citation, bibliographic coupling, etc. (4-6). History of Science is on the border line of Sciences and Social Sciences, therefore citations pertaining to the subject appear in both Science Citation Index (SCI) and Social Sciences Citation Index (SSCI). SCI and SSCI for the years 1985-1989 (five years) were scanned for the citations of authors who were found to be highly productive among the journals studied over a period of 1905 - 1986. SSCI hard volumes were scanned and citations of Highly Productive Authors pertaining to the articles appearing in the selected journals were jotted down on the slips. SCI CDROM version on PC-XT 286 were read and articles of the selected journals were searched out, followed by jotting down of computer citations on the data slips. Highly productive authors occurring as co-author(s) was/were also checked in citation index for their citations. Complete bibliographic details of the citations were noted from the source index of SSCI and SCI, and further citations related to the journals covered under study were taken from the reference list of the articles from the source index (2). Various parameters related to the citation were analysed and presented in the form of Tables and Graphs. Related citations of the selected journals and other Indian works were also jotted down from the source index of SSCI and SCI.

## 8.1 CITATION RANK OF THE SCIENTISTS

Highly productive authors were searched for their citation counts in SCI and SSCI, and the rank list of authors as shown in Table 43 was prepared accordingly. Out of the total of 20 highly productive authors, only 13 authors got citations. G.Sarton ranked as first with 21 citations. B.Datta and H.Gershenowitz scored 7 citations and 5 citations respectively, and were consequently placed second and third in the citation rank. Each of the four authors - N.Annandale, S.L.Hora, A.K.Bag, and S.Mahdihassan accounted for 3 citations and thus ranked fourth in the citation rank. Each of K.V.Sarma and R.C.Gupta scored 2 citations and was assigned fifth number in citation rank. Each of the four authors - I.H.Burkill, S.N.Sen, S.D.Sharma, and P.C.Sengupta, with one only citation secured last citation rank.

**Table 43**

**Citation Rank of the Scientists**

| Rank | Name of Scientist | Number of Citations |
|---|---|---|
| 1 | Sarton G | 21 |
| 2 | Datta B | 7 |
| 3 | Gershenowitz H | 5 |
| 4 | Annandale N | 3 |
| 4 | Hora S L | 3 |
| 4 | Bag A K | 3 |
| 4 | Mahdihassan S | 3 |
| 5 | Sarma K V | 2 |
| 5 | Gupta R C | 2 |
| 6 | Burkill I H | 1 |

| | | |
|---|---|---|
| 6 | Sen S N | 1 |
| 6 | Sengupta P C | 1 |
| 6 | Sharma S D | 1 |

## 8.2 CITATIONS IN SCI AND SSCI

Table 44 and Figure 32 present citation details such as cited year, citing year, citation availability of the authors. The earlier paper of N.Annandale 1906 was cited even in 1985. Similarly paper of G.Sarton published in 1913, also got citation in 1986. In addition, the earlier papers of S.L.Hora, B.Datta, P.C. Sengupta, I.H. Burkill were also cited during the period 1985-1989. This implies the historicity of basic research works done by N.Annandale, S.L.Hora, I.H. Burkill etc. In all, 50 papers got cited and received 53 citations. Out of which 24 and 14 citations were found to be exclusively in SSCI, and SCI respectively, whereas 15 citations were cited both in SSCI as well as in SCI as illustrated in the Figure 32.

## 8.3 YEARWISE CITATIONS

The maximum number of citations received in the year 1986 was 18 followed by 11 citations in 1988, 10 citations in 1985 as well as in 1987 individually, and 4 citations in 1989, as shown in Figure 33.

## 8.4 CITATION GAP

It was found in this study that though the literature on History of Science was published continuously from 1905 to 1986 but papers published earlier were cited later after stray gaps of years. For instance, the paper published in 1905, was cited immediately after a year in 1906, whereas the second citation to the literature appeared after a gap of 7 years in 1913, and the third citation to the History of Science appeared after a gap of 3 years in 1916. There were fifteen citation gaps shown by the cited years. Five blocks of different number of years observed continuity in citation occurrences. The highest citation gap was of 15 years from 1962 to 1977, followed by 10 years and 9 years of citation gap in 1943 to 1953 and 1953 to 1962 respectively.

The occurrences of citation gaps were more in the beginning in History of Science, first, second and third citation gaps occurred during the first eleven years of publication period covered in this research study (1905 to 1986). Fourth and fifth occurrences of citation gaps followed in twenty third year and twenty fifth year of publication respectively. The last occurrence of citation gap came in seventy eighth year of publication in the subject. Thus it was found, as is evident from Table 45 and Figure 34 that the trend in occurrences of citation gaps was going up.

## 8.5 BREAKUP OF CITATIONS

Out of the six journals taken into study, only three journals received the citations, besides some books on History of Science in India which also received citations as shown in Figure 35. The ISIS received 21 citations which was the highest among the journals studied. Indian Journal of History of Science (IJHS) and journals of the Asiatic Society of Bengal (ASB) received 11 citations and 7 citations respectively. The other publications which included books - monographs, manuscripts received only 14 citations.

## 8.6 PAPERS WITH CITATIONS AND SOURCE INDEX

Papers with citations related to highly productive authors are listed along with source index in this chapter. Source index gives the complete bibliographic details of the citations. While noting the bibliographic details, the cited items from the references were noted only for those items which are part of the literature of History of Science in India; these cited items are treated as 'related citations'. Precisely, the related citations are cited items related to the part of the literature of History of Science in India at the end of the citing item in the source index. There are 62 related citations. The source-wise breakup of related citations is given below:-

| Source item | Related citations |
|---|---|
| ISIS | 44 |
| IJHS | 2 |
| JASB | 1 |
| History of Ancient India | 2 |
| Others | 13 |

## Table 44

## Citations in Science Citation Index (SCI) and Social Sciences Citation Index (SSCI)

| Author | Cited papers | Cited years | Available in SCI | SSCI | Citations Number | total | Citing Year | Citations in 1985 | 1986 | 1987 | 1988 | 1989 |
|---|---|---|---|---|---|---|---|---|---|---|---|---|
| Annandale N | 3 | 1906 | Y | | 1 | 3 | 1985 | 1 | 0 | 0 | 1 | 1 |
| | | 1923 | Y | | 1 | | 1989 | | | | | |
| | | 1924 | Y | | 1 | | 1988 | | | | | |
| Bag A K | 3 | 1979 | Y | | 1 | 3 | 1986 | 0 | 1 | 0 | 2 | 0 |
| | | 1984 | | Y | 1 | | 1988 | | | | | |
| | | 1984 | | Y | 1 | | 1988 | | | | | |
| Burkill I H | 1 | 1916 | Y | | 1 | 1 | 1985 | 1 | 0 | 0 | 0 | |
| Datta B | 4 | 1935 | Y | Y | 2 | 7 | 1987, 1987 | 0 | 2 | 3 | 2 | 0 |
| | | 1938 | Y | Y | 2 | | 1987, 1988 | | | | | |
| | | 1962 | Y | | 2 | | 1986, 1988 | | | | | |
| | | 1983* | | Y | 1 | | 1986 | | | | | |
| Gershenowitz H | 5 | 1983 | | Y | 1 | 5 | 1986 | 0 | 3 | 0 | 2 | 0 |
| | | 1983 | | Y | 1 | | 1986 | | | | | |
| | | 1983 | | Y | 1 | | 1986 | | | | | |
| | | 1984 | | Y | 1 | | 1988 | | | | | |
| | | 1984 | | Y | 1 | | 1988 | | | | | |
| Gupta R C | 2 | 1981 | | Y | 1 | 2 | 1987 | 0 | 0 | 2 | 0 | 0 |
| | | 1983 | | Y | | | 1987 | | | | | |
| Hora S L | 3 | 1933 | | Y | 1 | 3 | 1986 | 1 | 2 | 0 | 0 | 0 |
| | | 1934 | Y | | 1 | | 1985 | | | | | |
| | | 1943 | Y | | 1 | | 1986 | | | | | |
| Mahdihassan S | 3 | 1962* | | Y | 1 | 3 | 1986 | 1 | 1 | 0 | 1 | 0 |
| | | 1981* | | Y | 1 | | 1985 | | | | | |
| | | 1984* | | Y | 1 | | 1988 | | | | | |
| Sarma K V | 2 | 1977 | Y | | 1 | 2 | 1986 | 0 | 1 | 0 | 1 | 0 |
| | | 1984 | | Y | 1 | | 1988 | | | | | |
| Sarton G | 21 | 1913 | Y | Y | 1 | 21 | 1986 | 5 | 8 | 5 | 0 | 3 |
| | | 1921 | Y | Y | 1 | | 1989 | | | | | |
| | | 1922 | | Y | 1 | | 1987 | | | | | |
| | | 1924 | Y | | 1 | | 1989 | | | | | |
| | | 1924 | | Y | 1 | | 1987 | | | | | |
| | | 1924 | Y | Y | 1 | | 1986 | | | | | |
| | | 1928 | Y | Y | 1 | | 1986 | | | | | |
| | | 1930 | Y | Y | 1 | | 1985 | | | | | |
| | | 1931 | Y | | 1 | | 1987 | | | | | |
| | | 1933 | | Y | 1 | | 1985 | | | | | |
| | | 1935 | Y | | 1 | | 1986 | | | | | |
| | | 1935 | | Y | 1 | | 1987 | | | | | |
| | | 1935 | Y | Y | 1 | | 1986 | | | | | |
| | | 1935 | Y | Y | 1 | | 1986 | | | | | |
| | | 1935 | | Y | 1 | | 1985 | | | | | |
| | | 1935 | Y | Y | 1 | | 1986 | | | | | |
| | | 1938 | Y | Y | 1 | | 1985 | | | | | |
| | | 1939 | Y | Y | 1 | | 1985 | | | | | |
| | | 1939 | | Y | 1 | | 1989 | | | | | |
| | | 1953 | Y | | 1 | | 1987 | | | | | |

| | | | | | | | | | | | | |
|---|---|---|---|---|---|---|---|---|---|---|---|---|
| | | 1953 | | Y | 1 | | 1986 | | | | | |
| Sen S N | 1 | 1983 | | Y | 1 | 1 | 1988 | 0 | 0 | 0 | 1 | 0 |
| Sengupta P C | 1 | 1941 | Y | Y | 1 | 1 | 1985 | 1 | 0 | 0 | 0 | 0 |
| Sharma S D | 1 | 1984* | | Y | 1 | 1 | 1988 | 0 | 0 | 0 | 1 | 0 |
| Total 13 | 50 | | | | | 53 | | 10 | 18 | 10 | 11 | 4 |

* Cited in Bibliography

Figure 32

Citations in Citation Indexes

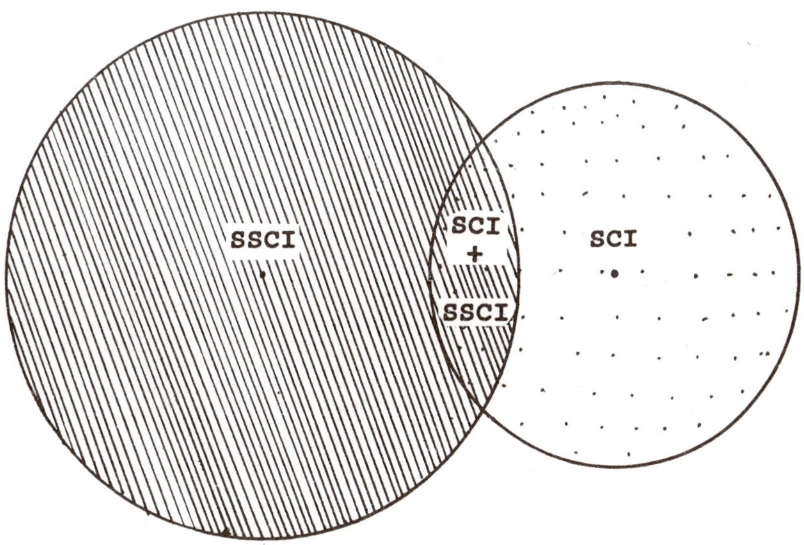

Table 45

Citation Gaps in the Cited Years

| Cited Year | Citations | Citation Gap in nth Year | nth Citation Gap |
|---|---|---|---|
| 1906 | 1 | 1 | 0 |
| 1913 | 1 | 8 | 1 |
| 1916 | 1 | 11 | 2 |
| 1921 | 1 | 16 | 3 |
| 1922 | 1 | 17 | 0 |
| 1923 | 1 | 18 | 0 |
| 1924 | 4 | 19 | 0 |
| 1928 | 1 | 23 | 4 |
| 1930 | 1 | 25 | 5 |
| 1931 | 1 | 26 | 0 |
| 1933 | 2 | 28 | 6 |
| 1934 | 1 | 29 | 0 |
| 1935 | 7 | 30 | 0 |
| 1938 | 2 | 33 | 7 |
| 1939 | 2 | 34 | 0 |
| 1941 | 1 | 36 | 8 |
| 1943 | 1 | 38 | 9 |
| 1953 | 2 | 48 | 10 |
| 1962 | 2 | 57 | 11 |
| 1977 | 1 | 72 | 12 |
| 1979 | 1 | 74 | 13 |
| 1981 | 2 | 76 | 14 |
| 1983 | 6 | 78 | 15 |
| 1984 | 7 | 79 | 0 |
| 1985 | 0 | 80 | 0 |
| 1986 | 0 | 81 | 0 |

Figure 33

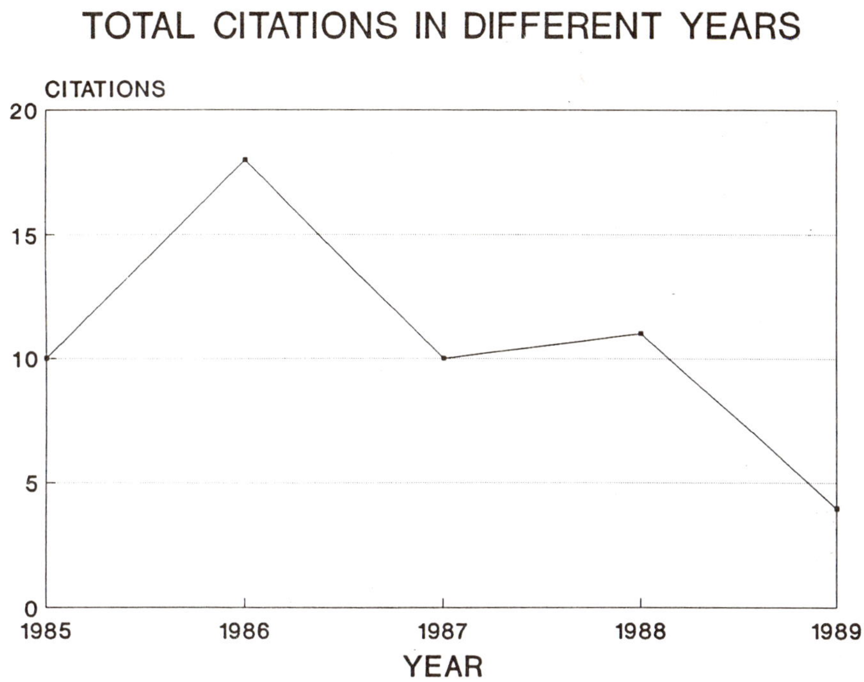


Figure 34

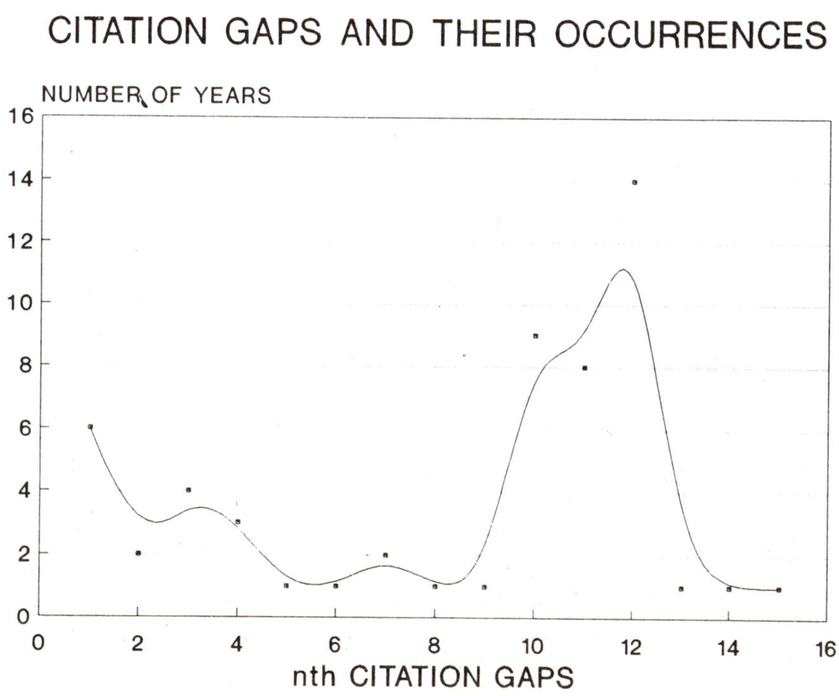

Figure 35

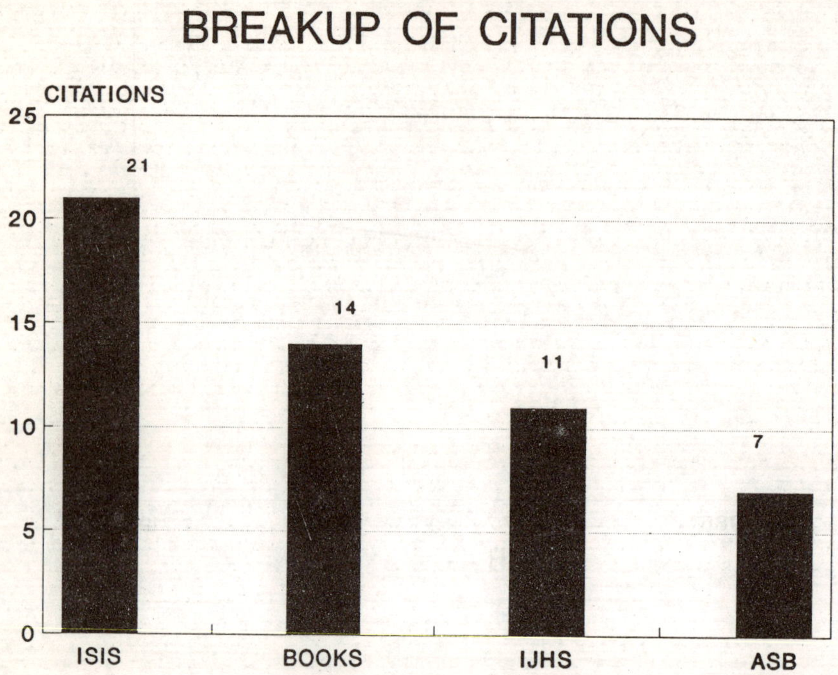
BREAKUP OF CITATIONS
CITATIONS
25
20
15
10
5
0
21
14
11
7
ISIS
BOOKS
IJHS
ASB

## PAPERS WITH CITATIONS

ANNANDALE N

| | | | | | |
|---|---|---|---|---|---|
| 1906 | Annandale N | J Asiat Soc Bengal | 2 | 289 | |
| | Tilak R | Zool Anzeiger | 215 | 231 | 85 |
| 1923 | Annandale N | J P Asiatic Soc Beng | 18 | 555 | |
| | Turner RD | Ophelia | 30 | 155 | 89 |
| 1924 | Annandale N | J P Asiatic Soc Beng | 19 | 399 | |
| | Davis GM | P A Nat SCI Philad | 138 | 466 | 86 |

BAG AK

| | | | | | |
|---|---|---|---|---|---|
| 1979 | Bag AK | Math Ancient Medieva | | 76 | |
| | Lam Ly | Arch Hist Ext SCI | 36 | 183 | 86 |
| 1984 | Bag AK | Indian J Hist SCI | 19 | 89 | |
| | Neu J | ISIS | R 78 | 5 | 87 |
| 1984 | Bag AK | Indian J Hist SCI | 19 | 397 | |
| | Neu J | ISIS | R 78 | 5 | 87 |

BURKILL IH

| | | | | | |
|---|---|---|---|---|---|
| 1916 | Burkill IH | P Asiat Soc Bengal | 12 | 239 | |
| | Nilsson LA | Plant Sys Evol | 150 | 223 | 8 |

DATTA B

| | | | | | |
|---|---|---|---|---|---|
| 1935 | Datta B | Hist Hindu Math | 1 | | |
| | Lam LY | Archiv Hist Ext SCI | 37 | 365 | 67 |
| | Joseph GG | Race Class | 28 | 13 | 87 |
| 1938 | Datta B | Hist Hindu Math | 2 | | |
| | Joseph GG | Race Class | 28 | 13 | 87 |
| | Shen KS | Arch Hist Ext SCI | 38 | 285 | 88 |
| 1962 | Datta B | Hist Hindu Math | | | |
| | Lam LY | Arch Hist Ext SCI | 36 | 183 | 86 |
| | Lam LY | Arch Hist Ext SCI | 38 | 101 | 88 |
| 1983 | Datta B | Indian J Hist SCI | 18 | 29 | |
| | Neu J | ISIS | R 76 | 5 | 85 |

GERSHENOWITZ H

| | | | | | |
|---|---|---|---|---|---|
| 1983 | Gersheno.H | Indian J Hist SCI | 18 | 1 | |
| | Neu J | ISIS | R 76 | 5 | 85 |
| 1983 | Gersheno.H | Indian J Hist SCI | 18 | 137 | |
| | Neu J | ISIS | R 76 | 5 | 85 |

| | | | | | | |
|---|---|---|---|---|---|---|
| 1983 | Gersheno.H | Indian J Hist SCI | | 18 | 154 | |
| | Neu J | ISIS | R | 76 | 5 | 85 |
| 1984 | Gersheno.H | Indian J Hist SCI | | 19 | 54 | |
| | Neu J | ISIS | R | 78 | 5 | 87 |
| 1985 | Gersheno. H | Indian J Hist SCI | | 19 | 261 | |
| | Neu J | ISIS | R | 78 | 5 | 87 |

GUPTA RC

| | | | | | | |
|---|---|---|---|---|---|---|
| 1981 | Gupta RC | Ganita Bharti | | 3 | 86 | |
| | Neu J | ISIS | R | 77 | 5 | 86 |
| 1983 | Gupta RC | Ganita Bharti | | 5 | 8 | |
| | Neu J | ISIS | R | 77 | 5 | 86 |

HORA SL

| | | | | | | |
|---|---|---|---|---|---|---|
| 1933 | Hora SL | J Asiatic Soc Bengal | | 29 | 1 | |
| | Arnold D | Past Present | | | 118 | 86 |
| 1934 | Hora SL | J P Asiatic Soc Bengal | | 29 | 271 | |
| | Alhassan JM | Marine Biol | | 88 | 117 | 85 |
| 1943 | Hora SL | J R Asiatic Soc Bengal | | 9 | 1 | |
| | Anurahda S | Rev Suiss Zool | | 93 | 291 | 86 |

MAHDIHASSAN S

| | | | | | | |
|---|---|---|---|---|---|---|
| 1962 | Mahdihas. S | Essays Hist Alchemy | | | | |
| | Neu J | ISIS | R | 76 | 5 | 85 |
| 1981 | Mahdihas. S | Indian J Hist SCI | | 16 | 223 | |
| | Cutcliff SH | Technol Cul | R | 26 | 353 | 85 |
| 1984 | Mahdihas. S | Indian J Hist SCI | | 19 | 74 | |
| | Neu J | ISIS | R | 78 | 5 | 87 |

SARMA KV

| | | | | | | |
|---|---|---|---|---|---|---|
| 1977 | Sarma KV | Panjab U Indological | | 10 | | |
| | Rajagopal CT | Arch Hist Ext SCI | | 35 | 91 | 86 |
| 1984 | Sarma KV | Vedanga Juotisa ITS | | | | |
| | Neu J | ISIS | R | 78 | 5 | 87 |

SARTON G

| | | | | | | |
|---|---|---|---|---|---|---|
| 1913 | Sarton G | ISIS | | 1 | 3 | |
| | Merton RK | ISIS | | 76 | 470 | 85 |
| 1921 | Sarton G | ISIS | | 4 | 25 | |
| | Richards E | Social ST S | | 19 | 697 | 89 |
| 1922 | Sarton G | ISIS | | 4 | 512 | |
| | Hon G | ANN SCI | R | 44 | 545 | 87 |
| 1924 | Sarton G | ISIS | | 6 | 26 | |

| | | | | | |
|---|---|---|---|---|---|
| | Richards E | Soc Stud SCI | 19 | 697 | 89 |
| 1924 | Sarton G | ISIS | 6 | 67 | |
| | Christie JR | Hist SCI | 25 | 99 | 87 |
| 1924 | Sarton G | ISIS | 6 | 74 | |
| | Reingold N | BR J Hist SCI | 19 | 243 | 86 |
| 1928 | Sarton G | ISIS | 10 | 67 | |
| | Glick TF | ISIS | 76 | 487 | 85 |
| 1930 | Sarton G | ISIS | 14 | 8 | |
| | Ober WB | AM J Dermatol | 6 | 541 | 84 |
| 1931 | Sarton G | ISIS | 16 | 315 | |
| | Moore JA | American Zoologist | 27 | 415 | 87 |
| 1933 | Sarton G | ISIS | 20 | 262 | |
| | Barmore FE | J Near East | 44 | 81 | 85 |
| 1935 | Sarton G | ISIS | 23 | | |
| | Sheynin OB | Arch Hist Ext SCI | 36 | 281 | 86 |
| 1935 | Sarton G | ISIS | 23 | 4 | |
| | Beirne P | AM J Sociol | 92 | 1140 | 87 |
| 1935 | Sarton G | ISIS | 23 | 6 | |
| | Stigler SM | J AM Stat A | 81 | 879 | 86 |
| 1935 | Sarton G | ISIS | 23 | 313 | |
| | Reingold N | BR J Hist S | 19 | 243 | 86 |
| 1935 | Sarton G | ISIS | 23 | 319 | |
| | Warner JH | Osiris | 1 | 37 | 85 |
| 1935 | Sarton G | ISIS | 24 | 107 | |
| | Merton RK | ISIS | 76 | 470 | 85 |
| 1935 | Sarton G | ISIS-Brugge | 29 | 99 | |
| | Kaunzner W | Archiv Hist Ext SCI | 32 | 1 | 85 |
| 1938 | Sarton G | ISIS | 30 | 495 | |
| | Rotton J | Psychol Bull | R 97 | 287 | 85 |
| 1939 | Sarton G | ISIS | | 31 | 118 |
| | Cohen IB | ISIS | 79 | 571 | 88 |
| 1953 | Sarton G | ISIS | 44 | 99 | |
| | Huang HT | Bioscience | 37 | 665 | 87 |
| 1953 | Sarton G | ISIS | 44 | 233 | |
| | Kertesz GA | ISIS | N 77 | 497 | 86 |
| SEN SN | | | | | |
| 1983 | Sen SN | Sulbasutras | | | |
| | Gupta RC | Centaurus | 31 | 114 | 88 |
| SENGUPTA PC | | | | | |
| 1941 | Sengupta PC | Brahamagupta Khandak | | | |
| | Abraham G | Arch Hist E | 34 | 221 | 85 |

SHARMA SD

| 1984 | Chhabra JG | Indian J Hist SCI | | 19 | 18 | |
|---|---|---|---|---|---|---|
| | Neu J | ISIS | R | 78 | 5 | 87 |

**Source Index**

(With Related Citations)

Abraham G. Theories for the motion of the sun in the Pancasiddhantika. Archive for *History of Exact Sciences*. 1985,34(30):221-230.12R.
Madras Christian Coll, Madras 600059, Tamil Nadu, India.

| Chaterjee B | Brahamagupta-Khandak | | 70 |
|---|---|---|---|
| Sastri TSK | Visheveshvaranand | 11 | 73 |
| Thibaut G | Pancasiddhantika-Var | | 1889 |

Alhassan JM, Criddle RS, Criddle KR, Thomson M, Summers B. Catfish epidermal secretions in response to threat or injury- A Novel defense response. *Marine Biology*. 1985, 88(2) : 117-123. 43R.
Kuwait UNIV, FAC SCI, Dept Biochem, POB 5969, Kuwait.

Anuradha S. Contributions to the study of Bagrid Fish. 19. Systematic position of Macrones-Halepensis-Colvillii Hora and Misra, 1943, with description of a new species (Siluriformes, Bagridae). *Revue Suisse De Zoologie*. 1986, 93(2) : 291-296. 6R.
Zoological Survey India, Indian Museum, fire Proof Spirit Bldg, 27 Jawaharlal Nehru Rd.,Calcutta 700016, India.

Arnold D. Cholera and colonialism in British India. *Past and present*. 1986 (113) : 118-151. 102R.
Univ Lancaster, Lancaster LA1 4YW, England.

Barmore FE, Turkish Mosque Orientation and the Secular Variation of the Magnetic-Declination. *Journal of Near Eastern Studies*. 1985, 44(2):81-98. 34R.
Univ Wisconsin, LA Crosse, WI 54601, USA.

| Renaud HPI | ISIS | 34 | 24 | 42 |
|---|---|---|---|---|
| Terrase H | ISIS | 24 | 109 | 35 |

Beirne P. Quetelet, Adolphe and the Origins of Positivist Criminology. American Journal of *Sociology*. 1987, 92(5): 1140-1169. 75R.
Univ So Marine, Dept Social, Portland, ME 04103, USA.

Christie JRR. Narrative and Rhetoric in Metzger, Helene Historiography of 18th-Century Chemistry. *History of Science*. 1987, 25(6) : 99-109. 24R.
Univ Leeds. Leeds LS2 9JT, W Yorkshire, England.

| Guerlac H | ISIS | 18 | 124 | 57 |
|---|---|---|---|---|
| Guerlac H | ISIS | 18 | 433 | 57 |
| Morris R | ISIS | 60 | 347 | 69 |
| Pelseneer J | ISIS | 15 | 351 | 31 |

Cohen IB. The Publication of Science, Technology and Society-Circumstances and Consequences. *ISIS*. 1988, 79(299):571-583. 39R.
Harvard Univ, Dept Hist SCI, SCI CTR 235, Cambridge, MA 02138.

| | | | | |
|---|---|---|---|---|
| (Annon) | ISIS | 29 | 303 | 38 |
| (Annon) | ISIS | 30 | 567 | 39 |
| (Annon) | ISIS | 33 | 181 | 41 |
| (Annon) | ISIS | 33 | 185 | 41 |
| Abraham GA | ISIS | 74 | 368 | 83 |
| Aharoni I | ISIS | 1 | 461 | 36 |
| Barner SB | ISIS | 6 | 65 | 36 |
| Birkhoff G | ISIS | 1 | 260 | 36 |
| Cohen IB | ISIS | 31 | 327 | 40 |
| Darby GOS | ISIS | 1 | 251 | 36 |
| Hykes OV | ISIS | 1 | 464 | 36 |
| Jones RF | ISIS | 31 | 439 | 40 |
| Merton RK | ISIS | 76 | 476 | 85 |

Cutcliffe SH,Roysdon CM, Mistiche JA. Current Bibliography in The History of Technology (1983) (Review). *Technology and Culture*. 1985,26(2):353-436. 999R.

Davis GM, Zhou YF, Zheng LC, Yang HM, Hoagland KE, Guo YH, Chen DJ, Chen PL. Anatomy and Systematics of Triculini (Prosobranchia, Pomatiopsidae, Triculinae), Fresh-Water Snails from Yunnan, China, with descriptions of New Species, Proceedings of the Academy of National Sciences of Philadelphia. 1986, 138(2):466-575. 26R.
Acad Nat SCI Philadephia, 19th & Pkwy, Philadelphia, PA 19103, USA.

Glick TF. Sarton, George and the Spanish Arabists. *ISIS*. 1985, 76(284):487-499. 38R.
Boston Univ. Dept Hist, Boston, MA 02215, USA.

| | | | | |
|---|---|---|---|---|
| Sarton G | ISIS | 10 | 67 | 28 |
| Glick TF | ISIS | 68 | 276 | 77 |

Gupta RC. New Indian Values of PI from the Manava Sulba Surta. Centaurus. 1988, 31(2) : 114-126. 11R.
Birla Inst Technol & SCI, PO Mesra, Ranchi 835215, Bihar, India.

| | | | | |
|---|---|---|---|---|
| Bhattacharya V | Baudhayana Sulbasutra | 70 | | 79 |
| Datta B | SCI Sulba | 21 | | 32 |

| | | | | |
|---|---|---|---|---|
| Khadilkar SD | Katyayana Sulba Sutra | 60 | | 74 |
| Kulkarni RP | Indian J Hist SCI | 13 | 32 | 78 |
| Sen SN | Sulbasutras | | | 83 |
| Shukla KS | Indian J Hist SCI | 15 | 151 | 80 |
| Thibaut G | Math Making Ancient | | 93 | 84 |
| Vangelder JM | Manava Srautasutra | | 300 | 60 |

Hon G. On Kepler Awareness of the Problem of Experimental Error. (Review). *Annals of Science*. 1987, 44(6):545-591.96R.

| | | | | |
|---|---|---|---|---|
| Aiton EJ | ISIS | 60 | 75 | 69 |
| Panofsky E | ISIS | 47 | 5 | 56 |
| Panofsky E | ISIS | 47 | 9 | 56 |
| Panofsky E | ISIS | 47 | 13 | 56 |
| Rosen E | ISIS | 47 | 79 | 56 |

Huang HT, Yang P. The Ancient Cultural Citrus Ant. *Bioscience*. 1987, 37(9):665-671. 19R.
Natl SCI FDN, Biochem, Washington, DC 20550, USA.

Joseph GG. Foundations of Eurocentrism In Mathematics. *Race Class*. 1987, 28(3):13-28. 22R.
Univ Manchester, Dept Econometr & Social Stat, Manchester M13 9PL, Lancs, England.

| | | | | |
|---|---|---|---|---|
| Srinivasi. CN | Hist Ancient Indian | | | 67 |

Kaunzner W.(GE) An early Latin Treatment of the AL-Khwarizmis Algebra in MS-Lyell-52 of The Bodleian-Library, Oxford. *Archive for History of Exact Sciences*. 1985, 32(1): 1-16. 49R.

| | | | | |
|---|---|---|---|---|
| Bortolotti E | ISIS Brugge | 29 | 410 | 38 |
| Kapp AG | ISIS | 22 | 152 | 34 |
| Thorndike L | ISIS Brugge | 13 | 81 | 29 |

Kertesz GA. Notes on ISIS-Von-Oken, 1817-1848. (Notes). *ISIS*. 1986, 77 (288):497-503. 35R.
Monash Univ. Dept Hist, Clayton. VIC 3168, Australia.

| | | | | |
|---|---|---|---|---|
| (Annon) | ISIS Oken | 37 | 1844 | |

LAM LY. The Conceptual Origins of our Numeral System and the Symbolic form of Algebra. *Archive for History of Exact Sciences*. 1986, 36(3):183-195. 42R.
Natl Univ Singapore, Dept Math, Singapore 0511, Singapore.

| | | | | |
|---|---|---|---|---|
| Bag AK | Math Ancient Medieva | | 76 | 79 |
| Kaye GR | J Asiat Soc Bengal | 3 | 475 | 07 |
| LAM LY | ISIS | 61 | 92 | 70 |
| LAM LY | ISIS | 63 | 370 | 72 |
| Smith DE | Hindu Arabic Numeral | | 41 | 11 |

LAM LY. Linkage-Exploring the similarities between the Chinese Rod Numeral System and Our Numeral System. *Archive for History of Exact Sciences*. 1987, 37(4) : 365-392. 19R.
Natl Univ Singapore, Dept Math, Singapore 0511, Singapore.

LAM LY. A Chines Genesis-Rewriting The History of Our Numeral System. *Archives For History of Exact Sciences*. 1988, 38(2) : 101-108. 9R.
Natl Univ Singapore, Dept Math, Singapore 0511, Singapore.

Merton RK. Sarton, George-Episodic Recollections by an Unruly Apprentice. *ISIS*. 1985, 76(284): 470-486. 26R.
Columbia Univ. New York, NY 10027. USA.

| | | | | |
|---|---|---|---|---|
| Cohen IB | ISIS | 75 | 13 | 84 |
| Merton RK | ISIS | 27 | 493 | 37 |
| Thackray A | ISIS | 63 | 473 | 72 |

Moore JA. Science as a Way of Knowing-Developmental Biology. *American Zoologist*. 1987,27(2):415-573. 395R.
Univ Calif Riverside, Dept Biol, Riverside, CA 92521, USA.

| | | | | |
|---|---|---|---|---|
| Baxter AL | ISIS | 68 | 363 | 77 |
| Omalley CD | ISIS | 47 | 117 | 56 |

Neu J. 110th Critical Bibliography of the History of Science and its Cultural Influences. (Review.) *ISIS*. 1985,76(285):5-240.999R.
Univ Wisconsin, Memorial Library, Madison, WI 53706, USA.

Neu J. 111th Critical Bibliography of the History of Science and its Cultural Influences. (Review). *ISIS*.1986,77(290):5-265.999R.

UnivWisconsin, Memorial Library, Madison, WI 53706, USA.

Neu J. Critical Bibliography 1987-112th Critical Bibliography of the History of Science and its Cultural Influences. (Review). *ISIS*. 1987, 78(295):5-244. 999R.
Univ Wisconsin, Memorial Library, Madison, WI 53706, USA.

Nilsson LA, Rason L, Randrianjohany E, Jonsson L. Pollination of Plectranthus-Vestitus (Lamiaceae) by Trap-Lining Hovering Bees in Madagascar. Plant Systematics and Evolution. 1985, 150(3-4):223-236. 21R.
Univ Uppsala, Inst Systematic BOT, POB 541, S-75121, Uppsala, Sweden.

Ober WB. Bottoms Up - The Fine-Arts and Flagellation.2.*American Journal of Dermatopathology*. 1984, 6(6) : 541-522. 24R.
Hackensack MED CTR, Dept Pathol, 30 Prospect AVE, Hackensack, NJ 07601, USA.

Rajagopal CT, Rangachari MS. On Medieval Kerala Mathematics. *Archive for History of Exact Sciences*. 1986, 35(2):91-99. 7R. Univ Madras, Ramanujan Inst Math, Madras 600005, India.

Reingold N. History of Science Today.1.Uniformity as Hidden Diversity - History of Science in the United-States, 1920-1940. *British Journal for the History of Science*. 1986, 19(63):243-262. 69R.
Smithsonian Inst, Washington, DC 20560, USA.

| | | | | |
|---|---|---|---|---|
| Abraham GA | ISIS | 74 | 368 | 84 |
| Thackray A | ISIS | 63 | 473 | 72 |
| Thackray A | ISIS | 66 | 445 | 75 |
| White L | ISIS | 75 | 171 | 84 |

Richards E, Schuster J. The Feminine Method as Myth and Accounting Resource - A Challenge to Gender Studies and *Social-Studies of Science*. *Social Studies of Science*. 1989, 19(4):697-720. 71R.
Autralia. Univ Wollongong, Dept SCI & Technol Studies. POB 1144, NSW 2500.

| | | | | |
|---|---|---|---|---|
| Sarton G | ISIS | 6 | 26 | 24 |

Rotton J, Kelly IW. Much ADO about the full moon - a Meta-Analysis of Lunar-Lunacy Research. *Psychological Bulletin*. 1985, 97(2):286-306. 156R.
Florida Int Univ, Dept Psychol, N-Miami, FL 33181, USA.

Shen KS. Historical Development of the Chinese Remainder Theorem. *Archive for History of Exact Sciences*. 1988, 38(4): 285-305. 10R.
Hangzhou Univ, Dept Math, Hangzhou, Peoples-R-China.

| | | | |
|---|---|---|---|
| Aryabhat | Aryabhatiya | 74 | 79 |
| Srihivasie. CN | Hist Ancient Indian | 101 | 67 |

Sheynin OB, Quetelet A. As a Statistician. (Review). *Archive for History of Exact Sciences*. 1986, 36(4):281-325. 132R.

Stigler SM. Craig, John and the Probability of History-from The Death of Christ to the Birth of Laplace. *Journal of the American Statistical Association*. 1986, 81(396):879-887. 12R.
Univ Chicago, Dept Stat, Chicago, IL 60637, USA.

Tilak R, Roy P. Description of a New Species of the Sub-Genus Rana-(PAA) Dubois from Chakrata Hills, District Dehra-Dun, Uttar Pradesh, India (Ranidae, Anura). *Zoologischer Anzeiger*. 1985, 215 (3-4): 231-239. 7R.
Zool Survey India. No Reg Stn, 13 Subash Rd, Dehra-Dun 248001, India.

Turner RD, Santhakumaran LN. The Genera Martesia and Lignopholas in the INDO-Pacific(Mollusca, Bivalvia, Pholadidae). *Ophellia*. 1989, 30(3): 155-186. 28R.
Harvard Univ, Museum Comparat Zool, Cambridge, MA 02138, USA. *Inst* Wood SCI & Technol, Bangalore, India.

Warner JH. Science In Medicine. *Osiris*. 1985, 1:37-58. 156R. Wellcome *Inst* Hist Med, London NW1 2BP, England.

| | | | | |
|---|---|---|---|---|
| Brieger GH | ISIS | 72 | 537 | 81 |
| Orr WJ | ISIS | 72 | 590 | 81 |
| Rosenberg C | ISIS | 74 | 356 | 83 |

## REFERENCES


1. E. Garfield, "Citation Indexes for Science," *Nature,* 122, No. 3159 (1955), 108-11.
2. E.Garfield, *"Science Citation Index : Journal Citation Reports,"* (1984).
3. Linda C. Smith, "Citation Analysis," *Library Trends,* 30, No.1 (1981), 83-106.
4. S. Cole and J.R. Cole, "Scientific Output and Recognition," *American Sociological Review,* 32, No.3 (1967), 377-90.
5. Pauline A. Scales, "Citation Analyses as Indicators of the Use of Serials: A Comparison of Ranked Title Lists Produced by Citation Counting and form Use Data," *Journal of Documentation,* 32, No.1 (1976), 17-25.
6. N. Wade, "Citaion Analysis : A New Tool for Science Administration," *Science,* 188, No.4187 (1975), 429-32.

# 9. FINDINGS, CONCLUSIONS AND SUGGESTIONS FOR FURTHER RESEARCH

In this study of literature on History of Science in India, published in six leading journals ranging from 1905 to 1986 has been subjected to scientometric analyses by applying the latest and leading techniques of scientometrics such as application of Lotka's Law to determine authors' productivity, collaborative pattern of authorship, highly productive authors and their pattern of contributions along with the mapping of subjects in their articles; the application of Zipf's Law to determine rank-frequency of occurrences of keywords and text value of articles as well as authors ; application of citation analysis to measure the impact of individual articles, authors and journals on the basis of citation counts. To undertake scientometric analysis, a machine-readable data base on History of Science was created out of the journals studied over the specific period of time. Growth of History of Science has been traced in the journals studied from the view point of literature, publication sources, and authors. In addition, a few hypotheses have also been tested on the basis of output of the above scientometric analyses. A precise and succinct resume of the conclusions and suggestions for further research in this study, follows in the following sections.

## 9.1 FINDINGS AND CONCLUSIONS

**9.1.1** Scattering of authors was measured and it was found that each of sixty percent of authors contributed only one contribution. The index constant of authors' productivity was found to be equal to 2.27, which approximates the inverse square value (i.e.,2) of Lotka's Law. The maximum deviation using Kolmogorov-Smirnov statistics test was found to be 0.035 at the 0.01 level of significance : this value being lesser than 0.534, upholds the applicability of Lotka's Law to the data of this study. Thus in simple terms it was found that if the frequency of the authors increases, the number of contributions decreases, whereas if the frequency of the authors decreases, the number of contributions increases.

**9.1.2** It was found that 86.87 percent of all the contributions were made by single authors, whereas authors in collaboration whether by two (10.45%), or three (1.27%) or four (0.19%) contributed only 11.91%, supporting the hypothesis that authors in History of Science mostly contribute individually.

**9.1.3** Twenty authors were found to be highly productive. In the decreasing order of productivity these are as follows : N.Annandale, S.L. Hora, I.H. Burkill, K.V. Sarma, S.R. Das, R.C. Gupta, H.Gershenowitz, G.Sarton, S. Chaudhuri, A.K.Bag, D.Hooper, D.C. Phillott, H.C. Dasgupta, B.Datta, B.Prashad, S.N. Sen, S.Mahdihassan, P.C. Sengupta, S.D.Sharma, K.Biswas. Peaks observed in the contributions curves of authors indicate the stages of their high productivity and that the pattern of their contributions never remains steady, thus supporting the hypothesis which states that contributions of the authors never remains steady.

**9.1.4** Over all journal preference followed by the highly productive authors was found to be JPASB > IJHS > JASBSC > IHQ > VIJ > ISIS > JASBL > JAS.

**9.1.5** Mapping of subjects revealed that besides concentrating mainly on three or four core subjects, the authors not only wrote on related subjects but some of them also touched the alien ones. For examples, N. Annandale, a zoologist wrote on Bismer Weighing Beam, and Loom Weaving Apparatus ; S.L. Hora, a fish expert wrote on scape goat, and birds; A.K. Bag, a mathematician also scanned technology. This supports the hypothesis that the authors penetrate into the areas other than tha areas of their normal pursuit.

**9.1.6** Rank-frequency of occurrences of keywords revealed that initially the number of word counts remained the same, although the frequency continuously changed. The value of the products of rank ***r*** of a word-type and frequency ***f*** of occurrences of the word-type fluctuated between 300-

400 and/or log values 2.38 to 2.6. The graphical presentation of rank *r* and frequency *f* on logarithmic scale turned out to be a straight line, upholding the applicability of Zipf's Law to the data on History of Science studied in this study. Thus the frequency of occurrences of a word, multiplied by its rank turned out to be near constant. The more the number of occurrences of a word, the higher its rank ; for instance, a word with maximum frequency of occurrences in a text, is ranked first.

**9.1.7** Mapping of keywords revealed that biological sciences has maximum number of keywords followed by astronomical sciences, mathematics, and technology.

**9.1.8** The authors' contributions during most prolific years were subjected to the determination of text value as the ratio of number of pages to the number of references. It was found that I.H. Burkill had the highest text value of 10, followed by D.Hooper with text value of 5, N.Annandale with text value of 2, S.L.Hora with text value of 1, and G. Sarton with text value of 0.9 and so on and so forth.

**9.1.9** In the wake of citation analysis of the total of twenty highly productive authors, it was found that only thirteen authors were cited ; G. Sarton ranked first with 21 citations, whereas B.Datta with 7 citations and H. Gershenowitz with 5 citations, ranked second and third respectively.

**9.1.10** Almost negligible citation gaps were found in the first half of the twentieth century as well as in the first half of 1980's, indicating the trend of continuity in citation occurrences which was found to be lacking during the period 1941 to 1979, during which as many as six citation occurrences occurred.

**9.1.11** The very fact that a paper of N.Annandale published as early as 1906 was cited in 1985, and a paper of G. Sarton published in 1913 was cited in 1986 along with the citations of the earlier papers of S.L.Hora, B.Datta, P.C.Sengupta, I.H.Burkill during the period 1985-1989, establishes the historicity of basic research works done by these authors. This supports the hypothesis that the citations in History of Science followed a phenomenon of continued citations.

**9.1.12** It was found that the journal 'ISIS' was not published from June 1914 to September 1919 during the period of first world war. It was further observed that the growth rate in History of Science fell in 1940's, when the country was under the stress and strain of freedom struggle as well as in the first half of 1960's, when the country had to face Chinese aggression as well as Indo-Pak war. Since 1980's there has been a sharp increase in growth rate. This supports the hypothesis that the depression in the growth of literature in the subject takes place during the period on political crisis and war conditions in the country.

## 9.2 SUGGESTIONS FOR FURTHER RESEARCH

**9.2.1** The growth of the literature in History of Science requires further investigation from the view points of exponential growth, obsolescence (negative growth), empirical formulations, mathematical modelling, etc.

**9.2.2** Journals which includes the articles on the subject, and figure in Critical Bibliography of ISIS have not been analysed scientometrically in this study, and consequently ought to be analysed separately.

**9.2.3** Studies of such type can also be carried out in the context of other national literatures so that a view of global History of Science may be obtained.

**9.2.4** Development of the subject can be further probed by examining the relationship of text pages with the references at levels of authors, articles, and journals. The text value and the occurrence of citations of papers can enable one to have a peep into the trends in different areas in the History of Science.

**9.2.5** Rank-frequency of occurrences of words need to be analysed further to have the much studied part of the literature, and to identify the less studied areas, for scientists in the History of Science for future studies.

**9.2.6** The study of citations as a basic phenomenon in scientific communication should invite attention, as the citation check is not only for all authors, but also for a wider range of years. Moreover the citation index is not all-comprehensive to cover all journals. More widely, the role of citations as indicators of knowledge transmission is an interesting area of investigation.

**9.2.7** The machine-readable data base on History of Science in India can be further built up, updated, and strengthened so that instant scientometric data values and graphical representations of the analysis could be performed.

## BIBLIOGRAPHY

Bibliography of select materials is listed in the following pages in accordance with the Turabian's format for bibliography given in his *Manual for Writers of Term Papers,Theses and Dissertations*. It is classified according to the types of materials/documents listed. Within each section, the entries are arranged according to the alphabetical order by author.

### BOOKS


1. Bradford, S.C. *Documentation.* London: Public Affairs Press, 1950.
2. Donohue, Joseph C. *Understanding Scientific Literatures: A Bibliometric Approach.* Cambridge: Massachusetts Institute of Technology Press, 1973.
3. Egghe L, and Rousseau R. *Introduction to Informetics : Quantitative Methods in Library, Documentation and Information Science.* New York: Elsevier Science Publishers, 1990.
4. Garfield, E., Sher, I.H. and Torpie R.J. *The Use of Citation Data in Writing the History of Science.* Philadelphia: Institute for Scientific Information, 1964.
5. Goffman, William, and Warren, Kenneth S. *Scientific Information System and the Principle of Selectivity.* New York: Praeger, 1980.
6. Guha, B. *Documentation and Information : Services, Techniques and Systems.* 2nd ed. Calcutta: World Press, 1984.
7. Guiter, H, and Aparov, M.V. *Studies in Zipf's Law.* Bochum: Brockmeyer, 1982.
8. Hjerppe, Roland. *A Bibliography of Bibliometrics and Citation Indexing and Analysis.* Stockholm: Royal Institute of Technology Library, 1980.
9. Hulme, E.W. *Statistical Bibliography in Relation to the Growth of Modern Civilization.* London: Grafton, 1923.
10. Lipetz, B.A. *The Measurement of Efficiency of Scientific Research.* Carlisle: Intermedia, Inc, 1965.
11. Meadows, A.J. *Communication in Science.* London: Butterworths, 1974.
12. Narin, F. *Evaluative Bibliometrics : The Use of Publication and Citation Analysis in the Evaluation of Scientific Activity.* New Jersey: Computer Horizons, 1976.
13. Nicholas, David, and Ritchie, Maureen. *Literature and Bibliometrics.* London: Bingley-Hamden. Con. Lineet Books, 1978.
14. Price. D.D. Solla. *Science, Technology and Society.* London: SAGE Publication.
15. ______________. *Little Science, Big Science.* New York: Columbia University Press, 1963.
16. ______________. *Science Since Babylon.* New Haven: Yale University Press, 1975.
17. Rao, I.K.R. *Quantitative Methods for Library and Information Science.* New Delhi: Wiley Eastern, 1983.
18. Shailendra, K. (jt.comp). *CSIR Research Output 1990: A Bibliometric Analysis.* New Delhi: Council of Scientific and Industrial Research, 1991.
19. Simpson, I.S. *Basic Statistics for Librarians.* 2nd ed. London: Clive Bingley Limited, 1983.

20. Turabian Kate L. *A Manual for Writers of Term Papers, Theses and Dissertations*. Chicago : University of Chicago Press, 1966.

21. U.S. National Science Foundation, National Science Board. *Science Indicators 1972*. Washington: Government Printing Office, 1973.

22. U.S. National Science Foundation, National Science Board. *Science Indicators 1974*. Washington: Government Printing Office, 1975.

23. Warren, Kenneth, S.(ed.). *Selectivity in Information System : Survival of the Fittest*. New York: Praeger, 1985.

24. Wolff, Howard K., and Pant, Prem R. *A Handbook for Social Science Research and Thesis Writing*. Kathmandu: Research Division Tribhuvan University, 1975.

25. Zipf, G.K. *Psycho-Biology of Language : An Introduction to Dynamic Biology*. Cambridge: MIT Press, 1965.

26. ________________. *Human Behavior and the Principle of Least Effort*. New York: Hafner Publishing Company, 1949.

27. Zunde, Pranas. *Empirical Laws for Information Systems Design*. Georgia: Georgia Institute of Technology, 1979.

**Articles and Periodicals**

28. Adair, W.C. "Citation Index for Scientific Literature," *American Documentation*, 6 (1955), 31.

29. Allan, D. Pratt. "A Measure of Class Concentration in Bibliometrics," *Journal of the American Society for Information Science*, 28 (1977), 285-92.

30. Allen, E.S. "Periodicals for Mathematicians," *Science*, 70 (1929), 529.

31. Arunachalam, S. "Citation Analysis as a Sociometric Tool," *Science Today*,3, No. 4 (1978), 14.

32. Arunachalam, S., and Markanday, S. "Science in Middle Level Countries : A Bibliometric Analysis of Scientific Journals of Australia, Canada, India and Israel," *Journal of Information Science*, 3, No.1 (1981), 13-26.

33. Arunachalam, S., and Srivastava, M.K. "Physics Research in Israel : A Preliminary Bibliometric Analysis," *Journal of Information Science*, 8 (1984), 185-95.

34. Aulakh, G.S., Jain, H.C., and Kochhar, V.B. "Bibliometric Study of Indian Phytochemical Literature-growth Pattern and Scatter," *Journal of Library and Information Science*, 8, No.1 (1983), 83.

35. Baker, D.B. "Chemical Literature Expands," *Chemical and Engineering News*, 44, No.23 (1966), 84-88.

36. Balog, C. "Multiple Authorship and Author Collaboration in Agricultural Research Publication," *Journal of Research Communication Studies*, 2 (1979), 159.

37. Basak, N.D., and Das, B.B. "Sactter of Indian Mathematicians Contributions in Primary Periodicals," *IASLIC Bulletin*, 28, No.2 (1983), 67-73.

38. Beyer, J.M., and Snipper, R. "Objective versus Subjective Indicators of Quality in Graduate Education," *Sociology of Education*, 47, 541-57.

39. Bhattacharyya, K.K. "Scattering Co-efficient and Rank of Periodical Literature on Glass and Ceramics," *IASLIC Bulletin*, 20, No.1 (1975), 1-6.

40. Bird, P.R. "Some Sampling Characteristics of Bibliometric Distribution," *Journal of Informetics*,

1, No.2 (1977), 69-80.

41. Bookstein, Abraham. "The Bibliometric Distributions," *Library Quarterly*, 46, No.4 (1976), 416-23.

42. Brookes, B.C. "Optimum P% Library of Scientific Periodicals," *Nature*, 232, No.5311, (1971).

43. Braun, T. "Some Comments on the Progress of Analytical Chemistry (1910-70)," *Talanta*, 23 (1976), 743.

44. Brodman, E. "Choosing Physiology Journals," *Bulletin of the Medical Library Association*, 32 (1944), 479-83.

45. Brookes, B.C. "The Derivation and Application of the Bradford-Zipf Distribution," *Journal of Documentation*, 24, No.4 (1968), 247-65.

46. _______________. "Aging in Scientific Literature," *Journal of Documentation*, 36 (1980), 164.

47. _______________. "Frequency-rank Distributions," *Journal of the American Society for Information Science*, 29, No.1 (1978), 5-13.

48. Brown, Pauline. "The Distribution of Articles in the Literature," *Australian Academic Research Libraries*, 8, No.1 (1977), 26-32.

49. Can, Fazli. "Similarity and Stability Analysis of the Two Partitioning Type Clustering Algorithms," *Journal of the American Society for Information Science*, 36, No.1 (1985), 3-14.

50. Carpenter, Mark P. "Similarity of Pratt's Measure of Class Concentration to the Gini Index," *Journal of the American Society for Information Science*, 30, No.2 (1979), 108-110.

51. Cason, H., and Lubotsky, M. "The Influence and Dependence of Psychological Journals on each other," *Psychological Bulletin*, 33 (1936), 95-103.

52. Chubin, D.E., and Moitra, S.D. "Content Analysis of References : Adjunct or Alternative to Citation Counting," *Social Studies of Science*, 5 (1975), 423.

53. Clarke, B.L. "Multiple Authorship Trends in Scientific Papers," *Science*, 143, No. 3608 (1964), 822-24.

54. Cole, J.R., and Cole, S. "The Ortega Hypothesis," *Science*, 178, No.4059 (1972), 368-75.

55. _______________. "Citation Analysis," *Science*, 183, No.4120 (1974), 32-33.

56. Cole, S., and Cole, J.R. "Scientific Output and Recognition: A Study on the Operation of the Reward System in Science," *American Sociological Review*, 32, No.3 (1967), 377-90.

57. Cole, F.J., and Eales, N.B. "The History of Comparative Anatomy," *Science Progress*, 11 (1917), 578-96.

58. Comroe, J.H. and Dripps, R.D. "Scientific Basis for Support of Biomedical Science," *Science*, 192 (1976), 105-11.

59. Crane, D. "The Gatekeepers of Science : Some Factors Affecting the Selection of Articles of Scientific Journals, " *American Sociologist*, 2 (1967), 195.

60. Crane, E.J. "Growth of Chemical Literature : Contributions of Certain Nations and the Effects of War," *Chemical and Engineering News*, 22, No.17 (1944), 1478-81.

61. Das Gupta, A.K. "Librametric Study," in *An Essay in Personal Bibliography - Ranganathan Fetschrift : A Bibliography of Writings on and by Dr S R Ranganathan* (Bombay: Asia Publishing), 2 (1967), 13-21.

62. Dobrov, G.M., and Mackay, A. "Not the Mysticism but the Science of Numbers," *Nature*, 219 (1968), 662.

63. Dutta, S., and Rajagopalan, T.S. "Literature Citations in Scientific and Technical Periodicals: A Survey," *Journal of Scientific and Industrial Research*, 17A (1958), 259-61.

64. Garfield, E. "Citation Analysis as a Tool in Journal Evaluation," *Science*, 178 (1972), 471-79.

65. ______________. "ISI is Studying the Structure of Science Through Co-citation Analysis," *Current Contents*, 7, No., Feb. 13 (1974), 5.

66. ______________. "Significant Journal of Science," *Nature*, 264 (1976), 609.

67. ______________."Citation Indexes for Science," *Science*, 122, No.3159 (1955), 108-11.

68. ______________. "Citation Indexing for Studying Science," *Nature*, 227, No.5259 (1970), 669-72.

69. ______________. "Is Citation Analysis a Legitimate Evaluation Tool?," *Scientometrics*, 1, No.4 (1979), 359-75.

70. Geller, Nancy L. "Lifetime-citation Rates : A Mathematical Model to Compare Scientists' Work," *Journal of the American Society for Information Science*, 32, No.1 (1981), 3-15.

71. ______________."On the Citation Influence Methodology of Pinski and Narin," *Information Processing and Management*, 14, No.2 (1978), 93-95.

72. Gera, J.C. "Social Science Citation Index : Its Efficacy for Information Retrieval," *Library Herald*, 15, No. 3-4 (1973/74) 132-45.

73. Ghosh, J.S. "Uncitedness of Articles in 'Nature', a Multidisciplinary Journal," *Information Processing and Management*, 11 (1975), 165-69.

74. Gilbert, G.N., and Woolgar, S. "The Quantitative Study of Science : An Examination of the Literature," *Science Studies*, 4 (1974), 279.

75. Goffman, W. "Mathematical Approach to the Spread of Scientific Ideas : The History of Mast Cell Research," *Nature*, 212 (1966), 449.

76. ______________. "A Mathematical Method for Analyzing the Growth of Scientific Discipline," *Journal of the Association for Computing Machinery*, 18, No.2 (1971), 173-85.

77. Goffman, W., and Harmon, G. "Mathematical Approach to the Prediction of Scientific Discovery," *Nature*, 229 (1971), 103.

78. Gordon, M. "Evaluating the Evaluators," *New Scientist*, (1977), 342.

79. Goudsmit, S.A. "Citation Analysis," *Science*, 183, No.4120 (1974), 28.

80. Gross, P.L.K., and Gross, E.M. "College Libraries and Chemical Education," *Science*, 66 (1927), 1229-34.

81. Guha, B., and Others. "Scattering of Articles by Indian Authors in Foreign Periodicals : A Sample Survey," *Annals of Library Science and Documentation*, 12, No.2 (1965), 76-86.

82. Gupta, D.K. "Chandrasekhar - Winner of 1983 Nobel Prize for Physics : A Citation Analysis of His Work," *Annals of Library Science and Documentation*, 30, No.3-4 (1983), 177-84.

83. Haitun, S.D. "Stationary Scientometric Distributions Part 1: Different Approximations," *Scientometrics*, 4, No.1 (1982), 5-25.

84. ______________. "Stationary Scientometric Distributions Part 2: Non-Gaussian Nature of Scientific Activities," *Scientometrics*, 4, No.2 (1982), 89-104.

85. ______________. "Stationary Scientometric Distributions Part 3: The Role of Zipf's Distribution," *Scientometrics*, 4, No.3 (1982), 181-84.

86. Hartsough, R., and Laffal, J. "Content Analysis of Scientific Writings," *The Journal of General Psychology*, 83 (1970), 193-206.

87. Hayes, Robert M. "Citation Statistics as a Measure of Faculty Research Productivity," *Journal of Education for Librarianship*, 23, No.3 (1983), 151-72.

88. Heine, M.H. "Indices of Literature Dispersion based on Qualitative Attributes," *Journal of Documentation*, 34, No.3 (1978), 159-61.

89. Henkle, H.H. "Periodical Literature of Biochemistry," *Medical Library Association Bulletin*, 27 (1938), 139-47.

90. Hirst, G. "Discipline Impact Factors: A Method for Determining Core Journal Lists," *Journal of the American Society for Information Science*, 29 (1978), 171.

91. ______________. "Computer Science Journals: An Iterated Citation Analysis," *IEEE Transactions PC*, 20 (1977), 233.

92. Hodder, P. "Limits to Collaborative Authorship in Science Publishing," *Journal of Research Communication Studies*, 2 (1979/1980), 169.

93. Hubet, J.J. "Bibliometric Models for Journal Productivity," *Social Indicators Research*, 4, No.4 (1977), 441-73.

94. Hurt, C.D. "An Examination of Differences Between Two Studies of Highly Cited Old Papers, "*Journal of the American Society for Information Science*, 31, No.3 (1980), 215-16.

95. Ikpahaandi, L. "An Overview of Bibliometrics: Its Measurements, Laws and their Applications, " *Libri*, 35 (1985), 163.

96. Inhaber, H., and Przednowek, K. "Quality of Research and the Nobel Prizes," *Social Studies of Science*, 6, No.1 (1976), 33-50.

97. Irvine, J., and Martin, B.R. "Evaluating Big Science : CERN's Past Performance and Future Prospects," *Scientometrics*, 7 (1985), 281.

98. Jain, T.C., and Goyal, S.P. "Study of Time lag in the Publication of Research Papers in some Selected Periodicals in Agricultural and Allied Sciences," *Annals of Library Science and Documentation*, 16, No.1 (1969), 11-14.

99. Kantor, Paul B. "A Note on Cumulative Advantage Distributions," *Journal of the American Society for Information Science*, 29, No.4 (1978), 202-04.

100. Kaula, P.N., and Singh, S.N. "Characteristics of Information Material : A Bibliometric Investigation in the Field of Social Science in India," *Herald of Library Science*, 19, No.4 (1980), 288-97.

101. Kendall, M.G. "The Bibliography of Operational Research," *Operational Research Quarterly*, 11, No.1-2 (1960), 31-36.

102. Kessler, M.M. "Bibliographic Coupling Between Scientific Papers, "*American Documentation*, 14, No.1 (1963), 10-25.

103. Kochen, M., and Tagliacozzo, R. "Matching Authors and Readers of Scientific Papers," *Information Storage and Retrieval*, 10, No.5-6 (1974), 197-210.

104. Kretschmer, Hildrun. "The Reflection of Lotka's Law in the Structure of Citations of a Journal," *Scientometrics*, 5, No.2 (1983), 85-92.

105. Lawani, Stephen M. "On the Heterogeneity and Clsssification of Author Self-citations," *Journal of the American Society for Information Science*, 33, No.5 (1982), 281-84.

106. Lehman, H.C. "Men's Creative Production Rate at Different Ages and in Different Countries," *Science Monthly*, 78 (1954), 321-26.

107. Lenk, Peter "Mapping of Fields Base on Nominations," *Journal of the American Society for Information Science*, 34, No.2 (1983), 115-22.

108. Long, J.S. "On Adjusting Productivity Measures for Multiple Authorship, "*Scientometrics*, 4, No. 5 (1982), 379-87.

109. Lotka, A.J. "The Frequency Distribution of Scientific Productivity," *Journal of the Washington Academy of Science*, 16 (1926), 317-23.

110. Maheswarappa, B.S., and Rao, K.S. "Journal Literature of Food Science and Technology : A Bibliometric Study," *Annals of Library Science and Documentation*, 29, No.3 (1982), 126-34.

111. Malwad, N.M. "Journal Ranking," *Science Today*, 13, No.4 (1978), 16.

112. Mandelbrot, B. "An Information Theory of the Statistical Structure of Language," *Proceedings of the Symposium on the Applications of Communication Theory*, (London: Butterworths, 1952), pp. 486-500.

113. Margolis, J. "Citation Indexing and Evaluation of Scientific Papers," *Science*, No.155 (1967), 1213-19.

114. Martin, B.R. and Irvine, J. "Assessing Basic Research: Some Partial Indicators of Scientific Progress in Radio Astronomy," *Research Policy*, 12, No.61 (1983), 95.

115. Martyn, John "Progress in Documentation : Citation Analysis," *Journal of Documentation*, 31, No.4 (1975), 290-97.

116. May, K.O. "Quantitative Growth of the Mathematical Literature," *Science*, 154, No. 3757 (1966), 1672-73.

117. Mccauly, J.E. "Multiple Authorship," *Science*, 141, No.3580 (1963), 579.

118. Mcgervey, J.D. "Citation Analysis," *Science*, 183, No.4120 (1974), 30-31.

119. Mckie, D. "The Scientific Periodicals from 1665 to 1978," *Philosophical Magazine (Commemoration Issue)*, (1948), 122.

120. Meadows, A.J., and O'Connor, J.G. "Bibliographical Statistics as a Guide to Growth Points in Science." *Science Studies*, 1 (1971), 95.

121. Merton, R.K. "The Matthew Effect in Science," *Science*, 159 (1968), 56.

122. Moravcsik, M.J. "A Progress Report on the Quantification of Science," *Journal of Scientific and Industrial Research*, 36 (1977), 195.

123. Nagpal, B.S., and others. "Current Trends in Research: Electronic Engineering." *RSPO, CSIR*, Series 2 (1966).

124. Naranan, S. "Power Law Relations in Science Bibliography: A Selfconsistent interpretation," *Journal of Documentation*, 27 (1971), 83.

125. Neelameghan, A. "Documentation of the History of Medicine in India," *Annals of Library Science*, 10, No.3-4 (1963), 116-42.

126. Neelameghan, A. Buche, V.V., and Gupta, B.S.S. "Distribution Pattern of the Duplication of Discoveries," *Nature*, 227 (1970), 1060.

127. Oppenheim, Charles. "Highly Cited Old Papers and the Reasons Why They Continue to be Cited," *Journal of the American Society for Information Science*, 29, No.5 (1978), 225-31.

128. Pao, Miranda Lee "Lotka's Test," *Collection Management*, 4, No. 1-2 (1982), 111-24.

129. Patel, N. "Quantitative and Collaborative Trends in American Sociology Research," *American Sociology*, 7 (1972), 5-6.

130. Potter, William Gray "Lotka's Law Revisited," *Library Trends*, 30, No.1 (1981), 21-39.

131. Price, D.D. Solla. "A Calculus of Science," *International Science and Technology*, 15 (1963), 37-43.

132. _______________. "Networks of Scientific Papers," *Science*, 149 (1965), 510.

133. _______________."Quantitative Measures of the Development of Science," *Archives Internationales d' Histoire des Sciences*, 14 (1951), 85-93.

134. _______________. "Collaboration in an Invisible College," *American Psychologists*, 21, No.11 (1966), 1011.

135. _______________. "A General Theory of Bibliometric and other Cumulative Advantages Processes," *Journal of the American Society for Information Science*, 27 No. 5-6 (1976), 292-306.

136. Pritchard, A "Statistical Bibliography or Bibliometry?," *Journal of Documentation*, 25, No.4 (1969), 348-49.

137. Radhakrishnan, T., and Kernizan, R. "Lotka's Law and Computer Science Literature," *Journal of the American Society for Information Science*, 30, No.1 (1979), 51-54.

138. Rahman, A., and Malik, S. "Current Trends in Research: Botany," *CSIR, RSPO*, Series No.1 (1966).

139. Rahman, A., Sen, N., and Sengupta, K.N. "Scientific Journals in India : A study of Their Characteristics," *CSIR RSIU, Survey Report*, No. 10, (1967).

140. Raisig, L.M. "Mathematical Evaluation of the Scientific Serial," *Science*, 131, No.3411 (1960),1417-19.

141. _______________. "Statistical Bibliography in the Health Sciences," *Bulletin Medical Library Association*, 50, No.3 (1962), 450-61.

142. Rajagopalan, T.S., and Others. "Reporting of Indian Mathematics Literature in Mathematical Review and Bulletin Signaletique," *Annals of Library Science and Documentation*, 12, No.3 (1965), 137-50.

143. Rangarajan, K.S., and Bhatnagar, P. "Mossbauer Effect Studies : Some Deduction from Bibliometric Effect Studies," *Annals of Library Science and Documentation*, 28, No.1-4 (1981), 32-8.

144. Rao, I.K.R. "Dispersion of Documents on Survey Analysis, Bradford and Pareto Distribution," *Library Science With a Slant to Documentation*, 9, No. 3 (1972), 396-403.

145. _______________. "Distribution of Scientific Productivity and Social Change," *Journal of the American Society for Information Science*, 31, No.2 (1980), 111-22.

146. Rothman, H. and Woodhead, M. "Publication Trends in Biological Control," *Nature*, 220 (1968), 1053.

147. Roy, Rustum "Approximating Total Citation Counts From Ist Author Counts and from Total Papers," *Scientometrics*, 5, No.2 (1983), 117-124.

148. Rudd, E. "The Effect of Alphabetical Order of Author Listing on the Carriers of Scientists," *Social Studies of Science*, 7 (1977), 269.

149. Saracevic, T, and Perk, L.L. "Ascertaining Activities in a Subject Area Through Bibliometric Analysis," *Journal of the American Society for the Information Science*, 24, No.2 (1973), 120-34.

150. Scales, Pauline A. "Citation Analyses as Indicators of the Use of Serials : A Comparison of Ranked Title Lists Produced by Citation Counting and from Use Data," *Journal of Documentation*, 36,No.3 (1976), 183-196.

151. Seetharama, S. "Survey Analysis : Development and Structure," *Library Science with a Slant to Documentation*, 9, No.2 (1972), 266-88.

152. Sen,B.K., "Indian Contributions in Nobel Lectures," *Annals of Library Science and Documentation*, 16, No.1 (1969), 35-6.

153. Sen, B.K., and Shailendra, K. "Evaluation of Recent Scientific Research Output by a Bibliometric Method," *Scientometrics*, 23, No.1 (1992), 31-46.

154. Sen, S.K., and Gan, S.K. "Mathematical Extension of the Idea of Bibliographic Coupling and Its Applications," *Annals of Library Science and Documentation*, 30 (1983), 78-82.

155. Sengupta, I.N. "Growth of Biochemical Literature," *Nature*, 244 (1973), 75-76.

156. _______________. "Recent Growth of Literature in Biochemistry and Change in the Ranking of Periodicals," *Journals of Documentation*, 29 (1973), 192-211.

157. _______________. "Weightage Formula to Rerank Biochemical Periodicals," *International Library Review*, 16, No.4 (1984), 447.

158. Shailendra, K., and Others. "Bibliometrics: A Tool for Scientific Evaluation," *ILA Bulletin*, 24, No.4 (1989), 194-207.

159. Shannon, C.E. "The Mathematical Theory of Communication," *Bell Telephone System Journal*, 27 (1948), 379-423, 623-58.

160. Sharma, K.D., Singhal, S and Gupta, B.M. "Electron Latice Scattering Technique for Citation Analysis and Evaluating the Impact of Research Papers," *Annals of Library Science and Documentation*, 27, No.1-4 (1980), 52-60.

161. Simon, H.A."On a Class of Skew Distribution Functions," *Biometrika*, 42 (1955), 425-40.

162. Small, H.G. "Cited Documents as Concept Symbols," *Social Studies of Science*, 8 (1978), 327.

163. _______________." Co-citation in the Scientific Literature : A New Measure of the Relationship Between Two Documents," *Journal of the American Society for Information Science*, 24 (1973), 265.

164. Smith, Linda C "Citation Analysis," *Library Trends*, 30, No.1 (1981), 83-106.

165. Subbaiah, R.Y. "Eighty Years of Grape Research in India (1901-82) : A Bibliometric Study," *Annals of Library Science and Documentation*, 31, No.1-2 (1984), 18-26.

166. Subramanyam, K. "Citation and Significance," *New Library World*, 76, No.905 (1975), 227-28.

167. _______________. "Bibliometric Studies of Research Collaboration - A Review," *Journal of Information Science*, 6, No.1 (1983), 33-8.

168. _______________. "Lotka's Law and the Literature of Computer Science," *IEEE Transactions*, 22, No.4 (1979), 187-89.

169. Swets, J.A. "Information Retrieval Systems," *Science*, 141 (1963), 245-50.

170. Tague, Jean. "The Success-Breeds-Success Phenomenon and Bibliometric Processes," *Journal of the American Society for Information Science*, 32, No.4 (1981), 280-286.

171. Thomas, E. "Computing Progress in Chemistry, "*Science*, 83 (1936), 159-61.

172. Verma, R.K. and Shailendra, K. "Publication Output in Fiber Optics," *Czechoslovak Journal of Physics*, Section B 36 (1986), 188-89.

173. Verma, R.K., and Shailendra, K. "An Analysis and Review of Bibliometric Literature," *First International Conference on Bibliometrics and Theoretical Aspects of Information Retrieval*,

(Belgium, 1987).

174. Verma, R.K., Sharma, Y.K. and Kharti, H.S.D. "Trends in Nuclear Research and Its Publication - An Analysis Based on 5 Years Coverage in ISA," *Annals of Library Science and Documentation*, 29, No.2 (1982), 64-9.

175. Vlachy, J. "On Publication Characteristics of Research Establishments," *Czechoslovakian Journal of Physics*, Section B 20, No.10 (1970), 1149-55.

176. Voos, Henery "Lotka and Information Science," *Journal of American Society for Information Science*, 25, No.4 (1974), 270-72.

177. Wade, N. "Citation Analysis : A New Tool for Science Administrators," *Science*, 188, No.4187 (1975), 429-32.

178. Westbrook, J.H. "Identifying Significant Research," *Science*, 132, No.343 (1960), 1229-34.

179. Wyllys, Ronald E. "Empirical and Theoretical Bases of Zipf's Law," *Library Trends*, 30, No.1 (1981), 53-64.

180. Xhignesse, L. and Osgood, C.E. "Bibliographic Citation Characteristics of the Psychological Journal Network in 1950 and 1960," *American Psychologists*, 22 (1967), 778-91.

181. Yaes, R.J. "Citation Analysis," *Science*, 183, No.4120 (1974), 31-32.

182. Ziman, J.M. "Information, Communication, Knowledge," *Nature*, 224 (1969), 318.

183. Zuckerman, H.A. "Nobel Laureates in Science : Patterns of Productivity, Collaboration, and Authorship," *Amercian Sociological Review*, 32, No.3 (1967),391-403.

184. ________________. "Patterns of Name Ordering Among Authors of Scientific Papers : A Study of Social Symbolism and its Ambiguity," *American Journal of Sociology*, 74 (1968), 276-91.

**Unpublished Material**

185. Barker, Dale Lockard. "Characteristics of the Scientific Literature Cited by Chemists of the Soviet Union." Unpublished Ph.D. dissertation, University of Illionis, 1966.

186. Deshmukh, Subhash. "Bio-medical Information: Application of Bibliometrics to Spinal Anesthesia." Unpublished AIS project, New Delhi: INSDOC, 1979.

187. Garg, Kailash Chand. "Application of Quantitative Methods to Information Science." Unpublished AIS project, New Delhi: INSDOC, 1981.

188. Matter, Gerorge A., and Broms, Henri. "The Myth of Garfield and Citation Indexing," Helsinki: School of Economics., 1982. (Working Paper.)

189. Mehdi, Baqir Raza. "History of Science: A Select List of Periodical Articles Covering the Period 1967-71," Unpublished AIS project, New Delhi: INSDOC, 1972.

190. Subramanyam, K. "Bibliometric Investigation of Computer Science Journal Literature," Unpublished Ph.D. dissertation, University of Pittsburg, 1975.

# PART II
# Database on History of Science

# HISTORY OF SCIENCE

## COMPLETE RECORDS IN YEARWISE ORDER

1.ANNANDALE N. Notes on Indian worm of the Genus Chaetogaster. JPASB. 1905, 1 : 117-20. Ref.2 Calcutta; Deputy Superintendent of the Indian Museum.

CONTENTS: 1. Description of chaetogaster bengalensis. 2. Bionomics of chaetogaster bengalensis.

2.ANNANDALE N. Additions to the collection of oriental snakes in the Indian Museum part 2 - specimens from the Andamans and Nicobars. JPASB. 1905, 1; 7 : 17-76. Ref.2 Calcutta; Deputy Superintendent of the Indian Museum.

CONTENTS: 1. 34 species have been discussed.

3.GORDON E M . Notes concerning the people of Mungeli Tahsil, Bilaspore District. JPASB. 1905, 1; 8 : 181-201. Ref. 2 KEYWORD: ANTHROPOLOGY; PREGNANCY.

CONTENTS: 1. 38 to 83 Loadings discussed. 2. Social behaviour, rites & marriage good faith , pregnancy, etc.

4.ANNANDALE N. Additions to the collection of oriental snakes in the Indian Museum, Part3. JPASB. 1905, 1; 8 : 208-14. Ref.3 Calcutta; Deputy Superintendent of the Indian Museum.

CONTENTS: 1. 15 Species discussed. 2. Habit & Habitat Diagnosis, Coloration & localities have been discussed.

5.DAS J M . Notes on the Kantabudiyas of Cuttack. JPASB. 1905, 1; 8 : 215. Anthropological secretary KEYWORD: ANTHROPOLOGY.

CONTENTS: 1. Marriage, food habits & their family titles discussed.

6.MOLONY E. Remarks on the Geology of the Gangetic Plain. JPASB. 1905, 1; 9 : 230-35.

CONTENTS: 1. Map of Ghazipur District of Ganges Khadir 2.Soil & river course.

7.ANNANDALE N. Notes on the species, external characters and habits of the Dugong. JPASB. 1905, 1; 9 : 238-43. Ref.4.

CONTENTS: 1. Anatomy of certain organs & structures in Dugong. 2. Dimensions, colour, integument, Trunk, limbs, Head & Habits.

8. GAGE AT. Hedyotis sisaparensis, a hitherto underscribed Indian species. JPASB. 1905, 1; 9 : 244. KEYWORD: PLANT.

CONTENTS: 1. Inflorescence, plant structures given.

9.ANNANDALE N. Ascaris halicoris Bird. JPASB. 1905, 1; 10 : 258-60. KEYWORD: BIRDS.

CONTENTS: 1. Birds measurement of various organs 2. Anatomy etc. figures given.

10.KING G; GAMBLE J.S. Materials for a flora of the Malayan Peninsula, no.16 1-91, no.17 93-273, no.18 275-386. JPASB. 1905, 74(Vol 1. Extra No.).

CONTENTS: 1. 88 orders have been described.

11. DAS S C. Origin of mankind (according to the Lamaic Mythology). JPASB. 1906, 2; 1 : 10-14. Ref. 25 KEYWORD: MAN; ORIGIN.

CONTENTS: 1. Kalpa, origin of five great race of jambudvipa.

12. MANN H H. . Proportion between the sexes in helopeltis Theivora waterhouse . JPASB. 1906, 2; 5: 177-81. KEYWORD: FAUNA.

CONTENTS: 1. Coloration & description of structure is given. 2. Observation from 1903 to 1906 in rainy days male & female.

13.HOSSACK W C. Preliminary note on the rats of Calcutta. JPASB. 1906, 2; 5 : 183-86. Calcutta; District Medical officer.

CONTENTS: 1. Key to rats of Calcutta. 2. measurements 3. Relative frequency of rats in Calcutta.

14. ANNANDALE N. Notes on the freshwater fauna of India. No. V. Some animals found associated with spongilla carteri in Calcutta. JPASB. 1906, 2; 5 : 187-96.

CONTENTS: 1. Feeding & larvae of animals etc. 2. Description of body structure.

15. ANNANDALE N; PAIVA C A. Notes on the freshwater fauna of India. No VI. Life history of an aquatic Weevil. JPASB. 1906, 2; 5 : 197-200.

CONTENTS: 1. Description of an aquatic Weevil - colour, Habits & Habitat.

16. ANNANDALE N. Notes on the freshwater fauna of India. No. VII. A new Goby from fresh and Brackish water in lower Bengal. JPASB. 1906, 2; 5 : 201-02. CONTENTS: 1. Diagnosis - length of spawning female.

17. ANNANDALE N. Contributions to oriental Herpetology. No. IV - notes on the Indian Tortoises. JPASB. 1906, 2 ; 6 : 203-06. Ref.2.

CONTENTS: 1. List of 42 Indian tortoises. 2. Skull, body, skeletons are mentioned.

18.ANNANDALE N. Notes on a rare Indo-pacific Bernacle. JPASB. 1906, 2; 6 : 207-08. Ref.2.

CONTENTS: 1. Conchoderma hunteri description.

19. LINSTOW V. Parasites from the Gharial. JPASB. 1906, 2; 7 : 269-71.

CONTENTS: 1. Mentioned the measurements of male & female bodies of Nematoda & Linguatulidae.

20. GURNEY R. Freshwater entomostraca in the collection of the Indian museum, Calcutta. JPASB. 1906, 2; 7 : 273-81. Ref. 21 KEYWORD: ENTOMOSTRACA.

CONTENTS: 1. 19 species are mentioned. 2. Describe locality measurement,etc.

21.ANNANDALE N. New Gecko from the eastern Himalayas. JPASB. 1906, 2; 7 : 287-88. KEYWORD: HERPETOLOGY.

CONTENTS: 1. Gymnodactylus himalayicus - diagnosis - Dimensions.

22. ANNANDALE N. Notes on the freshwater fauna of India. No. VIII. Some Himalayan Tadpoles. JPASB. 1906, 2; 7 : 289-92. Ref.2

CONTENTS: 1. Location & description of tadpoles (4 nos). 2. Habits of tadpoles.

23. WATSON E R. Preliminary note on the chemical examination of the milk and Butter-fat of the Indian Buffalo . JPASB. 1906, 2; 7 : 293-97. Ref. 5 Sibpur; Officiating professor of chemistry, Engineering College KEYWORD: MILK - CHEMISTRY; BUTTER FAT - CHEMISTRY.

CONTENTS: 1. Isolation & examination of sugar 2. Ratio of butyric acid caproic acid.

24. BURKILL 1.I.H. Parasite upon a parasite - viscum apparently V.articulatum, Burm., on Loranthus vestitus, wall., on quercus incana, Roxb. JPASB. 1906, 2; 7 : 299-301. Ref.9.

CONTENTS: 1. Host plants of Viscum articulatum.

25. BURKILL I H. Gentianacearum species asiaticus nova description. JPASB. 1906, 2; 7 : 309-27. KEYWORD: PLANT.

CONTENTS: 1. Described plants, folio, flores, calycis, corollae, stomina, ovarium etc. 2. 38 species described.

26.MOORE S L M. Swertiam novam japonicam Ex affinitate swertiae tetrapterae Maxim, descripseru. JPASB. 1906, 2; 7 : 329. KEYWORD: PLANT.

CONTENTS: 1. Herba caulis, folia, flores, calycis, corrollae, filaments, ovarium,etc.

27. STAPF O. Gentiana Hugelii, Griseb., redescribed. JPASB. 1906, 2; 8 : 337-38. KEYWORD: PLANT.

CONTENTS: 1. Described folia, flores, calyx, corolla, antherae, ovarium, capsula and semina,etc.

28. GAGE A T. Bulbophyllum Burkilli, a hitherto undescribed species from Burma. JPASB. 1906, 2; 8 : 343-44. KEYWORD: PLANT.

CONTENTS: 1. Description of inflorescences & rhizoma, etc.

29.PAIVA C A; ANNANDALE N. Notes on some rare and interesting insects added to the Indian museum collection during the year, 1905-06 . JPASB. 1906, 2; 8 : 345-58. London; Entomological Assistant, London Museum.

CONTENTS: 1. Location of species are mentioned.

30.BURKILL I H. Swertia angustifoloa, Ham. and its allies. JPASB. 1906, 2; 8 : 363-81.

CONTENTS: 1. Distribution of 9 species of swertia.

31. GANGULI P. Notes on the latitude of the Presidency college astronomical observatory. JPASB. 1906, 2; 9 : 383. Calcutta; Presidency college astronomical observatory KEYWORD: OBSERVATORY - PRESIDENCY COLLEGE.

CONTENTS: 1. Mentioned numerical value of latitude.

32.BOULENGER G A. Description of two Indian frogs. JPASB. 1906, 2; 9 : 385-86.

CONTENTS: 1. Rhacophorus taeniatus description of body structures 2. Ixalus annandalii description of body structures.

33. BURR M. Further note on earwigs (Dermaptera) in the Indian Museum, with the description of a New species. JPASB. 1906, 2; 9 : 387-90.

CONTENTS: 1. 5 genus are given. 2. Body measurements are given.

34. ANNANDALE N. Note on the habits of the earwigs Labidura lividipes, further note on earwigs in the Indian museum. JPASB. 1906, 2; 9 : 391.

CONTENTS: 1. Desription of labidura lividipes given, wings description.

35. ROGERS L. Short historical note on medical societies and medical journals in Calcutta. JPASB. 1906, 2; 9 : 393-97. KEYWORD: SOCIETIES-MEDICAL; JOURNAL-MEDICAL; INFORMATION SCIENCE.

CONTENTS: 1. 5 medical journals and Calcutta medical societies are mentioned.

36. BURKILL I H. Notes on the pollination of flowers in India. Note No. 1 - the pollination of Thunbergia grandiflora, Roxb, in Calcutta. JPASB. 1906, 2; 10 : 511-14. Ref.1.

CONTENTS: 1. Visitors for pollination mentioned. 2. Opening of flowers timing.

37. BURKILL I H. Notes on the pollination of flowers in India. Note No. 2 - the pollination of corehorus in Bengal and Assam. JPASB. 1906, 2; 10: 515-20.

CONTENTS: 1. Visitors at location of plants for pollination.

38. BURNKILL I H. Notes on the pollination of flowers in India. Note No. 3.The mechanism of six flowers of the north-west Himalayas. JPASB. 1906, 2; 10 : 521-25.

39.REGAN C T. Two new Cyprinoid fishes from the Helmand Basin. JPASB. 1906, 2; 1 : 8-9. KEYWORD: BASIN; HELMAND.

CONTENTS:1. Location& measurement of body.

40. ANNANDALE N. Notes on the freshwater fauna of India. No. III. Indian aquatic Cockroach and Beetle Larva. JPASB. 1906, 3; 4 : 105-07. Ref.5.

CONTENTS: 1. Habit & Habitat etc.

41. ANNANDALE N. Notes on the freshwater fauna of India. No. IV. Hydra Orientalis and its bionomical relations with other invertebrates. JPASB. 1906, 3; 4 : 109-16. Ref.9.

CONTENTS: 1.Hydra orientalis - Commensalism - chironomid Larva which feeds on H. orientails (enemies) - Prey (food of H. orientalis).

42.RAY J C. Hindu method of manufacturing spirit from Rice and its scientific explanation. JPASB. 1906, 3; 4 : 129-42. Ref 3 KEYWORD: ALCOHOL - RICE.

CONTENTS: 1. Kinds of liquor in earlier time. 2. Description of method - Bakhar - Caking - Brewing - Yeild of spirit - spent wash from the distillery.

43. WATSON E R. Silver dioxide and silver peroxynitrate. JPASB. 1906, 3; 4 : 143-54. Sibpur ; offg. Professor of Chemistry. Civil Engineering College.

CONTENTS: 1. History of silver dioxide. 2. Soulbe silver per-salts. 3. Preparation of silver peroxynitrate by electrolysis of aqueous silver nitrate solution.4. Silver dioxide - dil. sulphuric acid - Aqueous ammonia solution - Strong nitric acid.

44. ANNANDALE N. Notes on the freshwater fauna of India - a variety of spongilla lacustris from Brackish water in Bengal. JPASB. 1906, 3; 3 : 55-58. Ref.9.

CONTENTS: 1. 11 Indian spongillidae are mentioned.2. skeleton spicules & flesh spicules are described. 3. Life history of S. Carteri & S. deciplens are mentioned.

45. ANNANDALE N. Notes on the freshwater fauna of India . No.II .- The affinites of Hislopia. JPASB. 1906, 3; 3 : 59-63. Ref.4.

CONTENTS: 1. Occurrence in water. 2. Description of Hislopia & Norodonia.

46. HOOPER D. Instances of vegetable pottery. JPASB. 1906, 3; 3 : 65-67. Ref. 1 KEYWORD: POTTERY.

CONTENTS: 1. Vessels from the dried fruits e.g. bottle gourd, bel, coconut - uses. 2. Pulpy parts of fruits to make plastic clay & pots.3. Fruit aoula & its vessel 5 Root of asphodel,etc.

47. AIYAR T V R. Notes on some Sea Snakes caught at Madras. JPASB. 1906, 3; 3 : 69-72. Ref.2 KEYWORD: SNAKE SEA; SEA SNAKE.

CONTENTS: 1. Habits & Habitat of sea snakes.2. Description of Madras shore. 3. Venoms of sea snakes.

48. GAGE AT. Wormia mansoni: a hitherto undescribed species for Burma. JPASB. 1906, 3; 3 : 73. KEYWORD: PLANT.

CONTENTS: 1. Measurements of various parts in plants, features like ovule pollen structures mentioned.

49. ANNANDALE N. Tesludo baluchiorum, a new species. JPASB. 1906, 3; 3 : 75-76.

CONTENTS: 1. Diagnosis of T baluchiorum - locality, dimension.

50. MANN H H. Notes on the diet of tea garden coolies in upper Assam and its nutritive value. JPASB. 1907, 3; 2 : 103-08. KEYWORD: DIETARY.

CONTENTS: 1. Analysis of food component in the diet - rice, dal, spices, mustard oil, vegetable etc. 2. Protein and fuel value calories requirements of man are listed.

51. PHILLOTT D C. Notes on shanquar Falcon. JPASB. 1907, 3; 2 : 113-14. Ref.1 KEYWORD: FALCON - SHANQUAR.

CONTENTS: 1. History and evolution of term given.

52. PHILLOTT D C. Notes on common Raven - Corvus Corax. JPASB. 1907, 3; 2 : 115-18. Ref.11 KEYWORD: BIRDS.

CONTENTS: 1. Local variation in name of bird. 2. Body structure, eating habits.

53. SANYAL R B. Notes on the distribution of Macacus arctoides, Geoff. JPASB. 1907, 3; 2 : 119. KEYWORD: MONKEY.

CONTENTS: 1. Brown stump-tailed monkey location. 2. Colouration.

54. NEOGI P. Notes from the chemical laboratory of the Presidency College. Note No., I - a new method of preparing mercury iodide. JPASB. 1907, 3; 2 : 133-34. KEYWORD: CHEMISTRY.

CONTENTS: 1. Experiment given for preparation. 2. History of preparation.

55. NEOGI P. Notes from the chemical laboratory of Presidency College. Note No. II - nitro-ethane as a solvent of iodoform. JPASB. 1907, 3; 2 : 135. KEYWORD: CHEMISTRY.

CONTENTS: 1. Experiment for preparation given.

56. RAY PC. Note from the chemical laboratory of the Presidency college. Note No. III - on silver-mercuroso-mercury nitrate. JPASB. 1907, 3; 2 : 137-38. Ref.1 KEYWORD: CHEMISTRY.

CONTENTS: 1. Isomorphism in mercury. 2. Experiment and analysis of preparation.

57. ANDERSON A R S. Breynia Vredenburgi, an undescribed Echinoid from the Indian Ocean. JPASB. 1907, 3; 3 : 145-48. KEYWORD: ZOOLOGY.

CONTENTS: 1. Comparison of B. vredenburgi with B. australasiae. 2. Description of body structure.

58. BURKILL I H. Gentiana coronata, Royle. JPASB. 1907, 3; 3 : 149-68. Ref. 4 KEYWORD: BOTANY.

CONTENTS: 1. Distribution of plants. 2. Described inflorescence of 9 species.

59. ANNANDALE N. Notes on the freshwater fauna of India. No. IX. descriptions of new freshwater sponges from Calcutta, with a record of two known species from the Himalayas, and a list of the Indian forms. JPASB. 1907 , 3; 1 : 15-26. KEYWORD: FAUNA; SPONGES.

CONTENTS: 1. Description of spicules. 2. Genus Spongilla (10 species, genus Ephydatia (4 species),

Genus Trochospongilla (2 species) described. 3. List of freshwater sponges given.

60. HOOPER D. Well-waters from the Hadhramant, Arabia. JPASB. 1907, 3; 3 : 169-72. KEYWORD: ARAB.

CONTENTS: 1. Mineral analysis of water. 2. Fertility of soil.

61. PHILLOTT DC. Notes on the saker or Cherrug Falcon (F. Cherrug). JPASB. 1907, 3; 3 : 179-. Ref. 29 KEYWORD: FALCON.

CONTENTS: 1. Habitat and Habit. 2. Eating habits of bird.

62. CUNNINGHAM J A, MUKERJI S C. Notes from the chemical laboratory of the Presidency College. Note No. 4 - the electrical state of nascent gases : preliminary note . JPASB. 1907, 3; 3 : 193-97. Ref. 3 Presidency College. Calcutta KEYWORD: GASES - NASCENT.

CONTENTS: 1. Electrical phenomena with chemicals. 2. Measurement of positive and negative ions generated in gases.

63. CHAKRAVARTI N. Chronology of Indian authors, a supplement to M. Duff's chronology of India. JPASB. 1907, 3; 3 : 205-20. KEYWORD: INFORMATION SCIENCE; DUFF M - CHRONOLOGY.

CONTENTS: 1. List of authors, works and year, arranged yearwise.

64. DUTTA B B. Notes from the chemical laboratory of the Presidency College Note No.5-Reactions at low temperatures, Part I Aliphatic Iodochlorides. JPASB. 1907, 3; 5 : 229-33. Calcutta; Second Assistant, Chemical laboratory, Presidency College KEYWORD: REACTIONS-LOW TEMPERATURE; TEMPERATURE.

CONTENTS: 1. Amyliodide phenyl iodochloride.

65. TRAVERS M W. Note on the absorption of gases, vapours, and substances in solution by solids and amorphous substances. JPASB. 1907, 3; 5 : 235-39. Bangalore, Director, Indian Institute of Science.

CONTENTS: 1. Amorphous & solid substances. 2. Characters of the phenomena of absorption carbondioxide by charcoal, water by cotton & wool. 3. Absorption of substances from solution by solids and amorphous substances.

66. HOOPER D. Fats of Garcinia species. JPASB. 1907, 3; 5 : 257-59. KEYWORD: PLANT-FATS.

CONTENTS: 1. Ghee from murga seeds, Oil from gurgi seeds, Analysis of components. 2. Chois (Garcinia indica) oil & its component.

67. ANNANDALE N. Notes on the freshwater fauna of India. No. X. Hydra orientalis during the rain. JPASB. 1907, 3; 1 : 27-28. Ref. 1 KEYWORD: FAUNA.

CONTENTS: 1. Description of outer body structure.

68. ANNANDALE N. Notes on a specimen of Felis tristis, Milne-Edwards, in the Indian museum. JPASB. 1907, 3; 1 : 29-31. Ref. 1 KEYWORD: ZOOLOGY.

CONTENTS: 1. Body measurement, coloration.

69. BURKILL I H. Note on Swertia tongluensis, and on a new variety of Swertia purpurascens. JPASB. 1907, 3; 1 : 33-36. KEYWORD: BOTANY.

CONTENTS: 1. Habitat, Habits. 2. Description of the species.

70.LINSTOW V. Ascaris lobulator, schneider, ein parasit des darms von platanista gangetica. JPASB. 1907, 3; 1 : 37-38. Ref.5 KEYWORD: PARASITE.

CONTENTS: 1. Measurement of body parts.

71.PHILLOTT D C. Notes on the Shahin falcons (Falco peregrinator and F barbarus,Blanford). JPASB. 1907, 3; 5 : 389-93. Ref.8 Secretary, Board of Examiners KEYWORD: FALCON.

CONTENTS: 1. Nesting, coloration & Food habits etc. 2. Variation in local name.

72.PHILLOTT D C. Note on the lagar Falcon (Falco jugger). JPASB. 1907, 3; 1 : 39-45. Ref.21 KEYWORD: FALCON.

CONTENTS: 1, 1. Occurrence of lagar falcon 2. Habitat and habits.3. Eating habits.4. Catching of birds.

73.PHILLOTT D C.Note on the red-headed merlin (Aesalon Chicquera). JPASB. 1907, 3; 6 : 395-99. Ref.13 KEYWORD: BIRDS.

CONTENTS: 1. Food habits.2. Coloration & body structure.

74.BURKILL I H. BOSE G C.Abnormal branch of the mango (Mangifera indica, Linn). JPASB. 1907, 3; 6 : 427-32. KEYWORD: GROWTH-ABNORMAL MANGO.

CONTENTS: 1. Measurements of plant specimens. 2. Flowering axis & foliage leaves.3. Xylem from foliage & flowering side.

75.PHILLOTT D C. Method of catching wild fowl,herons and other water birds in the Punjab,Sindh and Kashmir. JPASB. 1907, 3; 6 : 433-39. Ref.1 KEYWORD: BIRDS-WATER.

CONTENTS: 1. Net for catching Herons, Ducks etc.2. Laying the net.3. Heron-catching in Kashmir. 4. Setting of nooses.

76.BURKILL I H.Fish trap from Northern Arakan. JPASB. 1907, 3; 7 : 441-42.

CONTENTS: 1. Description of Net/trap of bamboo.

77.MALLIK D N.Magnetic induction in spheroids. JPASB. 1907, 3; 7 : 443-57.

CONTENTS: 1. Magnetic induction in a magnetic substance as spheroid.2. Equation in detail is given.

78.KAYE G R. Notes on Indian mathematics Arithmetical notation. JPASB. 1907, 3; 7 : 473-508. Ref.39 Simla, Bureau of Education KEYWORD: NOTATION-ARITHMETICAL.

CONTENTS: 1. Greek,Christian & Sanskrit symbols. 2. Aryabhata, Brahmagupta,Bhaskare,Alliruni.3. Place-value notation.4. Modern notation-examinations.5. Woepcke statements.6. Arabian system.7. Indeterminate or Diophantive equations.8. Hoernle statements.

79.BURKILL I H. Notes on the Pollination of flowers in India-Note No.4.On Cotton in Behar. JPASB. 1907, 3; 7 : 517-26. Ref.5.

CONTENTS: 1. List of visitors enter the flowers.2. List of visitors to extra floral nectaries.3. number of individuals seen in inside the corolla, to the nectaries of bracts outside corolla.

80.PHILLOTT D C. Note on the Blue or common Heron (Ardea cinerea). JPASB. 1907, 3; 8 : 553-55. Ref.4 KEYWORD: BIRDS.

CONTENTS: 1. Vernacular names.2. Breeding of birds, food habits, coloration,etc.

81.BURKILL IH. Anguillicarpus-a new genus of the Cruciferae. JPASB. 1907, 3; 8 : 559-61. Ref.1.

CONTENTS: 1. Inflorescence-description,Habitat. 2. Differences of anguillicarpus & spirorhynchu

82 BURKILL IH. Variety of Ducrosia anethifolia, Boiss., from Baluchistan. JPASB. 1907, 3; 8 : 563-64. KEYWORD: FRUIT CONTENTS: 1. Description of fruit of D.anethifolia

83 BURKILL IH. Note on Impatiens balsamina, Linn., as a dye plant. JPASB. 1907, 3; 8 : 565-66. Ref 1 KEYWORD: PLANT-DYE; DYE

CONTENTS: 1. Distribution of plants 2. flowers for dyeing

84 GAGE AT. Case of lateral floral prolification of the inflorescence of the pine-apple(Ananas sativus) Schul.f.. JPASB. 1907, 3; 9 : 593.

85 PHILLOTT DC . Notes on sign-, gesture-, code-, and secret-language, etc., amongst the persians. JPASB. 1907, 3; 9 : 619-22. Ref 2 KEYWORD: FORENSIC SCIENCE

CONTENTS: 1. Silence, come here, yes, Astonishment, halt, go out, he is cracked etc.

86 FINLOW RS; BERGTHEIL. Method for producing immediate germination of Hard-coated seeds. JPASB. 1907, 3; 10 : 625. KEYWORD: SEEDS-HARD-COATED

CONTENTS: 1. Table is given for name of plant, germination capacity before & after Period of treatment

37 BURKILL IH; FINLOW RS. Three varieties of Corchorus capsularis,Linn Which are eaten. JPASB. 1907, 3; 10 : 633-38.

CONTENTS: 1. Habitat & Habit 2. Leaf structures

88 ANNANDALE N. Note on the freshwater fauna of India. No. XI. Preliminary note on the occurrence of Medusa (Irene ceylonensis, Browne) in the Brackish pool in the Ganges Delta and on the Hydroid stage of the species. JPASB. 1907, 3; 2 : 79-81. Ref 2

CONTENTS: 1. Diagnosis of species - body dimension 2. Vesicles 3. Description of body - Gonoplore (Medusa)

89 ANNANDALE N. Notes on the freshwater fauna of India. No. XII. the polyzoa occurring in Indian fresh and brackish pools. JPASB. 1907, 3; 2 : 83-93. Ref 12 KEYWORD: FAUNA

CONTENTS: 1. Location and description of body structure 2. 11 species described

90 SHIRAZI MK. List of 138 new words, chiefly European, that constantly occur in modern Persian newspapers : collected from the newspapers of the past six months. JPASB. 1907, 3; 1 : 9-13. KEYWORD: EUROPE; PERSIA; BIBLIOMETRICS; SCIENTOMETRICS

CONTENTS: 1. List of 138 words

91 KAYE GR. Notes on Indian mathematics No.2-Aryabhata. JPASB. 1908, 4; 3 : 111-41. Ref 49 Simla, Bureau of Education

CONTENTS: 1. Chronology of mathematician since B.C.600 to A.D.1120 2. Works of Aryabhata 3. Aryabhata & other mathematicians relations 3. Geometry by Aryabhata & trigonometry 4. Indeterminate equations

92 LEAKE HM. Studies in the experimental breeding of the Indian Cottons -an introductory note. JPASB. 1908, 4; (1) : 13-20. Ref 5 KEYWORD: COTTON-BREEDING

CONTENTS: 1. Leaf, value of leaf-factor 2. Cross-fertilization of the cotton in nature

93 PHILLOTT DC. Note on the Drum in Falconry . JPASB. 1908, 4; 4 : 159-64.

CONTENTS: 1. Method of beating the Drum 2. On beating the drum while in field

94 MUKHOPADHYAYA S. General theory of Osculating Conics. JPASB. 1908, 4; 4 : 167-78.

CONTENTS: 1. Geometrical & analytical methods for theory of osculating conics 2. Differential equations & expressions, relating to conics

95 BURKILL IH . Notes on the pollination of flowers in India. Note No.5-some autumn observations in the sikkim Himalaya. JPASB. 1908, 4; 4 : 179-95. Ref 6 KEYWORD: HIMALAYAS

CONTENTS: 1. Weather conditions of sikkim 2. Pendulous flowers, Horizontal flowers,erect flowers,Cleistogamic flowers 2. List of visitors to flowers.

96 BURKIL IH. Notes on the Pollination of flowers in India. Note No.6-the spring flora in the Simla Hills. JPASB. 1908, 4; 4 : 197-231. Ref 3

CONTENTS: 1. Description of flowers of viola serpens, viola patrinii,Arabia alpina & Crucifer 2. Pendent flowers, Horizontal regular & irregular flowers, erect & anemophilous flowers 3. List of visitors of flowers

97 DEY M. Some reciprocal relations of curves and surfaces. JPASB. 1908, 4; 4 : 239-45. Ref 1 Calcutta, Bengal National College

CONTENTS: 1. Method of elliptic inversion 2. Relation between the two loci

98 PHILLOTT DC. Note on the peregrine falcon (Falco peregrinus). JPASB. 1908, 4; 5 : 259-65. Ref 9

CONTENTS: 1. Location of peregrine, Habits & weight measurements 2. Food habits

99 KAYE GR. Use of the Abacus in ancient India. JPASB. 1908, 4; 6 : 293-97. Ref 6 KEYWORD: MATHEMATICS

CONTENTS: 1. Scholars statement about abacus 2. Numerical account of the sound 3. Uses of abacus

100 HOOPER D. Fat of the Himalayan Bear - Ursus torquatus,wagner. JPASB. 1908, 4; 2 : 33-34. KEYWORD: BEAR

CONTENTS: 1. fatty acids constants 2. fat constants,. Bears grease refining & analysis

101 NEOGI P. Reduction of Fehling's solution to metallic copper. A method of depositing a shining, mirror-like film of copper on glass vessels. JPASB. 1908, 4; 6 : 347-48. KEYWORD: MIRROR FILM

CONTENTS: 1. Faraday's depositing copper on glass 2. Method of copper deposition on Glass

102 DASGUPTA HC. Geological note on hill Tipperah (including the Lalmai range in Comillah District). JPASB. 1908, 4; 6 : 349-51. Ref 3 KEYWORD: GEOLOGY

CONTENTS: 1. Physical aspect, Fatikuli sandstone,Ferruginous concretions, fossil wood group, pyaro-land

103 HOOPER D. Oil of Lawsonia alba Lamp. JPASB. 1908, 4; 2 : 35 . KEYWORD: HENNA-OIL; PLANT-OIL

CONTENTS: 1. Henna (Mehndi) plant-uses 2. Analysis of seeds for oil carbohydrates,fibre,albuminoids etc

104 ANNANDALE N. Diagnosis of a living species of the genus Diplonema (Psyclodid Diptera). JPASB. 1908, 4; 6 : 353-54.

CONTENTS: 1. Body structure of male & female Diplonema superstes sp.nov.

105 DE M (MAHENDRANATH). Rationalization of algebraic equations. JPASB. 1908, 4; 7 : 355-62.

CONTENTS: 1. Cayley method of rationalization of algebraical equations 2. Details rationalization of algebraical equations given

106 MUKHOPADHYAYA S . Geometrical theory of a plane non-cyclic arc,finite as well as infinitesimal. JPASB. 1908, 4; 8 : 391-402. Ref 2 KEYWORD: GEOMETRY

CONTENTS: 1. Finite Arc-Definition & theorems 2. Infinitesimal arc-Definition & theorems

107 RAY PC. Retardation and acceleration is the dissolution of mercury in nitric acid in the presence of minute traces of ferric nitrate and manganous nitrate. JPASB. 1908, 4; 8 : 405. Ref 1 KEYWORD: NITRIC ACID-MERCURY

CONTENTS: 1. Details of experiment is given

108 LITTLE C . Note on the Calm region in the atmosphere above Calcutta, which,during the cold season, is at a height of 3,000 feet. JPASB. 1908, 4; 2 : 43-53.

CONTENTS: 1.Air movements of different heights 2. Meteorological description,curves & tables showing various positions of balloons

109 HIRST FC. Kosi river, and some lessons to be learnt from it. JPASB. 1908, 4; 9 : 463-87. KEYWORD: RIVER-KOSI

CONTENTS: 1. Banks of Kosi river 2. training of river, catchment area of Kosi 3. past,present & future history of river lying within the hills 4.Action of Kosi & measurements of bed,current silt, plains depression etc

110 BRUHL P. Proposals for a standard temperature for tropical countries. JPASB. 1908, 4; 9 : 489-96. KEYWORD: TEMPERATURE-STANDARD

CONTENTS: 1. Tables showing mean daily temperature of different cities,mean-minimum & maximum temperature & mean of the day temperature 3. Absolute maximum & minimum temperatures

111 MUKHOPADHYAYA S. General theory of Osculating conics (Second paper). JPASB. 1908, 4; 10 : 497-509. Ref 2

CONTENTS: 1. Higher affection of curvature 2. Equations in details

112 PHILLOTT DC; LAL PG. Hindustani-English vocabulary of Indian birds. JPASB. 1908, 4; 2 : 55-79. KEYWORD: BIRDS-DICTIONARY

CONTENTS: 1.Hindi, Urdu names are also given & reference to Jerdon

113 PHILLOTT DC . Eastern hoods for Hawks. JPASB. 1908, 4; 4 : 65-66.

CONTENTS: 1. Diagrams for different patterns of hawks-hood are given

114 BRUHL P. Recent plant immigrants. JPASB . 1908, 5; 11 : 603-56.

CONTENTS: 1. About 234 species are described

115 KING G; GAMBLE JS. Materials for a flora of the Malayan Peninsula No. 19 387-625 No. 20 626-728. JPASB. 1908, 74; pt2 : 387.

CONTENTS: 1. Three families apocynaceae, Asclepiadaceae, Loganiaceae & strychnos 2. 234

species describes +70 new species.No 20 3 families, 54 genera & 166 species described

116 KING G; GAMBLE JS. Materials for a flora of the Malayan Peninsula. JPASB . 1908(10), LXXIV : 730-908. Calcutta, Superintendent of Royal Botanic Garden, Indian forest department

CONTENTS: 1. 15 Genus are described with species 2. family-Gesneraceae in details

117 STEBBING EP. Loranthus Parasite of the moru and Ban Oaks (Quercus dilatata,Lindl.,and Quercus incana,Roxb). JPASB. 1909, 5; 7 : 189-95. Ref 12

CONTENTS: 1. Destructive effects of parasites-plant body parts 2. Description of Loranthus

118 DUTTA BB. Constituents of the roots of Arisaema concinnum,Schott, and A.Speciosum,Mart. JPASB. 1909, 5 ; 7 : 197-98.

CONTENTS: 1. Analyses of concinnum & A.speciosum-moisture Albuminoid,fat, Carbohydrates, fibre,Ash-N & Ca Silica & Phosphoric acod

119 LEAKE HM. Studies in the experimental breeding of Indian Cottons : an introduction notes. Part. 2. On buds and branching. JPASB. 1909, 5; 1 : 23-30. Ref 5

CONTENTS: 1. Buds - table indicating no. of plants from, generation pure type and on parents, accessory buds, position of buds 2. Breeding - lengthening of vegetative period 3. Economic importance, classification

120 RAY PC; GHOSH AC. Decomposition of Ammonium Platinichloride and Platinibromide under the influence of Heat. JPASB. 1909, 5; 8 : 263-67. Ref 1 Calcutta, Presidency College

CONTENTS: 1. Method, experiment & Chemical equations are given

121 SOMMERFELDT E. Notes on Mineralogy:Isomorphism between anthydrite and Barites. JPASB. 1909, 5; 8 : 269. KEYWORD: MINERALOGY-ANTHYDRITE; BARITES

CONTENTS: 1. Crystallographic symmetry of anthydrite & Barites

122 HOOPER D. Tamarisk Manna. JPASB. 1909, 5; 2 : 31-36. Ref 12 KEYWORD: HONEY

CONTENTS: 1. Tamarisk shrub and its honey or manna - perisia and Armenia description 2. Species of manna

123 HOOPER D. Secretion of Phromnia marginella, Olivier. JPASB. 1909, 5; 9 : 363-66.

CONTENTS: 1. Distribution of species Phromnia marginella 2. Habit and habitat 3. Analysis of secretion

124 INGLIS WA. Some of the problems set up by the rivers of Bengal. JPASB. 1909, 5; 10 : 393-405. KEYWORD: RIVERS - BENGAL

CONTENTS: 1. Course of rivers of Bengal 2. Constructions of bankers, control etc. on rivers of Bengal

125 DASGUPTA HC. Goniomya from the Cretaceous rocks of southern India. JPASB. 1909, 5; 2 : 57. Ref 3 KEYWORD: ROCK - CRETACEOUSFOSSIL

CONTENTS: 1. Fossil study of Goniomya

126 BURKILL IH. Drug astukhudus, nowadays Lavandula dentata, and not Lavandula stoechas. JPASB. 1909, 5; 3 : 67-71. Ref 4

CONTENTS: 1. Medicinal properties of drug L.dentata 2. Historical account of drug L.dentata

127 BURKILL IH. Coptis. JPASB. 1909, 5; 3 : 74-88. KEYWORD: DRUG-COPTIS

CONTENTS: 1. Coptis-drug,13 species in it 2. Distribution of genus coptis 3. Description of plant body,Analyses of coptis teeta inflorescence,leaves structure

128 BURKILL IH. First note on Cymbopogon Martini,stapf. JPASB. 1909, 5; 3 : 89-93. KEYWORD: OIL PLANT; PLANT-OIL

CONTENTS: 1. Distribution of oil plant-analysis 2. Description of plant body

129 BURKILL IH. Notes on the Pollination of flowers in India, Note No.7 a few observations made in the central provinces & Berar. JPASB. 1910, 6; 3 : 101-07.

CONTENTS: 1. Bee a visitor to cotton flower 2. list of visitors on flower 3. Observations from Asirgarh & Melghat

130 BRUNETTI E. Review of our knowledge of the oriental Diptera. JPASB. 1910, 6; 3 : 135-140. Ref 15

CONTENTS: 1. 9 genus are described 2. location of genus

131 DASGUPTA HC. Palaeontological notes on the Gangamopteris Beds of Khunmu (in Kashmir). JPASB. 1910 , 6; 4 : 179-80. Ref 7 KEYWORD: FISH-FOSSILS; FOSSILS-FISH

CONTENTS: 1. Palaeoniscid fish description 2. fish fossils

132 WATSON ER; GUPTA M; GANGULI SC. Chemical examination of the Butter-fat of the Indian Buffalo. JPASB. 1910, 6; 4 : 181-193. Bengal Government

CONTENTS: 1. Estimation of ratio of butyric and caproic acids by Henriques,process,Duclaux's method 2. Volatile fatty acids 3. Experimental data corresponding to the 16 curves

133 DASGUPTA HC. Probable indentity between clypeaster complanatus, Duncan and sladen,and clypeaster,duncanensis,Noetling. JPASB . 1910, 6; 1 : 19-21. KEYWORD: CLYPEASTER

CONTENTS: 1. Measurement of body structure 2. Ratios of dimensions

134 VREDENBURG EW. Chrondrodonta Boseri: a new species of fossil Lamelli-branches from the Hippurite-bearing beds of seistan. JPASB. 1910, 6; 5 : 255-58. Ref 2 KEYWORD: FOSSIL-GASTROPOD; GASTROPOD-GOSSIL

CONTENTS: 1. Gastropod & their development 2. Values in gastropod

135 HOOPER D. Medicinal Lizards. JPASB. 1910, 6; 6 : 301-303. KEYWORD: LIZARDS-MEDICINAL

CONTENTS: 1. Dried lizards for medicinal purposes 2. Examples of medicinal lizards

136 FERMOR LL. Note on a Palaeolithic implement made of manganese-ore. JPASB. 1910, 6; 8 : 381-83. Ref 2 Assistant superintendent, Geological Survey of India KEYWORD: INSTRUMENT-PALAEOLITHIC

CONTENTS: 1. Description of shape & size of implements

137 NEOGI, P; ADHIKARY BB. Chemical examination of Aurvedic metallic preparations. Part I. Shata-puta Lauha and shahashra-puta Lauha (Iron roasted a hundred and a thousand times). JPASB. 1910, 6; 8 : 385-391. Ref 1 KEYWORD: AYURVEDA - METALLIC PREPARATION

CONTENTS: 1. Method of preparation as given in Rasendra-shar-shangraha 2. Estimating free iron in the presence of ferrous and ferric 3. Qualitative and quantitative analysis

138 MAITRA SN. Experimental determination of the Electrochemical equivalent of Nickel. JPASB. 1910, 6; 2 : 39-55. Ref 5 Calcutta, Lecturer & demonstrator in Physics,Presidency College

CONTENTS: 1. Previous observation,preliminary experiments 2. Effects of purity of Salt

139 HARRISON EP. Note on the strength of sulphuric acid solution used in secondary cells in India. JPASB. 1910 , 6; 9 : 471-74. KEYWORD: BATTERIES

CONTENTS: 1. Secondary batteries - strength of dil acid and plates 2. Experiment to form secondary batteries 3. Calculation and value of density of solution in storage batteries

140 MUKERJEE SC. Preliminary note on the use of storage cells in Bengal. JPASB. 1910, 6; 9 : 475. KEYWORD: STORAGE BATTERIES; BATTERIES

CONTENTS: 1. Storage batteries-acids and chemical reaction

141 CHAUDHURI BL. Triacanthus weberi.sp.nov.. JPASB. 1910, 6; 9 : 497-501.

CONTENTS: 1. Fins, shape, teeth, colour and measurements of body

142 NEOGI P; ADHIKARY BB. Prepartion of Phenyl-nitro-methane by the interaction of mercurous nitrite and benzyl chloride. JPASB. 1910, 6; 10 : 503-505. KEYWORD: CHEMISTRY

CONTENTS: 1. Interaction between silver nitrate and benzylchloride 2. Interaction between mercurous nitrite and benzyl chloride

143 HOOPER D. Materia Medica Animalium India. JPASB. 1910, 6; 10 : 507-522.

CONTENTS: 1. 22 protozoa orders have been mentioned and species are indicated for medicine use

144 ASCOLI FD. Rivers of the delta. JPASB. 1910, 6; 10 : 543-56.

CONTENTS: 1. Sources of information - maps and papers, local information, observation of locality 2. Courses of rivers

145 NEOGI P; ADHIKARY BB. Reaction in presence of Nickle : (a) inability of nitrogen and hydrogen to combine in presence of iron and nickle. (b) reduction of the oxides of nitrogen, sulphur and phosphorus in presence of nickle . JPASB. 1910, 6; 11 : 593-97. KEYWORD: NITROGEN REDUCTION REACTION

CONTENTS: 1. Preparation of nitrogen free from nitric oxide 2. Nitrogen and hydrogen in presence of iron and nickle 3. Reduction of nitric oxide in presence of nickel 4. Reduction of sulphur dioxide in presence of nickle 5. Reduction of phosphorous pentaoxide in presence of nickle

146 DEY BB; SEN HK. Interaction of hydrazine sulphate with nitrites, and a new method for the determination of "Nitritic" Nitrogen. JPASB. 1911, 7; 4 : 109-14. KEYWORD: CHEMISTRY

CONTENTS: 1. Experiment with Barium nitrite solution, solution potassium nitrite, silver nitrite solution, tetra-methylammonium nitrite, benzylamine nitrite, tetra-methylammonium nitrite 2. Estimation of nitritic nitrogen

147 HOOPER D. Phosphorus in Indian food stuffs. JPASB. 1911, 7; 6 : 313-22. KEYWORD: FOOD-PHOSPHORUS

CONTENTS: 1. Beri-Beri diseases & lack of phosphorus 2. Analysis of paddy, rice in Bengal, Madras, Patna, Purulia etc 3.Examination of wheat & flour, Barley, Pulse etc.

148 BURKILL I H; FINLOW R S . Corchorus capsularis var oocarpus - a new variety of the common jute plant. JPASB. 1911, 7; 8 : 465-66. Ref 2

CONTENTS: 1. Description of fruit, Habitat & Habit

149 BURKILL I H. Polarity of the Bulbils of Dioscorea bulbifera, Linn. JPASB. 1911, 7; 8 : 467-69. KEYWORD: BOTANY

CONTENTS: 1. Bulbil & halved bulbils giving the position of shoots, adaxillary pole uppermost

150 BHIDE R K. New and revised species of Gramineae from Bombay. JPASB. 1911, 7; 8 : 513-20.

CONTENTS: 1. 9 species are described, inflorescence leaves, plant body etc. figures are given

151 JOSHI P T R; ROSE H A. Notes on the ethnography of the Bashahr state, Simla Hills, Punjab. JPASB. 1911, 7; 9 : 525-613. Private Secretary to the Raja of Bashahr. Punjab KEYWORD: ANTHROPOLOGY

CONTENTS: 1. Birth marriage, worship of God procedures 2. Songs, grades in society etc.

152 HOOPER D. Composition of India Yams. JPASB. 1911, 7; 3 : 57-62. KEYWORD: TUBER

CONTENTS: 1 Tubers analysis for their comparative food value 2. 40 Tuber species are studied

153 CHAUDHURI B L. Freshwater sting-Rays of the Ganges. JPASB. 1911, 7; 9 : 625-29. Ref 3 KEYWORD: FISH

CONTENTS: 1. Raia fluviatilis & Raia sancur habitat & habit 2. Measurement of body parts

154 HOOPER D. Asiatic milk-products. JPASB. 1911, 7; 3 : 63-67. Ref 5

CONTENTS: 1. Dahi, channa, dadhi analysed for fat, casein, sugar, lactic acid, water and others

155 KAYE G R. Brief bibliography of Hindu mathematics . JPASB. 1911, 7; 10 : 679-86. Ref 3 KEYWORD: INFORMATION SCIENCE

CONTENTS: 1. 12th c.A D Bhaskara onwards 2. 79 entries & Index to the bibliography authorwise

156 SMITH WW. Plantarum novarum in Herbario Horti Regii Calcuttensis cognitarum decas. JPASB. 1911, 7; 3 : 69-75. KEYWORD: FRUIT; FLORA; FLOWER

CONTENTS: 1. Description of fruit and flowers of different 9 species 2. Habit and Habitat are discussed

157 SMITH WW. New Gentian and two Swertias from the east Himalaya. JPASB. 1911, 7; 3 : 77-79. KEYWORD: FLORA CONTENTS: 1. 3 species described for inflorescence, habit and habitat

158 BURR M. Contribution to our knowledge of Indian Earwigs. JPASB. 1911, 7; 11 : 771-800.

CONTENTS: 1.Classification of Earwig 2.Description various species & characteristics of classes etc.

159 KAYE G R. References to Indian Mathematics in certain medieval works. JPASB. 1911, 7; 11 : 801-16.

CONTENTS: 1. Indian source general system of branches of mathematics 2. references to particular problem of Indian origin 3. references to modern system of arithmetical notation 'Indian'

160 BURKILL IH. Swertias chineses quatuor Novas, ex herbario G. Bonati, descripsit. JPASB. 1911, 7; 3 : 81-82. KEYWORD: CHINA

CONTENTS: 1. Description of species Swertia Bonatiana, S Duclouxii, S Rosea, S Patents 2. Inflorescence

161 WEST W. Descriptions of three new species of algae associated with Indian freshwater polyzoa. JPASB. 1911 , 7; 3 : 83-84. KEYWORD: ALGAE; FRESHWATER

CONTENTS: 1. 3 species described, habit, habitat

162 SMITH WW. Notes on Sterculia alata Roxb. var. irregularis, - a remarkable instance of leaf variation . JPASB. 1911, 7; 3 : 85-86. KEYWORD: BOTANY CONTENTS: 1. Description leaves, germination

163 RAY PC; RAKSHIT JN. Methylamine nitrite (Methylammoniun nitrite). JPASB. 1911, 7; 4 : 91.

CONTENTS: 1. Preparation of methylamine nitrite

164 GAGE AT. Materials for a flora of the Malayan Peninsula, no 26. JASB. 1912, 36; 75 : PI-V469-527.

CONTENTS: 1. family Euphorbiaceae.22. 7 families 33 genera, 189 species, 78 species new

165 GAMBLE JS. Materials for a flora of the Malayan peninsula No-22 1-204,23 205-278,24 279-391,25 393-468. JPASB. 1912, 75; Pt. 1 : 1-468. Indian forest department

CONTENTS: 1. Four volumes of this work, in 21 parts have already appeared

166 RAY PC; RASIK LD. Allylammonium nitrite. JPASB. 1912, 8; 3 : 101-02. Calcutta; Presidency College, Chemical laboratory

CONTENTS: 1. Preparation of allylammonium nitrite & properties

167 RAY PC, RASIK LD. Isomeric allylamines.. JPASB. 1912, 8; 3 : 103.

CONTENTS: 1. Hydrolysis of allyl mustard oil 2. Preparation of allylamine

168 RAKSHIT JN. Preliminary note on sodium diacetamide, CH3.CO/CH3.CO-N.Na. JPASB. 1912, 8; 3 : 105. Calcutta; Chemical laboratory, Presidency College

CONTENTS: 1. Method of preparation of acetamide metallic derivatives of acid amides

169 RAKSHIT JN; RAY PC. Piperazinium nitrite. JPASB. 1912, 8; 3 : 107.

CONTENTS: 1. Nitrite of alkylammonium bases 2. Preparation of piperazinium nitrite

170 MALLIK DN. Note on the secular cooling of the earth and a problem in conduction of heat. JPASB. 1912, 8; 6 : 223-26. Calcutta; Professor, Presidency college KEYWORD: EARTH-SECULAR COOLING; HEAT-CONDUCTION, EARTH

CONTENTS: 1. Determination of probable age of earth 2. Radium & probable age of the earth, Mathematical formulation

171 MUKERJI BL. Date of Varaha Mihira's Birth. JPASB. 1912 , 8; 7 : 275-78.

CONTENTS: 1. Thibant examination of varaha mihira's works & Divedi S

172 SIRCAR AC. Possible Chemical method of distinguishing between seasoned and unseasoned teakwood. JPASB. 1912, 8; 7 : 303-09. Dacca; Chemical Laboratory, Dacca College

KEYWORD: WOOD-TEAK, SEASONED, UNSEASONED

CONTENTS: 1. Method based on difference in resins composition 2. method based on isolation of crystalline substance

173 BHIDE RK. Two more species of Gramineae from Bombay. JPASB. 1912, 8; 8 : 311-14.

CONTENTS: 1. Sporobolus scabrifolius, Bhide chloris quinquesetica, Bhide

174 FERMOR LL. Preliminary note on the origin of meteorites. JPASB. 1912, 8; 9 : 315-24. Ref 11

CONTENTS: 1. Formation of garnets in the earth's crust & origin of meteorites 2. Meteoritic hypothesis concerning formation of solar system.

175 HOOPER D. Ash of the plantain (MusaSapientum Linn). JPASB. 1912, 8; 10 : 363-66. Ref 1

CONTENTS: 1. Ash of fig leaves for bleaching silk to white 2. uses of plantain 3. Chemical Analysis of ash of plantain

176 BHADURI K. Compound of sodium cuprous thiosulphate and acetylene cuprous acetylide. JPASB. 1912, 8; 10 : 367-68.

CONTENTS: 1. Grammatical analysis & preparation of compounds

177 RAY PR; SEN HK. Note on the interaction of Hydrazines with Ferri-cyanides. JPASB. 1912, 8; 10 : 369-70.

CONTENTS: 1. Oxidation of hydrazines 2. Method of evaluating hydrazine

178 RAY PC; RASIK LD. Isomeric allylamines (second communication). JPASB. 1912, 8; 10 : 371-72. Calcutta; Chemical Laboratory, Presidency College

CONTENTS: 1. Allylamine from Kahlbaum & redistillation

179 ANNANDALE N. Recent advances in our knowledge of the freshwater fauna of India . JPASB. 1912, 8; 2 : 39-53. Ref 28

CONTENTS: 1. Biological relations between freshwater animals 2. seasonal changes in life-cycle of invertebrates 3. Effect of environment on plastic organisms (fresh-water sponges)

180 DATTA RL; SEN H. New series of the double sulphate of Barium with sulphates of the substituted ammonium bases, pt.I. JPASB. 1912, 8; 10 : 399-400. Calcutta; Chemical Laboratory, Presidency College KEYWORD: SULPHATE-DOUBLE

CONTENTS: 1. Pyridine Barium Sulphate 2. Quinoline Barium Sulphate 3. Piperazine barium Sulphate

181 DASGUPTA HC. Crystallized slag from Kulti. JPASB. 1912 , 8; 10 : 401-02. Ref 5 KEYWORD: SLAG - CRYSTALLIZED

CONTENTS: 1. Black, Blue & light grey slag analysis for crystals

182 GRAVELY FH. Preliminary account of a revised classification of Indo-Australian passalidae. JPASB. 1912, 8; 10 : 403-07. Ref 5 Assistant superintendent Indian Museum

CONTENTS: 1. Characteristics of families & sub-families of passalidae

183 BENTLEY CA. Melaria according to the newer researches. JPASB. 1912, 8 : CXXIV.

CONTENTS: 1. Method of quinine & mosquito-destruction 2. Educating tye living beings

184 GAGE AT. Materials for a flora of the Malayan Peninsula, no 26. JASB. 1912-36; 75 : 469-527(I-V).

CONTENTS: 1. family Euphorbiaceae.22. 7 families 33 genera, 189 species, 78 species new etc.

185 ANNANDALE N. Polyzoa of the lake of Tiberias. JPASB. 1913; 6 : 9. Ref 9 KEYWORD: LAKE - TIBERIAS; POLYZOA; TIBERIAS - LAKE

CONTENTS: 1. Comparison of Fredericella sultana forma typica, indica, Jordanica, Duplessisi. 2. Plumatella auricomis sp. Nov - Habitat & Habits

186 RAKSHIT JN. Action of stannic chloride on phenylhydrazine. JPASB. 1913, 9; 3 : 131-35. Calcutta; Chemical Laboratory, Presidency college

CONTENTS: 1. Preparation of phenylhydrazine 2. decomposition of phenylhydrazine-experiment

187 HOSTEN RH. Earliest Jesuit Printing in India. JPASB. 1913, 9; 4 : 149-68. Ref 72 KEYWORD: PRINTING-JESUIT

CONTENTS: 1. Block-printing, printing with Movable type 2. Goa history of jesuit & printing 3. Tmil, Badaga, malabar printing 4. Chaldean printing

188 ANNANDALE N. Introduction to a report on the biology of the lake of Tiberias. JPASB. 1913, 9; 1 : 17-23. Ref 2

CONTENTS: 1. Map & location of Tiberias & lake of Tiberias 2. Climate of Tiberias 3. Vegetation & Zooplankton of lake

189 BURR M. Indian dermaptera collected by A.D. Imms. JPASB. 1913, 9; 5 : 183-87. KEYWORD: IMMS AD

CONTENTS: 1. Pygidicranidae, Labiduridae, Labidae, forficulidae 25 species given with their locations

190 CHIBBER HM. Variations in the flowers of Limnanthemum indicum, on Thwaites. JPASB. 1913, 9; 5 : 191-93. Ahmedabad; Acting professor of Botony, Gujarat College

CONTENTS: 1. 457 flowers examined & inf. given in table about style & stigma length, calyx, corolla size

191 MUKHERJEE H; DUTTA RL. Double mercuri-periodides of substituted ammonium bases tetrapropylammonium mercuri-periodide. JPASB. 1913, 9; 6 : 205-06.

CONTENTS: 1. Method preparation & estimation are given

192 DATTA RL. Action of nitrosyl chloride on secondary amines methyl-benzyinitrosamine and ethylbenzyinitrosamine. JPASB. 1913, 9; 6 : 207-08.

CONTENTS: 1. Method of preparation of methylbenzyinitrosamine & ethylbenzyinintrosamine 2. Action of nitrosyl chloride on two secondary amines

193 ANNANDALE N. Leeches of the lake of Tiberias. JPASB. 1913, 9; 6 : 211-14. Ref 4

CONTENTS: 1. 3 families given, Body structure described

194 CARPENTER GH. New springtail from Galilee. JPASB. 1913 , 9; 6 : 215-17. Ref 2 KEYWORD: ZOOLOGY

CONTENTS: 1. Order Collembola & Cyphoderus genneserae, sp. nov are described, locality & teeth, body segments etc

195 LAIDLAW FF. Note on the dragonflies of Syria and the Jordan valley. JPASB. 1913, 9; 6 : 219-20. KEYWORD: FLIES-DRAGON

CONTENTS: 1. species of Agrionidae & grouping are mentioned

196 ANNANDALE N. Note on a Sponge-larva from the Lake of Tiberias. JPASB. 1913, 9; 6 : 221-22.

CONTENTS: 1. Larva of Nudospongilla mappa. colouration, body cells & size

197 RAUSSELET CF. Note on Rotifers from Galilee. JPASB. 1913, 9; 6 : 229-30.

CONTENTS: 1. Species taken near the surface of lake Tiberias 2. Species reared in Calcutta from dried mud.

198 GURNEY R. Entomostraca from the lake of Tiberias. JPASB. 1913, 9; 6 : 231-32. KEYWORD: PLANKTON; TIBERIAS - LAKE; LAKE - TIBERIAS

CONTENTS: 1. Plankton of Tiberias, location in water.

199 GHOSH E. Internal anatomy of the blind Prawn of Galilee (typhlocaris galilea, calman). JPASB. 1913, 9; 6 : 233-39. Ref 4

CONTENTS: 1. Digestive system, vascular system, Reproductive system and Nervous system are described with figures

200 ANNANDALE N; KEMP S. Crustaca decapoda of the lake of Tiberias. JPASB. 1913, 9; 6 : 241-58. Ref 11

CONTENTS: 1. Measurement of crabs 2. 3 species of crustaca Decapoda are described 3. 3 families are described

201 CHRISTIE WAK. Composition of the water of the lake of Tiberias. JPASB. 1913, 9; 1 : 25-29. Ref 4

CONTENTS: 1. Inorganic constituents of lake water 2. comparison of water of Caspian sea Tiberias 3. Salinity of lake

202 DIGLBY E. Nor'westers and Monsoon prediction. JPASB. 1913, 9; 8-9 : 305-12.

CONTENTS: 1. Nor'west occurring in hot weather & following monsoon and formation of minor storms 3. Anemographic records & observations 4. formation of storm 5. Not a cyclone but an oblique air slit 6. monsoon prediction

203 ANNANDALE N. Notes on the fishes,batrachia and reptiles of the lake of Tiberias. JPASB. 1913, 9; 1 : 31-41. Ref 10

CONTENTS: 1. Analysis of the fish-fauna 2. List of fish of lake, description of fish 3. Batrachia & reptiles -Aquatic species (5 species) -Terrestrial reptiles (4 species)

204 CLEGHORN ML. Notes on the Pollination of Colocasia antiquorum. JPASB. 1913, 9; 8-9 : 313-15.

CONTENTS: 1. Spadix & spathe functions in cross-pollination 2. Four distinct parts of spadix

205 SEWELL RBS. Notes on the biological work of the R.I.M.S.S. "Investigator" during survey seasonsons, 1910-11 and 1911-12. JPASB. 1913, 9; 8-9 : 329-90. Ref 24 Calcutta; Surgeon-Naturalist,Marine survey of India KEYWORD: BIOLOGY

CONTENTS: 1. Shore collecting -Mainland stations -pisces, insecta, Hirudinea, Coelenterata, Porifera, crustacea etc. 2. Mid-water Trawling 3. observations on the surface planktom 4. Succession of Planktonic organisms

206 ANNANDALE N; BROWN JG; GRAVELY FH. Limestone caves of Burma and the Malay Peninsula. JPASB. 1913, 9; 10 : 391-424. Ref 13

CONTENTS: 1. Geology of cave-bearing limestones 2. Fauna of limestone caves 3. Notes on day tablets

207 DUTTA RL. Preparation and decomposition of monochloro-and dichlorobenzylamines. JPASB. 1913, 9; 10 : 425-26. KEYWORD: MONOCHLOROBENZYLAMINES; DICHLOROBENZYLAMINES

CONTENTS: 1. Interaction of dichlorocarbamide with aliphatic & aromatic amines 2. Method of obtaining chloramines 3. Spontaneous decomposition under water of monochlorobenzylamine.

208 BRUNETTI E. Noxious diptera from Galilee. JPASB. 1913, 9; 1 : 43-45. KEYWORD: DIPTERA-NOXIOUS

CONTENTS: 1. family Muscidae, Hippoboscidae discription

209 WHITEHOUSE RH. Planarians of the lake of Tiberias. JPASB. 1913, 9; 11 : 459-63. Ref 1 Belfast; Queen's university

CONTENTS: 1. Planaria tiberiensis n.sp. habitat & habits, planaria Salina, n.sp, P. Barroisi n.sp

210 PRESTON HB. Molluscan faunal list of the lake of Tiberias, with description of new species. JPASB. 1913 , 9; 11 : 465-75.

CONTENTS: 1. 43 species identified & location given 2. 8 families given

211 EDWARDS FW. Tipulidae and Culicidae from the lake of Tiberias and Damascus. JPASB. 1913, 9; 1 : 47-51.

CONTENTS: 1. 3 & 7 species of Tipulidae & Culicidae described. Head, Thorax, Abdomen & occurrence & other remarks

212 HORVATH G. Aquatic and semi-aquatic rhynchota from the lake of Tiberias and its immediate vicinity. JPASB. 1913, 9; 11 : 477-80.

CONTENTS: 1. 13 semi-aquatic species are described & location given 2. 8 aquatic species are described & location given

213 STEPHENSON J. Aquatic Oligochaeta from the Lake of Tiberias. JPASB. 1913, 9; 1 : 53-56.

CONTENTS: 1. Two species described body segments structure is given

214 ANNANDALE N. Account of the sponges of lake of Tiberias, with observations on certain genera of spongillidae. JPASB. 1913, 9; 1 : 57-88. Ref 36

CONTENTS: 1. Systematic account of the collection 2. Biology & distribution of the sponges of the lake of Tiberias 3. Classification of the potamolepidinae and of some obscure genera of spongillinae

215 RAKSHIT JN. Double compound of mercuric oxide with acetone. JPASB. 1913, 9 : IXXXVIII. CONTENTS: 1. Method of preparation given

216 JANA S. New compound of ethylacetoacetate with mercuric oxide. JPASB. 1913, 9 : XCV.

CONTENTS: 1. Method of preparation given

217 GHOSH E. Reproductive system of Atopos Simroth. JPASB. 1914, 10; 4 : 111-13. Ref 5

CONTENTS: 1. Various parts of reproductive system of Atopos are described 2. Hermaphrodite gland

218 RAMASWAMI MS. Note on leaf variation in Heptapleurum Venulosum seam. JPASB. 1914, 10; 4 : 115.

CONTENTS: 1. Leaflets on leaf & chief line variations

219 SMITH JM. Shous or Bighorned deer of Tibet. JPASB. 1914, 10; 4 : 117-19.

CONTENTS: 1. Antlers of Cervus wallichii description given

220 RAMASWAMI MS. New species of Diospyros from the Tinnevelly Hills. JPASB. 1914, 10; 2 : 47-50. Calcutta; officiating curator of the Herbarium, Royal Botanic Garden KEYWORD: BOTANY

CONTENTS: 1. Description of plants, measurements of various parts.

221 PRAIN D; BURKILL IH. Synopsis of the Dioscoreas of the old world, Africa excluded, with descriptions of new species and of varieties.. JPASB. 1914, 10; 1 : 5-41. Ref 5 KEYWORD: BOTANY

CONTENTS: 1. Key to the species 2. Enumeration with diagnosis -8 sections description of plant parts

222 CHATTERJEE G. Demonstration apparatus for determining Young's Modulus. JPASB. 1914, 10; 2 : 51-56.

CONTENTS: 1. Description of apparatus 2. Uniformity of magnification 3. Experiment to determine the load-elongation curve of copper given

223 RAMASWAMI MS. Studied on the leaf structure of Zoysia pungens wild. JPASB. 1914, 10; 2 : 57-63.

CONTENTS: 1. Comparative anatomy of leaves of Madras city 2. soil, temperature, Light, Air, Epidermis, Motor cells, Stomata, vascular bundles & Bundle-sheaths, mechanical tissue & chlorophyll-containing tissue of leaves are described

224 GRAVELY FH. Evolution and distribution of Indian spiders belonging to the subfamily Aviculariinae. JPASB. 1914, 10(10) : 411-20. Ref 7 Assistant superintendent, Indian museum

CONTENTS: 1. Distribution Indian spiders-Five group opf aviculariinae are discussed

225 CLEGHORN ML. Note on the floral mechanism of Typhonium trilobatum. JPASB. 1914, 10(10) : 421-24. KEYWORD: FLOWER

CONTENTS: 1. Tuberous rooted plant-habit & habitat 2. Spadix and spathe 4. trap-mechanism of Ghet-kachu (T.trilobatum)

226 ANNANDALE N; SHASTRI MH. Relics of the worship of mud-turtles (Trionychidae) in India and Burma. JPASB. 1914, 10(5) : 131-38. Ref 4 KEYWORD: TURTLES-MUD

CONTENTS: 1. Mud-turtles kept living in shrines at the present day 2. Some instances of the use of mud-turtles in worship and iconography in Northern India 3. Description of mud-turtles species

227 SOUTHWELL T. Short account of our present knowledge of the Cestode fauna of British India and Ceylon. JPASB. 1914, 10(5) : 139-45. Dy.Director of fisheries, Bengal,Behar and Orissa KEYWORD: FAUNA

CONTENTS: 1. Segmentation of strobia,genital pores castoda 2. Final host,larval names & intermediate host of cestoda

228 SCHULTEN C. Hot springs in Raj Darbhanga, Khargpore hills, distrit Monghyr. JPASB. 1914, 10(5) : 147-52. KEYWORD: SPRINGS-HOT

CONTENTS: 1. 5 hot springs are mentioned alongwith temperature & chemical analysis 2. Rameswara koond, Lachni Koond, Bhimfandh,Richikoond and Haveli Khargpore-hot springs

229 GRAVELY FH. Evolution and distribution of certain Indo-Australian passalid Coleoptera. JPASB. 1914, 10(6) : 201-09. Ref 2 Assistant Superintendent, Indian Museum KEYWORD: AUSTRALIA

CONTENTS: 1. Evolution & distribution of beetles 2. Modifications in classification of coleoptera

230 LLOYD RE. Presence and absence of the gall-bladder in certain rodents. JPASB. 1914, 10(6) : 211-14. Ref 1 Calcutta; Professor of Biology, Medical College

CONTENTS: 1. Utility of gall-bladder 2. 19 species of rodents are examined

231 CALDWELL KS. Improved method of using Oil gas. JPASB. 1914, 10(6) : 215-21. Ref 1 KEYWORD: OILGAS-USE

CONTENTS: 1. Oil gas combustion 2. Calorific value of gas mixture 3. method of mixing air with gas 4. Analysis of oil gas

232 JACKSON VH; MUKERJEE AT. Improvements in measurements with quadrant electrometers. JPASB. 1914, 10(6) : 227-40. Ref 5 KEYWORD: ELECTROMETERS-QUADRANT

CONTENTS: 1. Improvements in accuracy of measurement 2. Improvements in insulation 3. Increase of accuracy obtained in electrometer

233 HAYDEN HH. Note on the application of the principle of isostatic compensation to the conditions prevailing beneath the Indo-Gangetic alluvium. JPASB. 1914, 10(7) : 275-80. Ref 1

CONTENTS: 1. Origin of Himalaya 2. Rift hypothesis

234 DUTT BC. SEN SN. Action of nitric Oxide on metallic peroxides suspendel in water.Part I. JPASB. 1914, 10(8) : 287-91. Scotish Churches College Laboratory

CONTENTS: 1. Action of gas in peroxide 2. Preparation of nitric oxide, lead oxide & Barium oxide& their reactions in presence of water

235 MANN HH; KANITKAR NV. Notes on the fat of Garcinia indica, the so-called kokam butter. JPASB. 1914, 10(8) : 293-94. Ref 3 KEYWORD: BUTTER-KOKAM

CONTENTS: 1. Analysis of butter sample prepared in lab. & bought in bazar 2. Non-volatile faty acids & lower volatile faty acids

236 D'ORCHYMONT A. Hydrophilidae from the lake of Tiberias. JPASB. 1914, 10(9) : 357-60. Ref 2

CONTENTS: 1. Male & famale species are described body structure 2. 6 species are described

237 TATTERSALL,WM. Amphipoda anmd isopoda from the lake of Tiberias. JPASB. 1914, 10(9) : 361-67. Ref 7 U.S.A.Manchester,Manchester Museum

CONTENTS: 1. 3 species of amphipoda & 4 species of isopoda are described-localities,gnathopod & pleopod

238 KIEFFER JJ. Chironomides du lac de Tiberiade. JPASB. 1914, 10(9) : 369-72. KEYWORD: LAKE,TIBERIAS

CONTENTS: 1. Six species are described-body structure & measurement

239 IYENGAR MOP. Observations of the defoliation of some Madras trees. JPASB. 1915,

11(1) : 19-25. Ref 7 Saidapet, Lecturer, Teachers College

CONTENTS: 1. Odina group 2. New & Old leaves 3. Effect of sea-breeze on trees

240 FYSON PF. Note on the flora of the south Indian Highlands. JPASB. 1915, 11(1) : 27-28. KEYWORD: HIGHLANDS-SOUTH INDIA

CONTENTS: 1. Nilgiri & Pulney hills-phanerogamic species

241 BURNS W; PRAYAG SH. Grafting the mango inflorescence. JPASB. 1915, 11(1) : 7-8. Bombay; Economic Botanist, Bombay, Bombay Agricultural Department

CONTENTS: 1. Mango inflorescence-wholly or partly vegetative 2. 3 figures are given

242 CHOWDHURI BL. Weighing beam called Bisa danga in Orissa, with short notes on some weights and measure still current among the rural population of that Division. JPASB. 1915, 11(1) : 9-16.

CONTENTS: 1. Wood-stick beam 2. Weights & measures - palas,tolas,Addha 3. multiples of Addha

243 ANNANDALE N. Report on the biology of the Lake of Tiberias ,fifth series the distribution and origin of the fauna of the Jorden river-system with special reference to that of the lake of Tiberias. JPASB. 1915 , 11(10&11) : 437-76. Ref 72

CONTENTS: 1. List of agnatic fauna of Tiberias Basin 2. Distribution of fauna of Tiberias Basin 3. Origin of fauna of Jordan system

244 HORNELL J. Recent Pearl fishery in Palk Bay with biological notes upon pearl Oysters. JPASB. 1915, 11(5-6) : 153-64. Madras,Superintendent, Pearl & Chank fisheries to the Government of Madras

CONTENTS: 1. Pearl fishery 2. Limitations of pearl Oyster Habitat in palk Bay 3. Pearl Oyster spat 4. New potential pearl-inducing cestode larva

245 SMEETH WF. Geological history of Southern India. JPASB. 1915, 11(5-6) : 141-51. Ref 1 Mysore, State Geologist of Mysore KEYWORD: GEOLOGY

CONTENTS: 1. Rock formation in Mysore 2. Carboniferous, Cretaceous periods 3. Quartzites rocks 4. limestone, gneiss (Peninsular,Champion)

246 MACMOHAN PS. Action of Light on silver chloride. JPASB. 1915, 11(7-8) : 179-85.

CONTENTS: 1. Oxygen requirement in darkening of silver chloride-experiment 2. Effect of Heat on darkened substance

247 MITRA SC. North Indian folk-medicine for Hydrophobia and Scorpion sting. JPASB. 1915, 11(7-8) : 217-30. Ref 14 KEYWORD: MEDICINE

CONTENTS: 1. Original text & translation are given for mantras,cure of Hydrophobia,Incantation for exercising away the venom of scorpion,Incantation for charming water wherewith to exercise away venom from a person bitten by a dog or a jackal,Incantation for exercising away the venom of dog & jackal

248 DASGUPTA HC. Palaeontological notes from Hazara. JPASB. 1915, 11(9) : 253-57. Ref 17 KEYWORD: PALAEONTOLOGY

CONTENTS: 1. Triassic-Trias limestone & quartzite 2. Jurassic-spiti shales 3.Gieumal sandstone 4.Tertiary beds

249 EVERSHED J. Sunspots and prominences. JPASB. 1915, 11(9) : 293-95.

CONTENTS: 1. Dark patches on sun's disc-photographs study,magnetic field,various positions 2. Three sunspots photographs

250 MITRA SC. North Indian Charms for securing immunity from the virus of scorpion stings. JPASB. 1915, 11(9) : 305-10. Ref 15 Chapra, Pleader, Judge's court

CONTENTS: 1. Components of charms,touching blossoming mango tree,act of waving hands over affected parts 2.Instances of charms by yogis

251 RAMASWAMI MS. New species of Tephrosia,from sind - Tephrosia Falciformis, Ramaswami. JPASB. 1916, 12 : 125-27. Calcutta; Botanical survey of India, Royal Botanic Garden

CONTENTS: 1. Habitat & Measurements of various plant structures 2. Pods structure

252 PRASAD B. Seasonal conditions governing the pond life in the Punjab. JPASB. 1916, 12 : 141-45. Lahore; Alfred, Patiala Research Student, Punjab University Government College KEYWORD: FAUNA

CONTENTS: 1. Periodic physical change in natural freshwater fauna 2. 6 forms selected for study

253 GIBSON CS; SIMONSEN JL. Note on the constituents of the bark of the Hymenodyctyon excelsum. JPASB. 1916, 12 : 161-63. KEYWORD: DRUG

CONTENTS: 1. Bark - medicinal purposes 2. Hymenodyctyon Excelsum - bark as tonic, febrifuge and astringent 3. Extract of fluid from bark - experiment

254 BURKILL IH. Notes on the pollination of flowers in India - note No. 8, Miscellanea. JPASB. 1916, 12 : 239-65. Ref 9

CONTENTS: 1. Birds visiting flowers 2. Sphingid flowers 3. Butterfly flower 4. Bee-flower 5. Irregular flowers suited for Apis and other small bees 6. Malvaceae, Araceae 7. Ant-patrols and extrafloral nectaries

255 BURKILL IH. Note on the Terai forests between the Gandak and the Teesta. JPASB. 1916, 12 : 267-72. KEYWORD: FORESTS - TERAI

CONTENTS: 1. River course, trade routes, natural vegetation 2. Rulers role in Gandak and Teesta areas

256 MITRA SC. Some Indian ceremonies for disease - transference. JPASB. 1917, 13 : 13-21. Ref 12

CONTENTS: 1. Matani Rath Kahadvi - disease - transference-Bombay : act of performance 2. Telgu near sea-coast - in epidemic;DT 3. Cholera DT - at Coconada in south India 4. Disease - driving expedition

257 GIBSON CS; SIMONSEN JL. Stereochemistry of Alanine derivatives. JPASB. 1917, 13 : 191-93.

CONTENTS: 1. Crystallographic study of racemic and resolved compounds 2. Effects of change in conc. of group R. etc.

258 PHELPS WH. Notes on the habits of Cyrtophora citricola. JPASB. 1917, 13 : 23-26. KEYWORD: SILK

CONTENTS: 1. Body markings and structures 2. Cocoons and spinning

259 FYSON PF. Note on a malformation of common pine-apple (Ananas sativus, Schult.). JPASB. 1917, 13 : 253. KEYWORD: PLANT - ABNORMAL, PINE-APPLE

CONTENTS: 1. Aggregate flowers - inflorescence of pine-apple 2. Cause of malformation of pine-apple with plates

260 FYSON PF. Notes on the ecology of the Nilgiri and Pulney highland plateaus. JPASB. 1917, 13 : 255-59. Madras; Presidency College KEYWORD: FLORA

CONTENTS: 1. Flora of Nilgiri and Pulney Plateaus 2. Ecological conditions at plateaus 3. Xerophytic conditions, driest situation - underlying rocks 4. Summer rains area 5.Shade-loving plants - orchids etc.

261 MITRA SC. Secrecy and silence in north Indian agricultural-ceremonies. JPASB. 1917, 13 : 27-42. Ref 40

CONTENTS: 1. Sowing seeds 2. Threshing harvested crops 3. Winnowing threshed-out grains 4. Heaping up cleaned grains 5. Measuring grains

262 CHATTERJI SC. Detection of Lactic and Glycolic acids and a suggestion regarding the constitution of Morphine. JPASB. 1917, 13 : 281-85. Gauhati; Cotton College

CONTENTS: 1. Morphine formulae - molecular structure 2. Colouration in detection

263 JACOB KC. Observations on pollination in Alysicarpus. JPASB. 1917, 13 : 287-92. Ref 7 Coimbatore; Agricultural College CONTENTS: 1. Weeds-alysicarpus 2. Flora structure and mode of pollination 3. Explosive mechanism by Sun's action by Keel bending

264 DASGUPTA HC. Occurrence of limburgite in British Baluchistan. JPASB. 1917, 13 : 293-98. Ref 30 KEYWORD: PALAEONTOLOGY

CONTENTS: 1. Palaeontological collections - trap rock 2. Senonian group - rock is described (limburgite) 3. Chemical analysis of rocks 4. Molecular ratios of constituents of limburgitic

265 RAO YR. Notes on some south Indian cecidomyiids causing galls in grasses. JPASB. 1917, 13 : 299-306. KEYWORD: GALLS - GRASSES

CONTENTS: 1. Silver-shoot (gall) caused by fly 2. Species of grasses showing galls formation 3. Life history of fly 4. 4 figures are given

266 ANNANDALE N. Bismer weighing beam from the Darjeeling district. JPASB. 1918, 14 : 243-44. Ref 1

CONTENTS: 1. Description of beam function

267 MEHTA KC. Observations and experiments on the rust on Launaea aspleniifolia DC. JPASB. 1918, 14 : 247-52. Ref 3 Agra; Agra College KEYWORD: PARASITE

CONTENTS: 1. Cause of rust, anatomy of parasite 2. Effects of rust on Launaea aspleniifolio 3. Phenomenon connected with disease

268 BHATTACHARYA DR. Notes on the anatomy of a double monstrosity in the Chick. JPASB. 1918, 14 : 333-37. KEYWORD: ABNORMAL - ANATOMY CHICK

CONTENTS: 1. External features, endoskeleton, body-cavity and digestive organs, respiratory organs, heart and blood vessels, nervous system etc.

269 FISHER C. Preliminary note on the flora of the anaimalais. JPASB. 1918, 14 : 379-88. Indian Forest Service

CONTENTS: 1. Anaimalai Hills - geographical description 2. Nilgiri goat, snakes and tribes of area 3. Botanical investigation 4. Synoptical table of phanerogamia and cryptogamia

270 RAKSHIT JN. Isolation of Porphyroxine. JPASB. 1918, 14 : 407-08.

CONTENTS: 1. Preparation of porphyroxine

271 COTTER G DE P. Geotectonics of the tertiary Irrawaddy basin. JPASB. 1918, 14 : 409-20. Ref 17 Geological Survey of India

CONTENTS: 1. Burma area - eocene time 2. Yenangyat - singu area 3. Geoanticline and sinking geosyncline 4. Sea retreated to south 5. Ancient coast-line 6. Evidences of recent upheaval 7. Age of anticlines of Irrawaddy basin

272 SAHA MN. New theorem in elasticity. JPASB. 1918, 14 : 421-23. Ref 2 Calcutta; Mathematical Physics, University College of Science

CONTENTS: 1. Equations of motion of elastic system 2. Applications of theorem

273 SAHA MN; CHAKRAVARTI S. Pressure of light. JPASB. 1918 , 14 : 425-32. Ref 6 Calcutta; University College of Science KEYWORD: LIGHT - PRESSURE

CONTENTS: 1. Description of apparatus for pressure of light 2. Measurement of energy

274 VREDENBURG EW. Considerations regarding a possible relationship between the charnockites and the Dharwars. JPASB. 1918, 14 : 433-48. Ref 1 KEYWORD: ROCKS; GEOLOGY

CONTENTS: 1. Rocks apparently intermediate in their characters between Dharwars and Charnockites 2. Stratigraphical relationships of Charnockites 3. Association of charnockite and Khondalite series 4. Dyke-rocks apparently connected with charnockites and Dharwars

275 VREDENBURG E. Note on the occurrence of Dolium variegatum Lamarck at Maskat , with considerations on its geographical distribution at the present day and in former geological times. JPASB. 1918, 14 : 449-52. KEYWORD: FAUNA

CONTENTS: 1.Fauna of Australia - distribution through geology, history of fauna

276 ANNANDALE N; PRASAD B. Note on the taxonomic position of the genus camptoceras, Benson and of Lithotis japonica, Preston (Mollusca Pulmonata). JPASB. 1918, 14 : 457-62. Ref 14 Zoological Survey of India

CONTENTS: 1. Radular teeth of camptoceras, habits 2. Geographical range, 4 species are described

277 CLEGHORN ML. Note on the vitality and longevity of silkworm moths during the cold and rainy seasons in Bengal. JPASB. 1919, 15 : 101-05.

CONTENTS: 1. Diagrams showing vitality of silkworm moths in relation to seasons in Bengal 2. Species of longevity and immunity from disease of silkworm moths- conditions

278 VREDENBURG E. Occurrence of Cypraea nivosa, Broderip, in the Mergui Archipelago. JPASB. 1919, 15 : 137-42.

CONTENTS: 1. Description and distribution of Cypraea nivosa Broderip 2. Variability in species of maritius and Mergui-measurements 3. Comparison with Cypraea vitellus

279 VREDENBURG E. Two albino varieties of Cypraea erosa, Linnaeus. JPASB. 1919, 15 : 143-45. KEYWORD: ZOOLOGY

CONTENTS: 1. Two varities of Cypraea erosa Linnaeus - Dimensions, occurrence, comparison with other forms

280 VREDENBURG E. Occurrence of Cypraea piriformis, Gray in the Mergui Archipelago. JPASB. 1919, 15 : 147. Ref 1 KEYWORD: ZOOLOGY

CONTENTS: 1. Location & measurements of species.

281 SOUTHWELL T; PRASAD B. Observations on the intra-uterine embryos of elasmobranchs. JPASB. 1919, 15 : 149-52. Ref 1 KEYWORD: EMBRYOS - INTRA-UTERINE; FISH

CONTENTS: 1. Oviparous ans viviparous in Elasmobranchs 2. Placenta, its cord, appendages and branchail filaments

282 KAYE GR. Ancient Hindu spherical astronomy. JPASB. 1919, 15 : 153-89. Ref 15 KEYWORD: ASTRONOMY - SPHERICAL; TIME

CONTENTS: 1. Spherical trigonometry 2. Ascensional difference 3. Lagna (piont of ecliptic on horizon) 4. Length of day 5. Planets mean motions 6. Eclipses lunar and solar 7. Parallax 8. Tables of cycles, years, months, days, latitudes etc.

283 SEN NN. Interaction of phosphorus halides and arsenious and arsenic compounds. JPASB. 1919, 15 : 263-65.

CONTENTS: 1. Method of interaction, reactions given

284 ANNANDALE N; CARTER HG. Notes on the vegetation of Seistan. JPASB. 1919, 15 : 267-97. Ref 1 Zoological Survey of India Botanical Survey of India KEYWORD: AFGHANISTAN; BALUCHISTAN

CONTENTS: 1. Vegetation of stony desert, banks of saline streams, alluvium, sand dumes, Hamun-i-Helmand 2. Systematics list of plants in Seistan

285 BARTER E. Radiation pressure : the fallacy in Larmor's proof. JPASB. 1919, 15 : 299-302.

CONTENTS: *pages are missing

286 KENOYER LA. Notes on vallisneria. JPASB. 1919, 15 : 303-04. Allahabad; Allahabad School of Agriculture KEYWORD: PLANT - AQUATIC

CONTENTS: 1. Habitat and habit of vallisneria s. 2.Comparison of European form, American form and Indian form

287 CHATTERJI NN. Rationalisation of algebraic equations. JPASB. 1919, 15 : 305-07. KEYWORD: EQUATIONS - ALGEBRAIC

CONTENTS: 1. Detailed method of rationalisation of algebraic equations

288 HAINES HH. Some new species of plant from Behar and Orissa. JPASB. 1919, 15 : 309-17.

CONTENTS: 1. 12 new species are desrcibed - location, inflorescence, other structures

289 RAI H. Note on nitrogen. A new method of preparation. JPASB. 1919, 15 : 319-20.

CONTENTS: 1. Electrolysis of ammonium chloride in aqueous solution 2. Experiment given for nitrogen preparation

290 RAI H; DUNNICLIFF HB. Purification of Indian Sesame (Til) oil. JPASB. 1919, 15 : 321-28. Lahore; Chemical Laboratory, Government College KEYWORD: OIL PLANT

CONTENTS: 1. Constituents of Til oil 2. Effects of sunlight 3. Effects of air and sunlight combined 4. Effects of sulphuric acid and caustic soda

291 PILGRIM GE. Suggestions concerning the history of the drainage of northern India arising out of a study of the Siwalik Boulder Conglomerates. JPASB. 1919, 15 : 81-99. Ref 20 Geological Survey of India

CONTENTS: 1. Pleistocene boulder beds of Siwalik series - course of route 2. Tributary of river, draining Eastern Himalayas 3. V-shape of Himalayan rivers 4. Gangetic alluvium 5. Maps given showing distribution of land and water in Eocene, miocene, pliocene and pleistocene

292 RAY PC. Chemical knowledge of the Hindus of old. ISIS. 1919, 2 : 322-25. Ref 8 KEYWORD: IRON PILLAR; METALLURGY

CONTENTS: 1. Blue colour of flames (copper) 2. Metallic preparation 3. Zinc 4. Hardness of steel 5. Iron pillar - kutab near Delhi, Irongirdle - Puri, Ornamental gates - Somnath, Iron-gun-Nurvar

293 KAYE GR. Indian mathematics. ISIS. 1919, 2 : 326-56. Ref 32 KEYWORD: PIE; CHINA

CONTENTS: 1. Value of pie in table since BC 1700 to AD 1150 2. Chinese and Arabic mathematics in relation to India 3. Discussion of three periods - Sulvasutra period C 200 AD, - Astronomical period C 400-600 AD, - Hindu mathematical period 600-1200 AD 4. Modern place-value system 5. Numerical notations 6. -Indeterminate equation problems (list of formulas given), -rational right angle triangle, -Pure geometry

294 JACKSON VH; MUKERJEE AT. Utility of desicants in electrostatic measurements. JPASB. 1920, 16 : 1-11. Ref 13

CONTENTS: 1. Calcium chloride 2. Metallic sodium, P2O5, quicklime, H2SO4 - solution electrical insulation

295 JACKSON VH; MUKERJEE AT. Improvements in measurements with quadrant electrometers.Part. II. Simplified arrangements for accurate and continuous work. JPASB. 1920, 16 : 13-22. Ref 4 KEYWORD: ELECTROMETERS-QUADRANT

CONTENTS: 1. Method of working and general arrangements 2. Description of instrument

296 DUNNICLIFF HB. Automatic control of the separation of a liquid into fractions limited by specified densities. JPASB. 1920, 16 : 159-64. Aravankadu; Cordite Factory, South India KEYWORD: DISTILLATION - FRACTIONAL

CONTENTS: 1. Acids and guncotton - description of apparatus

297 ANNANDALE N. Loom used by the Gaodar Herdsmen of Seistan. JPASB. 1920, 16 : 23-25. KEYWORD: WEAVING

CONTENTS: 1. Weaving - loom, apparatus description and figures given

298 ANNANDALE N; PRASHAD B. Further notes on the genus Camptoceras (Mollusca Pulmonata). JPASB. 1920, 16 : 27-33. Ref 3

CONTENTS: 1. Earlier work on the species - camptoceras hirasei, walker C. subspinosum sp. nov. 2. Radular teeth of camptoceras 3. Measurements of shells 4. Taxonomic position of genus camptoceras

299 BHATTACHARYA DR; DAS BK. Note on persistent oviducts and abnormal testes in a male Rana tigrina. JPASB. 1920, 16 : 293-96. Allahabad; Muir Central College KEYWORD: ANIMAL - ABNORMAL GROWTH, TESTES, OVIDUCT

CONTENTS: 1. Disposition and character of oviducts 2. Disposition, relative size and minute structure of right and left testes

300 DASGUPTAHC. Short note on the Cretaceous echinoid cyrtoma, m'clelland. JPASB. 1920, 16 : 297-300. Ref 24

CONTENTS: 1. 5 species of cretacous fossils - location and description

301 BRAHMACHARI UN. Preparation of Urea-antimonyl tartrate,a new compound. JPASB. 1920, 16 : 339-40.

CONTENTS: 1. Solubility cure & analysis of C,H,N & Sb in the salt

302 MATTHAI G. Preliminary observations on cocoon-formation by the common Lahore Leech Limnatis (Poecilobdella) Granulosa (Sav.). JPASB. 1920, 16 : 341-46. KEYWORD: LEECH-COCOON

CONTENTS: 1. 34 specimens observered for cocoon formation 2. figures given

303 BOSE SR. Records of agaricaceae from Bengal. JPASB. 1920, 16 : 347-54. Calcutta; Professor of Botany, Carmichael Medical College

CONTENTS: 1. Distribution & Habitat 2. Pileus, stalk, gills, margin, spore, cystidia are described 12 species

304 PRASHAD B. Tigari - a primitive type of boat used in eastern Bengal. JPASB. 1920, 16 : 35-38. Director of fisheries, Bengal, Bihar and Orissa KEYWORD: ARTISIAN

CONTENTS: 1. Construction of Tigari (Boat) - use and principle

305 ANNANDALE N. Opening address in a discussion on the value of bodily measurements in distinguishing human races. JPASB. 1920, 16 : 41-56. Ref 7 Zoological Survey of India KEYWORD: ANTHROPOLOGY

CONTENTS: 1. Anthropometric measurement of human being

306 KAYE GR. Hindu Astronomical Deities. JPASB. 1920, 16 : 57-75. Ref 88

CONTENTS: 1. Vedic deities, post-vedic 2. Astronomical myths 3. Medieval solar cults 4. Geographical distribution 5. Medieval rituals, manual symbols, planetary worship - influences, Iconography

307 SARTON G. Synthetic Bibliography: with special reference to the History of Science. ISIS. 1920, 3; 8 : 159-70. Ref 6 KEYWORD: BIBLIOGRAPHY - SYNTHETIC; INFORMATION SCIENCE; BIBLIOGRAPHY - ENCYCLOPAEDIC; BIBLIOGRAPHY - ANALYTIC

CONTENTS: 1. Definitions - analytic bibliography, - synthetic bibliography 2. How to prepare a good bibliography of the History of Science 3. How to prepare a good encyclopaedic bibliography 4. How to prepare a good analytic bibliography 5. Insufficiency of analytic bibliography 6. Synthetic bibliography 7. Synthetic bibliography of the History of Science 8. Classification scheme

308 SUBRAMANIAM LS. Genus Cerebella in India. JPASB. 1921, 17 : 205-08. Delhi Pusa, Assistant to Imperial mycologist,Agricultural Research Institute

CONTENTS: 1. Cerebella found associated with Gramineae 2. 8 Species are described various structures

309 SINGH BK. Recent advances in stereochemistry. JPASB. 1921, 17 : 213-30. Patna. Patna College KEYWORD: STEREOCHEMISTRY

CONTENTS: 1. Optical activity & Chemical constitution 2. Optical activity in homologous series 3. effect of conjugated unsaturation 4. Effect of position-isomerism on optical activity

310 BHATTACHARYA DR. Notes on the venous system of Varanus bengalensis. JPASB. 1921, 17 : 257-61. Ref 12 Allahabad, Zoological Department, Muir Central College

CONTENTS: 1. Anterior Abdominal vein 2. Epigastric veins 3. Umbilical vein 4. Post-caval vein 5. Hepatic portal system 6. Renal Portal vein

311 MITRA P. Prehistoric writing in India and Europe. JPASB. 1921, 17 : 279-85. Ref 22 Calcutta, Lecturer, Calcutta University KEYWORD: WRITING-EUROPE; INDIA

CONTENTS: 1. Indian Brahmi characters

312 KATE GR. Rgveda: a note on the arrangement of the books. JPASB. 1921, 17 : 325-27. KEYWORD: INFORMATION SCIENCE

CONTENTS: 1. ten mandalas (cycles) or books & their grouping

313 HORA SL. Homology of the Weberian Ossicles. JPASB. 1922, 18 : 1-4. Ref 3 Asstt Superintendent, Zoological survey of India KEYWORD: FISH

CONTENTS: 1. Variation in structure of tripus-Labeo rohita 2. Bone structure figures are given

314 DOVER C. Resume of recent progress in our knowledge of the Indian wasps and bees. JPASB. 1922, 18 : 17-23. Ref 11

CONTENTS: 1. 10 family are described-species mentioned 2.Distribution

315 HORNELL J. Pearl formation in the Indian pearl oyster. JPASB. 1922, 18 : 213-19.

CONTENTS: 1. Cestode larval from systs found in fissnus of pearl-oyster 2. Stayes of pearl oyster 3. Secreting fissue 4. Classes of pearls,decalcification,muscle pearls etc

316 HARRISON EP; SEN NN. Automatic "make and break" Key for actuating the heating and high potential circuits of a Coolidge X-ray tube. JPASB. 1922, 18 : 221-24. KEYWORD: X-RAY TUBE

CONTENTS: 1. Circuits and diagram of apparatus to run coolidge tube

317 DOVER C; RAO HS. Note on the Diplopterous wasps in the collection of the Indian Museum. JPASB. 1922, 18 : 235-49. Ref 13

CONTENTS: 1. Description of seven new forms of vespidae 2. Taxonomy of vespidae 3. Description of female, male measurements Habitat, location etc

318 CHATTERJI NN. Rationalisation of algebraic equations. JPASB. 1922, 18 : 251-52. Ref 1 KEYWORD: EQUATION-ALGEBRAIC

CONTENTS: 1. Detail method of equation is given

319 HORA SL. Modification of the swim-bladder in hill-stream fishes. JPASB. 1922, 18 : 5-7. Asstt. Sup. Zoological survey of India.

CONTENTS: 1. Elimination of chambers, Types of bladders

320 ANNANDALE N; DOVER C. Advances in our knowledge of the fauna of the fresh and brackish water of India, with a bibliography for the years 1912-1922. JPASB. 1922, 18 : 527-55. KEYWORD: FRESHWATER FAUNA

CONTENTS:

321 ANNANDALE N. Bivalve molluscs injuring brickwork in the Calcutta docks. JPASB. 1922, 18 : 555-57. Zoological survey of India

CONTENTS: 1. Martesia fluminalis-Damage described 2. 4 species are described for Damage

322 ROBINSON HC; KLOSS CB. Remarks on Mr C Stuart Baker's new volume on the birds in the "Fauna of British India". JPASB. 1922, 18 : 559-69.

CONTENTS: 1. Review of the book-list of addenda & corrigenda

323 CLEGHORN ML. Observations on the bat-flowers of the mohwa (Bassia latifolia). JPASB. 1922, 18 : 571-76. KEYWORD: BOTANY

CONTENTS: 1. Swarms of bats & Mohwa tree-inflorescence, pollination floral mechanism etc. given 2. Flowers figuers are given

324 BRUHL P; BISWAS KP. New species of cylindrospermum from Bengal - Cylindrospermum

doryphorum, Bruhl et Biswas. JPASB. 1922, 18 : 577-80.

CONTENTS: 1. Vegetative cells measurements,reproductive cells measurements 2. Heterocysts of C.doryphorum spn.

325 PRASHAD B. Observations on the luminosity of some animals in the genetic delta. JPASB. 1922, 18 : 581-85. Ref 5 Calcutta Zoological survey of India KEYWORD: ZOOLOGY

CONTENTS: 1. Bombay Duck-Harpodon nephereus-fish,prawn,Ctenophores description about body of structures

326 CHOPRA B. Preliminary note on isopoda of the family Bopyridae parasitic on Indian Decapoda Macrura. JPASB. 1922, 18 : 69-71. Research Assistant, Zoological Survey of India

CONTENTS: 1. Distribution of isopoda

327 WOODLAND WNF. Renal Portal system (Renal venous meshwork) and kidney excretion is vertebrata. JPASB. 1922, 18 : 85-193. Ref 48 Allahabad, Senior professor of Zoology, Muir Central College KEYWORD: VERTEBRATA-RENAL PORTAL SYSTEM

CONTENTS: 1. Two system of sinusoids in frog kidney 2. Kidney secretion with & without renal afferent vein supply 3. Repetition of experiments of gurwitsch and interpretation of results theoretical considerations 4. Tubules venus encapsulated glomeruli

328 HORA SL. Observations on the Oral apparatus of the tadpoles of megalophrys parva Boulenger. JPASB. 1922, 18 : 9-15. Ref 21

CONTENTS: 1. Tadpoles on surface,in mid-water, on bottom 2. dissection of oral apparatus

329 PRASHAD B. Revision of Kobelt's nomenclature of the India ampullariidae. JPASB. 1922, 18 : 91. Ref 5 Zoological survey of India

CONTENTS: 1. Species are described - distribution

330 THAPAR GS. Arterial system of the lizard varanus bengalensis (Daud) with notes on Uromastix and Hemidactylus. JPASB. 1923, 19 : 1-13. Ref 18 Lucknow, Zoological Department, Lucknow University

CONTENTS: 1. Heart & main arteries in V.bengalensis 2. Arterial arches 3. Innominate & its branches 4. Origin & distribution of ligament & mesenteric artery 5. Dorsal aorta 6. Notes on arterial system of Uromastix hardwickii(Gray) and Hemidactylus flaviviridis Ruppel

331 ANNANDALE N; HORA SL. Fish, recent and fossil- a review of some recent Americans work. JPASB. 1923, 19 : 101-03. KEYWORD: AMERICA

CONTENTS: 1. Bibliography of fishes 2. Genera of fishes 3. Classification of fishes including families & genera as far as known,abstract review of these

332 HORA SL. Certain local names of the fishes of the genus Garra. JPASB. 1923, 19 : 105-09. Ref 5 Officiating superintendent,Zoological survey of India

CONTENTS: 1. Distribution of fish & their local names in India & meaning of local names-25 local names of fish of genus Garra

333 DUTTA SK. Peculiar disposition of the liver and the kidney in the fish genera Clarias and Saccobranchus. JPASB. 1923, 19 : 111-20. Ref 19 Allahabad Research scholar, Zoology Department, University of Allahabad

CONTENTS: 1. Material & technique 2. Position & Structure of liver & kidney 3. Histology of liver & Kidney

334 GUDGER EW. Sources of the material for Hamilton Buchanan's fishes of the Ganges,the fate of his collections, drawings and notes, and the use made of his date. JPASB. 1923, 19 : 121-36. Ref 21 USA, New York, Associate in lcthyology, American Museum of Natural History

CONTENTS: 1. Sources of Buchanan's material 2. fate of Buchanan's collections, notes & drawings 3. uses made of Buchanan's data

335 HORA SL. Adhesive apparatus on the toes of certain Geckos and tree-frogs. JPASB. 1923, 19 : 137-45 Ref 10 Officiating Superintendent,Zoological survey of India

CONTENTS: 1. Adhesive property of Gecko's foot 2. Finger-disc of Hyla annectens (tree-frog)

336 LAW SC. Observation on the breeding of some common birds in the vicinity of Calcutta. JPASB. 1923, 19 : 147-51.

CONTENTS: 1. Nest of birds in neighbourhood of Calcutta 2. Measurements of nests-4 species

337 SINGH H. Anatomy and Bionomics of the red cotton bug Dysdercus cingulatus (Fabr). JPASB. 1923, 19 : 15-42. Ref 30 Lyallpur, Assistant Professor of Entomology, Punjab Agricultural College

CONTENTS: 1. Habits & general account 2. Anatomy-external (Head, thorax, Abdomen) 2. Anatomy-internal (Digestive system,Nervous system) 4. Reproduction and life-history 5. Diagrams are given

338 GHATAK JC. Conception of the Indian astronomers concerning the precession of the equinoxes. JPASB. 1923, 19 : 311-21.

CONTENTS: 1. Surya-siddhanta, Soma-siddhanta, Vasistha-siddhanta, Brihat-samhita, Laghu-manasa, Jyotirvidabharanam, Adbhuta-sagara, Karana-Kutuhalam, Graha-laghava, Jatakarnaoa,Siddhanta-rahasya,Siddhanta-Darp ana-equinox es

339 GREGORY JW; GREGORY CJ. Zoological results of the percy sladen trust expedition to Yunnan under the leadership of professor J.W.Gregory F.R.S.(1922) etc. Introduction. JPASB. 1923, 19 : 383-84.

340 ANNANDALE N. Land Molluscs. JPASB. 1923, 19 : 385-97. Zoological survey of India

CONTENTS: 1. Description of collection 2. Zoogeography 3.Land & freshwater shells

341 ANNANDALE N. Zoological Results Aquatic gostropod molluscs. JPASB. 1923, 19 : 399-422. Ref 11

CONTENTS: 1. Chinese viviparidae,-shells 2. 5 familes described

342 PRASHAD B. Zoological Result Bivalve Molluscs. JPASB. 1923, 19 : 423-28. Ref 8 Calcutta, Zoological Survey of India

CONTENTS: 1. 9 species described, Habit & Habitat, 2. Location etc

343 TATTERSALL WM. Zoological Results Amphipod Crustacea. JPASB. 1923, 19 : 429-35. Cardiff,Professor of Zoology, University College

CONTENTS: 1. 2 species Occurrence & described to Head, eyes,side plates,pleon,Antenna,Gnathopods,peracopods & uropods etc

344 KEMP S. Zoological Results Decapod Crustaceae. JPASB. 1923, 19 : 437-45. Ref 6 Zoological Survey of India

CONTENTS: 1. Species described, bodyly measurements,legs,carapace,rostrum etc.

345 FRASER FC. Zoological Results Odonata. JPASB. 1923, 19 : 447-64.

CONTENTS: 1. 31 species studied-earlier studies mentioned 2. Head,thoroax,prothorax,legs,wings,Abdomen etc described

346 BROWN JC. Occurrence of Ostrea gryphoides schlotheim. JPASB. 1923, 19 : 75-80. Ref 6 Superintendent, Geological survey of India

CONTENTS: 1. Oysters in clive street 2. Character of associated fauna 3. Origin of Oyster bed

347 DASGUPTA HC. Fossil Pectinidae from Hathab, Bhavanagar state (Kathiawar). JPASB. 1923, 19 : 81-86. Ref 23

CONTENTS: 1. Distribution & description of fossils- 5 species 2. Diagrams are given

348 RAU, LR. Age of the Utatur marine transgression. JPASB. 1923, 19 : 87-89. Ref 7 Bangalore, Lecturer in Geology, Central College

CONTENTS: 1. fossils of the area, Cretaceous deposits-Ammonites

349 STAMP LD, LORD L. Preliminary note on the ecology of part of the riverine tract of Burma. JPASB. 1923, 19 : 91-100. Ref 10

CONTENTS: 1. Ecological factors 2. Vegetation and edaphic control 3. Description of plant formations 4. Agriculture, forestry & Ecology-relations

350 DAVIES LM. Notes on the Geology of Kohat, with reference to the homotaxial position of the salt Marl at Bahadur Khel. JPASB. 1924, 20 : 207-24. Ref 8

CONTENTS: 1. Nummulitic series at kohal 2. Nummulitic series at Bahadur Khel 3. Rocks at various position 4. Correlation of Kohat Bed & taxial position 5. Bahadur Khel Gypsum

351 YAJNIK NA; KOHLI SJ. Radioactive of some Indian minerals. JPASB. 1924, 20 : 225-38. Ref 1 Lahore; Department of chemistry, Forman Christian College

CONTENTS: 1. Radium content of (8) minerals 2. Methods for examination of radiations from radioactive bodies 3. Ionising action of the rays on the surrounding gas experiment given in detail 4. Examination of minerals

352 DASGUPTA HC. Occurrence of Scylla serrata forskal in the upper testing beds of Hathab, Bhavanagar(Kathiawar). JPASB. 1924, 20 : 239-41. Ref 11 KEYWORD: FOSSIL

CONTENTS: 1. Description of fossils-crab

353 WALKER GT. Wings of gliding birds. JPASB. 1924, 20 : 243-46. Ref 1

CONTENTS: 1. Anterior of wings-measurements of wings 2. Vulture's wings

354 LAW SC. Local names of some birds of the Manbhum district . JPASB. 1924, 20 : 247-50. Calcutta

CONTENTS: 1. 50 local names of birds given alongwith scientific names & their association

355 LAW SC. Parus major cinereus breeding in the ,24-Perganahs. JPASB. 1924, 20 : 251-52. Ref 1 KEYWORD: BIRDS

CONTENTS: 1. Nests & eggs of birds 2. Description of birds

356 LAW SC. Kalidasa and the migration of birds I. JPASB. 1924, 20 : 253-60. KEYWORD: BIRDS-MIGRATION; MIGRATION-BIRDS

CONTENTS: 1. Raj hansas - nature, reactions to weather conditins etc. hankering, anxiety & restlessness

357 LAW SC. Kalidasa and the migration of birds II. JPASB. 1924, 20 : 261-73. KEYWORD: BIRDS-MIGRATION; MIGRATION-BIRDS

CONTENTS: 1. Hansas & Rajhansas - seasons & their behaviour 2. Pink-footed goose, Bar-headed goose, Common Flamingo etc.

358 LELE SH. Studies in Bombay fish I Revision of the genus Dupane (Cuv. and val). JPASB. 1924, 20 : 275-88. Ref 13 Bombay; Department of Zoology, Royal Institute of science KEYWORD: FISH

CONTENTS: 1. Historical resume 2. Morphology 3. Anatomy 4.Description of two species

359 DOVER C. Further notes on the Indian diplopterous wasps. . 1924, 20 : 289-305.

CONTENTS: 1. systematic position of Indian wasps 2. 37 species are described, Habitat & habits, face region

360 GEORGE CJ. Root-sucking aphids of Coimbatore. JPASB. 1924, 20 : 307-11.

CONTENTS: 1. Arts & aphids phenomena 2. 10 species are described for body structures, Habitat & Habits 3. National enemies of this species of aphids

361 UVAROV BP. Orthoptera (except Blattidae) collected by Prof. Gregory's expedition to Yunnan. JPASB. 1924, 20 : 313-25. Ref 1

CONTENTS: 1. 31 species of orthoptera 2. 8 text figure are given

362 HANITSCH R. Blattidae collected by Prof. Gregory's expedition to Yunnan. JPASB. 1924, 20 : 337-38. KEYWORD: ZOOLOGY

CONTENTS: 1. 3 species are described - colouration, habitat, habits, body segment, length etc.

363 RAO HS. Note on a brackish-water actinian from Madras Zoological Survey of India. JPASB. 1924, 20 : 339-47. Ref 4 KEYWORD: ZOOLOGY

CONTENTS: 1. Pedal disc, autonomy and regeneration, tentacles and oral disc contraction 2. Fertile mesentery - structure 3. Distribution of actinian

364 CHAUDHURI H. Study of a disease of the garden peas (Pisum sativum) due to sclerotium rolfsii. JPASB. 1924 , 20 : 349-54. Lahore; Department of Botany, Panjab University KEYWORD: PLANT DISEASE

CONTENTS: 1. Relation of sclerotia formation to food material 2. Isolation of fungus from host 3. Nature of infection 4. Growth and temperature 5. Effect of darkness on sclerotium formation 5. Thermal death point and control

365 CHAUDHURI H. Oedogonium nagii, sp. nov.. JPASB. 1924, 20 : 355-58. Ref 10 Lahore; Department of Botany, Panjab University KEYWORD: BOTANY

CONTENTS: 1. Oedogonium filaments 2. Sexual reproduction, description - vegetative plants 3. Measurements of structures 4. A sexual reproduction

366 BISWAS K. Subaerial algae of Berkuda island, in the Chilka Lake, Ganjam district, Madras Pres.. JPASB. 1924, 20 : 359-. KEYWORD: ALGAE - SUBAERIAL; LAKE - CHILKA

CONTENTS: 1. Grouping of algae - soil roof, wall and tree algae 2. Distribution and structure of 9 species

367 NAMASIVAYAM D. Liesegang rings. JPASB. 1924, 20 : 367-69.

CONTENTS: 1. Agar-agar gel in capillary-tubes and test tubes - experiments 2. Colloids formation 3. Gelatin film on glass plates - experiments 4. Mechanism of formation of rings or stratification

368 RAY S. Earth's electric field and the vertical potential gradient. JPASB. 1924, 20 : 371-74. Ref 2 KEYWORD: EARTH - ELECTRIC FIELD; EARTH - VERTICAL POTENTIAL GRADIENT

CONTENTS: 1. Curve of electric potential gradient with height above earth's surface 2. Diurnal variation of potential gradient

369 CHHIBBER HL. Microscopic study of the old copper slags at Amba Mata and Kumbaria, Danta State, N. Gujarat, India. JPASB. 1924, 20 : 375-82. Ref 10 KEYWORD: SLAGS - COPPER

CONTENTS: 1. Megascopic characters of slag 2. Microscopic characters of slag 3. Crystal figures are shown

370 RIBEIRO S. Coleoptera of the family Paussidae. JPASB. 1924, 20 : 383-. Ref 27 Zoological Survey of India

CONTENTS: 1. Structure of Paussidae 2. Classification and zoo-geographical region of paussidae

371 AYYAR ASR. Note on Ardhanarisvara. JPASB. 1924, 20 : 79-80. KEYWORD: HERMAPHRODITE

CONTENTS: 1. South Indian Amazon & Egyptian influence

372 SHAHIDULLAH M. Etymologies of kubha, Icagh-, Icagh-, Icagh-, gevaya, and Laghulo (in the inscriptions of Asoka. IHQ. 1925, 1; (1) : 100-2. Ref 4

373 ACHARYYA PK. Buddha as an Architect. IHQ. 1925, 1; (2) : 234-44. Ref 43 CONTENTS: 1. Town-planning 2. Rooms 3. Gates 3. Hot-air baths 5. Superstructure (Woodwork or brickwork) 6. Furniture articles

374 SRINIVASACHARI CS. Report of the Archaeological department, Hyderabad State. IHQ. 1925, 1; (3) : 565-66. KEYWORD: PAINTINGS - PRESERVATIONS

CONTENTS: 1. Preservation of painting 2. Methods and chemicals.

375 DAS SR. Aspects of the History of Hindu astronomy. IHQ. 1925, 1; (4) : 657-76.

CONTENTS: 1. Vedic times 2. Surya Siddhanta 3. Motions of necessary bodies 4. Relation with Chinese, Egyptian and Greeks astronomy 5. Babylonian system

376 DEB HK. When Kurus fought Pandavas. JPASB. 1925, 21 : 211-20. Ref 13 KEYWORD: CHRONOLOGY

CONTENTS: 1. Great war - Kurus and Pandavas dates of its occurrence by Aryabhata, Varahamihira etc.

377 DEB HK. Five-yearly yuga and the Saptarshi cycle. JPASB. 1925, 21 : 221-28. Ref 8 KEYWORD: ASTRONOMY

CONTENTS: 1. Formula related to Saptarshi reckoning 2. Scheme of quinquennial yuga 3. Earlier reference to yuga and Saptarshi cycle. Newer and older

378 CHAKRAVARTI PC. Patanjali : as he reveals himself in the Mahabhasya. IHQ. 1926, 1; (4) : 738-. Ref 37

CONTENTS: 1. Scientific theories in the Mahabhasya 2. Knowledge of medical science 3. Society 4. Sanskrit 5. Geographical knowledge

379 BASU PC. Art and Philosophy in Hindu temple building. IHQ. 1926, 2; (2) : 225-31.

KEYWORD: BUILDING - TEMPLE; BUILDING CONSTRUCTION

CONTENTS: 1. Buddhistic system

380 CHAKRAVARTI PC. Patanjali: as he reveals himself in the Mahabhasya. IHQ. 1926, 2; (2) : 262-89. Ref 80

CONTENTS: 1. In Mahabhasya 2.Mystical personage 3. Author of varttika on Ayurveda 4. Parentage & birth place 5. Knowledge of vedic literature

381 CHAKRAVARTI PC. Patanjali: as he reveals himself in the Mahabhasya. IHQ. 1926, 2; (2) : 464-94. Ref 111 KEYWORD: TIME

CONTENTS: 1. Poetry knowledge 2. God 3. Philospher 4. Substance 5. Time

382 HORA SL. Manuscripts drawings of fish in the library of ASB.III. Fish drawings among the zoological drawings in the collection of Lieut. col. Sir Alexander Burnes (1805-1841) by Dr PB Lord. JPASB. 1926, 22 : 117-25. KEYWORD: LORD PB

CONTENTS: 1. 19 plates description is given in table 2. Reptiles - 6 plates, mammals - 12 plates, birds 104 plates, fish - 19 plates

383 PRASHAD B. Dates of publication of PM Heude's memoris on the molluscs of China. JPASB. 1926, 22 : 127. Ref 2 Calcutta; Zoological Survey of India, Indian Museum KEYWORD: MOLLUSCS; CHINA; HEUDE PM

384 PRASHAD B. Dates of publication of Hanley and Theobald's "Conchologia Indica". JPASB. 1926, 22 : 129-30. Calcutta; Zoological Survey of India, Indian Museum KEYWORD: MOLLUSCS

CONTENTS: 1. Most comprehensive illustrated work on molluscs 2. Dates of issue and contents of various parts

385 HORA SL. Albino Magur Clarias batrachus Linn. JPASB. 1926, 22 : 131-32. Ref 2 Officiating Superintendent, Zoological Survey of India

CONTENTS: 1. Body colour and patches

386 HORA SL. Goat employed as scapegoat in the Bilaspore district, central on a provinces . JPASB. 1926, 22 : 137.

387 SEWELL RBS. Investigations regarding an epidemic of fish mortality in the tank in the Indian Museum compound. JPASB. 1926, 22 : 177-201. Ref 19 Director, Zoological Survey of India KEYWORD: FISH-EPIDEMIC

CONTENTS: 1. Meteorological condition prevailing from 9-20 Feb. 1926 2. General condition of tank 3. Epidemic 4. Post-morton of fish 5. Conditions of water in Tank on day of epidemic 6. Causation of epidemic 7. General effect of mortality

388 CHOPRA B. Note on fish-mortality in Indian river, in Upper Burma. JPASB. 1926, 22 : 203-10. Ref 3 Calcutta; Zoological Survey of India

CONTENTS: 1. Water current at different positions 2. Mortality cause

389 DASGUPTA HC. Palaeontological notes on the panchet beds at Deoli, near Asansol. JPASB. 1926, 22 : 215-17. Ref 6 KEYWORD: PALAEONTOLOGY

CONTENTS: 1. Description of fossils -Corapace of branchyurous crab, -Stegocephalian cranium, -reptilian coracoid

390 CHIBBER HL. Extinct iron industry of the neighbourhood of mount Popa, Upper Burma, with notes on the microscopic study of the slags. JPASB.

1926, 22 : 219-32. Ref 22 KEYWORD: INDUSTRY - IRON

CONTENTS: 1. Geology, ore-materials 2. Mineralogy of concretions and origin 3. Microscopic old iron-slags -megascopic characters, -crystals structures plates are given

391 DATTA B. Hindu value of pie. JPASB. 1926, 22 : 25-42. Ref 84 Calcutta; University College of Science

CONTENTS: 1. Early values-Sulvasutras 2. Values of pie : 3.1416, /~10, 22/7 and their origin 3. Dimensions of earth and value of pie deduced therefrom 4. Dimensions according to Aryabhata 5. Greek influence in Aryabhata value of pie 6. Date of Aryabhata's value of pie 7. Later hindu values - closer approximations 8. Pie values in series

392 LAW SC. Little-noticed habits of some birds of the district Parganas-24. JPASB. 1926, 22 : 411-20. Ref 12

CONTENTS: 1. Nests of Centropus Sinesis (Coucal) 2. Nests of Merops virides (Bee-cater) -sandy earth house -measurements

393 RIBEIRO S. Note on a simulid larva found associated with a may-fly Nymph. JPASB. 1926, 22 : 69-70. Entomological Assistant, Zoological Survey of India KEYWORD: INSECTS

CONTENTS: 1. Photograph of may-fly Nymph - habitat and habits

394 HORA SL. Short note on the distribution and habits of the bivalva Balwantia Soleniformis (Benson). JPASB. 1926, 22 : 71-76. Ref 6 Officiating Superintendent, Zoological Survey of India

CONTENTS: 1. Distribution 2. Habits 3. Boring mechanism 4. Instrument used for collecting B. Soleniformis

395 HORA SL. Note on a Hermaphrodite loach. JPASB. 1926, 22 : 77-80. Ref 6 Officating Superintendent, Zoological Survey of India

CONTENTS: 1. Nemachilus montanus (loach) anterior parts 2. Deformed caudal fin 3. Measurements of body structures

396 HORA SL. Peculiar fishing implement from the Kangra valley, Punjab. JPASB. 1926, 22 : 81-84. Ref 3 Officiating Superintendent, Zoological Survey of India

CONTENTS: 1. Kalerni and types of knots and fishing implements 2. Running fish-noose (Aikeghi) 3. Types of materials used

397 HORA SL. Occurrence of the polyzoon, plumetella fruticosa in running water in the Kangra valley, Punjab. JPASB. 1926, 22 : 85-86. Ref 1 KEYWORD: RUNNING WATER

CONTENTS: 1. Polyzoon and their colony in running water-location

398 MUKHERJEE D. Digestive and reproductive systems of the male ant Dorylus labiatus Schuck. JPASB. 1926, 22 : 87-92. KEYWORD: ANTS

399 HORA SL. Manuscripts drawings of fish in the library of the ASB. I. Fish drawings in the Mackenzie collection. JPASB. 1926, 22 : 93-98. Ref 7

CONTENTS: 1. 20 fish drawings only name mentioned alongwith vernacular name, scientific name and locality

400 HORA SL. Manuscripts drawings of fish in the library of ASB. II. Fish drawings in Buchanan-Hamilton's zoological drawings. JPASB. 1926, 22 : 99-115. Ref 11

CONTENTS: 1. 4 volumes of Buchanan's zoological drawings 2. Two of these contain fish drawings - described 3. Letters description to obtains after 2 vols of drawings

401 MORELAND WH. Mogul unit of measurement. JRAS. 1927 : 102-03.

CONTENTS: 1. Linear square measurement 2. Ain-i-Akbari : yard length in mogul period

402 GEORGE CJ. South Indian aphididae. JPASB. 1927, 23 : 1-12. KEYWORD: APHIDIDAE

CONTENTS: 1. 34 species are mentioned, earlier studies mentioned 2. Brief description of species are given 3. Factors controlling multiplications 4. Wing production 5. Association with Ants

403 CHHIBBER HL. Origin and mineral constitution of the late tertiary fossil wood of Burma. JPASB. 1927, 23 : 13-26. Ref 34 Rangoon; Head of Deptt. of Geology and Geography, University College, University of Rangoon KEYWORD: GEOLOGY

CONTENTS: 1. Mineralogical constitution of fossil wood 2. Physical characteristics 3. Nature of fossil wood 4. Specific gravity determinations 5. Age of beds containing fossils wood and conditions favourable to its formation 6. Puckering and contortion in fossil wood

404 DATTA B. Hindu method of testing arithmetical operations. JPASB. 1927, 23 : 261-67. Ref 18 Calcutta; University College of Science

CONTENTS: 1. Hindu method of Proof -Aryabhata II (c. 950 AD) 2. Arabic method 3. Origin of proof by nine

405 BANERJI RD. Indian affinities of Ainu pottery. JPASB. 1927, 23 : 269-72. Ref 14 KEYWORD: POTTERY - AINU

CONTENTS: 1. Suspension vessel from Japan 2. Glazed suspension vessel 3. Painted mininature suspension vessel with ring 4. Drinking vessel, wine-cooler 5. Suspension bottle

406 MAHALANOBIS PC. Analysis of race-mimture in Bengal. JPASB. 1927, 23 : 301-33. KEYWORD: ANTHROPOLOGY

CONTENTS: 1. Anglo-Indians in Calcutta their analysis 2. Geographical and cultural classification 3. Geographical and cultural factors 4. Analysis of caste system 5. Nasal, cephalic, Indian stature etc.

407 KRISHAN S; CHAUDHURI H. Preliminary report on injection experiments with special reference to the production of alkaloids and general metabolism in plants. JPASB. 1927, 23 : 335-38. Lahore; University Chemical Laboratories KEYWORD: BOTANY; PLANT

CONTENTS: 1. 7 species of alkaloid plants are taken for experiment 2. Method and procedure 3. Stem of plants are injected near soil level

408 HORA SL. Lunar periodicity in reproduction of insects. JPASB. 1927, 23 : 339-41. Ref 2 KEYWORD: REPRODUCTION - INSECTS; MAY-FLY

CONTENTS: 1. May flies, aquatic insects

409 LAW SC. Pericrocotus speciosus (Lath) occurring in Ranchi district, Chota Nagpur. JPASB. 1927, 23 : 343-44. KEYWORD: BIRDS

CONTENTS: 1. Distribution and measurements of body structure

410 HORA SL. Further note on the manuscript drawings of fish in the Mackenzie collection. JPASB. 1927, 23 : 345-46. Edinburgh; Zoological Dept. , University of Edinburgh

CONTENTS: 1. 2 Vols of India office library 2. Ist volume contains fish drawing and other 3. 2nd volume contains "natural history and botanical drawings"

411 RATNAKAR JD. Historical stone horse in the Lucknow Museum. JPASB. 1927, 23 : 369-79.

Benares; Shivalaghat KEYWORD: HORSE - STONE

CONTENTS: 1. Inscription study and religious views

412 PRASHAD B. Dates of publication of "The fishes of India" by Dr. Francis Day. JPASB. 1927, 23 : 399-401. Ref 4

CONTENTS: 1. Main work published in four parts between 1875-1878

413 DAS SR. Precession and liberation of the equinoxes in Hindu astronomy. JPASB. 1927, 23 : 403-13. Ref 24

CONTENTS: 1. Soma Siddhanta rule for precession 2. Brahma Siddanta - precession 3. Surya Siddhanta - precession 4. Vasistha Siddhanta - precession 5. Vrddha Siddhanta - precession 6. Laghu Manas - precession 7. Mahasiddhanta of Aryabhata II (875 Saka) - precession 8. Bhaskara Siddhanta Siromani - precession 9. Liberation and revolution theory - annual rate

414 HORA SL. Remarks on Gunther-Day controversy regarding the specific validity of Hamilton Buchanan's cyprinus chagunio. JPASB. 1927, 23 : 415-17. KEYWORD: ICHTHYOLOGY

CONTENTS: 1. Relations between Barbus beavani, Gthr. and Cyprinus Chagunio Ham. Buch.

415 SOHONI VV. Bibliography of meteorological papers in the publications of the Asiatic Society of Bengal, 1788-1928. JPASB. 1927, 23 : 453. KEYWORD: INFORMATION SCIENCE; METEOROLOGY

CONTENTS: 1. 405 papers listed chronologically 2. List of 405 papers published in 7 publications of ASB 3. Author, subject and geographical indices are given

416 BANERJEE MN. On metals and metallurgy in ancient India. IHQ. 1927, 3; (1) : 121-23. Ref 104

CONTENTS: 1. Metallurgy in Rg-veda

417 CHATTOPADHYAYA K. Patanjali and his knowledge of Science. IHQ. 1927, 3; (1) : 181-82.

CONTENTS:

418 DAS SR. Origin and development of numerals. IHQ. 1927, 3; (2) : 356-75. Ref 36

CONTENTS: 1. Numerals in arabs 2. Numerals in Europe 3. Hindu origin of numerals and the gradual evolution of the modern forms 4. Numerals with space value in literature

419 SHARMA VV. Method of plastering walls for painting (Sudhalepavidhanam). IHQ. 1927, 3; (1) : 53-59.

CONTENTS: 1. Outlines of the pictures 2. Brushes etc.

420 DATTA B. Present mode of expressing numbers. IHQ. 1927, 3; (3) : 530-40. Ref 11 KEYWORD: NUMBERS - EXPRESSING

CONTENTS: 1. Principal of expressing numbers (i) Writing out the number name in full (ii) by means of symbols or sign 2. Scripts of numbers 3. Hindu system 4. Greek system 5. Babylonians

421 DAS SR. Cosmographical theories of the Hindu Astronomeris. IHQ. 1927, 3; (3) : 599-603. Ref 9

CONTENTS: 1. Objects and their operations 2. Circumference of universe/Brahmanda

422 BANERJEE MN. On metals and Metallurgy in ancient India. IHQ. 1927, 3; (4) : 793-802. Ref 33

CONTENTS: 1. Gold 2. Silver 3. Ayas 4. Washing of Gold 5. Blowing 6. Smelting 7. Welding

423 GHOSH B. Age of Manusamhita. IHQ. 1927, 3; (4) : 808. Ref 5 KEYWORD: DOCUMENT AGE

424 DAS SR. Origin and development of Numerals. IHQ. 1927, 3; (1) : 97-120. Ref 48

CONTENTS: 1. Meaning and use of numerals 2. Hindu-Arabic Question 3. Hindu numerals origin and development 4. Zero 5. Place value

425 BHADURI JL. Note on a double chick embryo. JPASB. 1927, 33 : 449-52. Ref 21 Calcutta; Zoology Deptt. university College of Science Ballygunge KEYWORD: EMBRYOLOGY; ANIMAL - DEFORMITY

CONTENTS: 1. Description of specimen - anterior portion of double chick embryo 2. Figures are given

426 RANGACHARI K. Town-planning and house-building in ancient India (Silpasastras). IHQ. 1927, 4; (4) : 813-36. KEYWORD: BUILDING CONSTRUCTION

CONTENTS: 1. Underground tunnel (Ajanta and Ellora) 2. Selection of land 3. Classification of towns

427 COYAJEE JC. Astronomy and astrology in the Bahram Yasht. JPASB. 1928, 24 : 223-33.

CONTENTS: 1. Method of study adopted by Bahram Yasht 2. Antiquity, and locality of formation of scheme in Yasht 3. Yasht present day zodiacal scheme 4. Identification of individual zodiacal constellations 5. Application of our method to their Yashts

428 BOGDANOV L. Afghan weight and measures. JPASB. 1928, 24 : 419-35. Ref 3

CONTENTS: 1. Law introducing metric system in Afghan - measurements and weights, lenght, surface, capacity, liquids, temperature

429 DAS SR. Lunar and solar eclipses in Hindu astronomy. JPASB. 1928, 24 : 437-54. Ref 23 KEYWORD: ECLIPSE MOON; ECLIPSE SOLAR; MOTION - SUN, MOON, PLANETS

CONTENTS: 1. Motion of sun, moon, diametres of sun and moon 2. Eclipse 3. Valana used in projection of eclipse

430 DASGUPTA HC. Batrachian and reptilian remains found in the panchet beds at Deoli, Bengal. JPASB. 1928, 24 : 473-79. Ref 35

CONTENTS: 1. Labyrinthodont skull 2. Rhynchocephalian vertebra 3. Carnivorous dinosaurian tooth 4. Age of Panchet beds

431 HORA SL. Habitat and systematic position of two imperfectly known loaches from Afghanistan. JPASB. 1928, 24 : 481-84.

CONTENTS: 1. Nemachilus - genus, 2 species are described - body structure

432 BHADURI JL. Case of hermaphroditism in a common frog Rana tigrina Daud, with a note on the classification of hermaphroditic cases. JPASB. 1928, 24 : 485-99. Ref 34

CONTENTS: 1. Description of specimen Rana tigrina - external characters, urino genitial organ 2. Classification of hermaphroditism - classification systems

433 RANGACHARI K. Town-planning and house-building in ancient India (Silpasastras). IHQ. 1928, 4; (1) : 102. Ref 4 KEYWORD: BUILDING CONSTRUCTION

CONTENTS: 1. House of various castes, location and building 2. Houses of various uses their design and location

434 DAS SR. Astronomical instruments of the Hindus. IHQ. 1928, 4; (2) : 256-69. Ref 53 KEYWORD: INSTRUMENTS - ASTRONOMICAL

CONTENTS: 1. Moon, Ecliptic 2. Gnomon 3. Water-clocks 4. Astrolabe .5. Nadi - valaya yantra 6. Instruments in astronomical texts

435 DAS SR. Hindu Calendar. IHQ. 1928, 4; (3) : 483-511. Ref 103 KEYWORD: TIME

CONTENTS: 1. Time and seasons 2. Year, month, day, date 3. Measure of time 4. System of nomenclature .5. Divisions of time

436 DAS SR. Seasons and the year-beginning of the Hindus. IHQ. 1928, 4; (4) : 653-66. Ref 42

CONTENTS: 1. Sun's annual motion 2. Division of the seasons 3. System of nomenclature

437 DAS SR. Alleged Greek influence on Hindu astronomy. IHQ. 1928, 4; (1) : 68-77.

CONTENTS: 1. Vedas astronomy in and in astronomical works 2. Apollonious ideas

438 GANGULI S. Notes on Indian Mathematics: a criticism of George Rusby Kaye's interpretation. ISIS. 1929, 12 : 132-45. Ref 36 KEYWORD: KAYE GR; PIE

CONTENTS: 1. Aryabhata's Mahasiddhants 2. Value of pie

439 MAJUMDAR DN. Race and adaptability. JPASB. 1929, 25 : 149-56. Ref 1 KEYWORD: ANTHROPOLOGY

CONTENTS: 1. Korwas - primitive tribes of India 2. Customs, food, adaptability

440 BASU PC. Anthropometry of the Bhuiyas of Mayurbhanj. JPASB. 1929, 25 : 157-63.

CONTENTS: 1. Cephalic, Nasal indices 2. Length-height Index 3. Statistical analysis

441 BONNELL B. Geophilid centipedes from the bed of the cooum riverMadras. . 1929, 25 : 181-84.

CONTENTS: 1. Localities 2. Habitat & problems 3. Centipedes & experiment with water-movements & air capacity

442 SINHA JC. Indo-American trade, past and present. JPASB. 1929, 25 : 201-21. Ref 35 KEYWORD: AMERICAN

CONTENTS: 1. Beginning of Indo-American trade. 2. Jay Treaty 3. Effect of the Treaty 4. Trade with Bengal 5. Import into India 6. Decline of trade & its causes 7. Growth of trade from 1905-06 8. Other countries competition

443 SOHONI VV. Meteorological normals of Calcutta. JPASB. 1929, 25 : 237-83. Ref 3

CONTENTS: 1. Historical introduction 2. Previous data 3. Blanford's description of climate of calcutta 4. Tables of Normals-pressure, temperature, humidity, cloud 5. Some abnormals records of Calcutta wind etc.

444 BRAHMACHARI UN; DASGUPTA JM. contribution to the chemistry of certain new aromatic antimonials. JPASB. 1929, 25 : 301-06. Ref 11

CONTENTS: 1. Seven compounds their composition, properties are described

445 CHOPRA B. Drop-door fishing trap used in the Myitkyina District, on a upper Burma. JPASB. 1929, 25 : 311-13. Ref 2 KEYWORD: TRAP-FISHING DROP-DOOR; FISH TRAP

CONTENTS: 1. Measurement of Kya myon (crude trap) like rat-trap - description method of catching fish

446 AYYAR KVK. Karoura of Ptolemy. IHQ. 1929, 5; (1) : 119-24. Ref 6

CONTENTS: 1. Karur or town 2. Ptolemy's geographical knowledge

447 CHOWDHURI JN. Condition of education and architecture in the Bahmani kingdom. IHQ. 1929, 5; (1) : 124-28. Ref 1

CONTENTS: 1. Kulburya (a palce) mosque 2. Kulburgh bazar 3. Mahmud Gawan (mathematician) 4. Allanddin II (Science lover)

448 GHOSH E. Was the equation of time known to the vedic sages?. IHQ. 1929, 5; (1) : 136-37. KEYWORD: TIME - EQUATION

CONTENTS: 1. Sun and Etasa (horse-God) 2. Allegorical story

449 POUSSIN L; DELA V. Some notes on the Tattvasamgraha. IHQ. 1929, 5; (2) : 354-55.

CONTENTS: 1. Only reference quoted 2. I by Embar Krishnamacharya

450 BANERJEE MN. Iron and steel in the Rgvedic age. IHQ. 1929, 5; (3) : 432-40. Ref 23 KEYWORD: RGVEDA

CONTENTS: 1. Dried medicinal plants and birds wings 2. Iron with medicinal plants fused to form steel 3. Fire hammering etc. 4. Ore

451 DATTA B. Scope and development of the Hindu Ganita. IHQ. 1929, 5; (3) : 479-512. Ref 105 KEYWORD: KNOWLEDGE TRANSMISSION; MATHEMATICS - DEVELOPMENT

CONTENTS: 1. Science of calculation 2. Ancient classification 3. Value and importance of Ganita 4. Ganita in Hindu education 5. Greek influence 6. Greek astronomy in India 7. Astronomy effect on the Ganita 8. Aryabhata and Mahavira 9. Sridhara 10. Brahmagupta 11. Aryabahta II 12. Hindu geometry 13. Early Hindu Ganita 14. Sulva-sutras and later Hindu mathematics 15. Grahaganita or jyotise (Science of calculation of the planets) 16. Patiganita and Dhulikarma (astronomical calculations) Arithmetic (geometry and mensuration together etc.)

452 BAGCHI PC. On some Tantrik texts studied in ancient kambuja. IHQ. 1929, 5; (4) : 754-69. Ref 16 KEYWORD: MYSTICISM

CONTENTS: 1. Tantras classification 2. Authors and texts

453 DAS SR. Notes on Indian astronomy. ISIS. 1930, 14 : 388-402. KEYWORD: SURYA SIDDHANTA; PARALLAX

CONTENTS: 1. Suryasiddhant, Motion of planets, parallax, etc.

454 KRISHNASWAMI GV. Reform of the Indian Calendars. ISIS. 1930, 14 : 403-10. Ref 15 KEYWORD: CALENDAR

CONTENTS: 1. Equinoxes, sayana, nirayana 2. Chronolgical description of calendar 3. Year, ayana, month, tithi & nakshatra

455 DATTA B. Hindu names for the rectilinear geometrical figures . JPASB. 1930, 26 : 283-90. Ref 56 KEYWORD: GEOMETRICAL-RECTILINEAR FIGURES

CONTENTS: 1. Name of rectilinear figures according to number of sides 2. Different naming systems - compound ending 3. Classification of triangles 4. Classification of quadrilaterals

456 PRASHAD B. Reflections on Zoological research in India. JPASB. 1930, 26 : 291-300. Ref 26 KEYWORD: RESEARCH-ZOOLOGICAL

CONTENTS: 1. Ways & obstacles in original zoological research 2. Experience of research in Molluscs-authors views

457 BHADURI JL. Notes on the arterial system of the common Indian Toad Bufo melanostictus schneid. JPASB. 1930, 26 : 301-15. Ref 11

CONTENTS: 1. Material & method 2. Description of arterial system 3. Carotid arch 4. Systematic or Aortic arches 5. Pulmo-cutaneous arch

458 SEN AC. Studies on Indian Ichneumonidae. JPASB. 1930, 26 : 317-38. Ref 22

CONTENTS: 1. Material & method 2. General morphology of ichneumonidae 3. Morphology of xanthopimpla pedator 4. Pro-thorax 5. Meso-thorax 6. Meta-thorax, legs, wings, abdomen, genitalia

459 CHAKRAVARIT C. Meteorological proverbs of the people of Bengal. JPASB. 1930, 26 : 371-77. Ref 4

CONTENTS: 1. 20 proverbs are given in original text with translation

460 BLATTER B. New species of Indian plants. JPASB. 1930, 26 : 339-66.

CONTENTS: 1. Plants from Bombay, Madras & Mount Abu 2. 13 families are described, Habitat, Location, Habits & structures measurements etc

461 DASGUPTA HC. New Theropod Dinosaur (Orthogoniosaurus Matleyi, n. gen. et n. sp.) from the Lameta beds of Jubbulpore. JPASB. 1930, 26 : 367-69. Ref 13 KEYWORD: DINOSAUR-THEROPOD

CONTENTS: 1. Description, systematic position are given

462 BRAHMACHARI UN; DASGUPTA JM. Synthesis of a few antimonials of therapeutic interest. JPASB. 1930, 26 : 413. KEYWORD: THERAPY; MEDICINE; DRUG

463 CHAUDHURI H; SINGH G. Wither-tip disease of citrus plants. JPASB. 1930, 26 : 523-32. Ref 5 KEYWORD: PLANT DISEASE; FUNGUS

CONTENTS: 1. Morphology of fingus 2. Study of fungus in culture; isolation 3. Growth in different media 4. Effect of different concentrations on growth 5. Isolation of different strains 6. Dissemination & infection

464 BISWAS K. Notes on the organisms in the filtered water of Calcutta. JPASB. 1930, 26 : 533-40. Ref 41 Calcutta; Sibpur, Royal Botanic Garden KEYWORD: WATER

CONTENTS: 1. Chemical examination & biological examination are performed 2. 5 species studied

465 MAHALANOBIS PC. Tests and measures of group divergence. Part I: Theoretical formulae. JPASB. 1930, 26 : 541-88. Ref 17 KEYWORD: STATISTICS-DIVERGENCE

CONTENTS: 1. Tests of divergence 2. Measures of divergence 3. Coefficient of Divergence in Means 4. Results of sampling experiments 5. Principle of eqipartition of variance & a coefficient of familial differentiation 6. Coefficient of divergence in variability

466 CHAKRAVARTI C. Antiquity of Tantricism. IHQ. 1930, 6; (1) : 114-26. Ref 56 KEYWORD: MYSTICISM

CONTENTS: 1. Tantricism - its universal character - its prevalence among primitive peoples 2. Destroying the enemies by destroying an image of him

467 CHAKRAVARTI PC. Naval warfare in ancient India. IHQ. 1930, 6; (4) : 645-64. Ref 54 KEYWORD: NAVY

CONTENTS: 1. India shipping and maritime activities 2. Early traces in literature 3. From 4c B.C-7c A.D. 4. Navy and naval operations in Bengal 5. In Punjab and Sindh 6. In Tamil land

468 DATTA B(B). Two Bhaskaras. IHQ. 1930, 6; (4) : 727-36. Ref 36

CONTENTS: 1. Biography and his work date 2. Bibliographical description of this works 3. Commentaries on his works

469 GANGULI R. Cultivation in ancient India. IHQ. 1930, 6; (4) : 737-46. Ref 19 KEYWORD: AGRICULTURE

CONTENTS: 1. Rgveda on Agriculture. Atharvaveda 2. Soil and its classification 3. Meteorological observations leading to forecasting of rain-fall 4. Keeping of cattle needed for cultivation

470 BAGCHI PC. On some Tantrik texts studies in ancient kambuja. IHQ. 1930, 6; (1) : 97-107. Ref 14 KEYWORD: MYSTICISM

CONTENTS: 1. Agamas and their influence, classification

471 THOMPSON RC. Assyrian prescription for diseases of ear . JRAS. 1931, 1 : 1-25. Ref 109 KEYWORD: EARS - DISEASES; DISEASES - EARS

CONTENTS: 1. Given more than 50 prescriptions and care about ears etc.

472 WESENDONK OGV (FOREIGN). Kalavada and the Zervanite system. JRAS. 1931, 1 : 53-109. Ref 213 KEYWORD: TIME; IRAN

CONTENTS: 1. Indian conceptions of time (various scholars views on the problem, allusion to the Kalavada) 2. Iranian conceptions of time (Iranian sources, christain references)3. Mesopotamian conceptions of time 4. Greek and Hellensitic speculations on time 5. Development of Zervanism 6. Iranian contact with India

473 TUCCI G. Notes on the Nyaya-pravesa by Sankarasvamin. JRAS. 1931, 2 : 381-413. Ref 12 KEYWORD: LOGIC

CONTENTS: 1. Indian logic text 2. Chinese interpretation 3. Commentary by Kuei-chi on the Nyaya-Pravesa

474 BHADURI JL. Observation on the course of the facial vein and the formation of the external jugular vein in some Indian frogs and toads. JPASB. 1931, 27 : 101-10. Ref 13

CONTENTS: 1. Material and method 2. Description of anterior veins in Bufo melanostictus 3. Description of anterior veins of Rana tigrina

475 SEN AC. Genitalia of the common Indian cockroach-Periplaneta americana Linn. JPASB. 1931, 27 : 111-18. Ref 8

CONTENTS: 1. Abdomen & terminal abdominal structures 2. Cerci 3. Genitalia - (i)Male - (ii) Female 4. Styli

476 ACHARJIE MN. Short note of the red-billed Chough, Pyrrhocorax Linn. JPASB. 1931, 27 : 119-21.

CONTENTS: 1. Distribution & measurements of species 2. Culmen size, wings, bill from nostril, tarsus

477 HORA SL. Buchanan's ichthyological manuscript entitled 'Piscium Bengalae inferioris delineationes. JPASB. 1931, 27 : 123-35. Ref 4 KEYWORD: ICHTHYOLOGY; BUCHANAN-HAMILTON

CONTENTS: 1. 81 species are described; scientific & local names in gangetic fishes etc. 2. Manuscript 3. Buchanan's early interest in fishes of Ganges 4.

Buchanan's investigation of fishes 5. Buchana's fish drawings

478 HORA SL; MUKERJI DD. Further notes on Hamilton Buchanan's Cyprinus chagunio. JPASB. 1931, 27 : 137-39. Ref 9 KEYWORD: FISH

CONTENTS: 1. Head, anal fin of male & female Barbus chagunio 2. location of fish

479 CHERIAN MC. South Indian Acarina . JPASB. 1931, 27 : 141-47. KEYWORD: PESTS-CROPS

CONTENTS: 1. 7 families are described 2. species have description of Habitat, Host-plants etc

480 SINHA JC. Jute in early British days. JPASB. 1931, 27 : 149-56. Ref 15

CONTENTS: 1. uses of jute 2. local names of jute 3. Variety, vernacular name & description of plant given 4. 5 drawings of jute plant are given

481 SAHNI B; RAO. Jurassic Plants from the Rajmahal Hills. JPASB. 1931, 27 : 183-208. Ref 40 KEYWORD: FOSSIL

CONTENTS: 1. Localities, with table showing distribution of species 2. Description -Equisetales -filicales -Cycadophyta -Coniferales -incertae 3. 35 fossils plants photos are given

482 DEB HK. Hindu calendar. and the earlier siddhantas. JPASB. 1931, 27 : 271-83. Ref 8

CONTENTS: 1. History 2. Yuga, solar days 3. quinquennial yuga 4. five yearly yuga etc.

483 LAW SC. Pahariya names of some birds of Darjeeling. JPASB. 1931, 27 : 313-16. Calcutta

CONTENTS: 1. 35 local pahariya names & scientific names are given

484 NARAYANASWAMI V. Provenance of early Malayan plant collections. JPASB. 1931, 27 : 327-477.

CONTENTS: 1. Diary of Kunstler's wanderings in Malay Peninsula between 188-1886 2. Volume I-V of flora of Malay Peninsula by HN Ridley with page reference 3. Index is also given reference to vol & page no.

485 MUKERJI DD. Observations on the burrowing toad, Cacopus globulosus gunther . JPASB. 1931, 27 : 97-100. Ref 11

CONTENTS: 1. Comparison of C. Globulosus with Dr. Anderson's two Specimens C.Systoma 2. Habitat & Habits

486 IVANOW W. Ismailitic work by Nasiru'd-din Tusi. JRAS. 1931, 3 : 527-564. Ref 79 KEYWORD: PERSIAN

CONTENTS: 1. Work : Raudatu't taslim has chapteer (tasawwurs) 2. Tasawwurs - existence of things, living nature : biological, physical and psychial processes, universal physical substances, human, biological and other processes, logic, qualities of things

487 PRZYLUSKI J (FOREIGN). Varuna, god of the sea and the sky. JRAS. 1931, 3 : 613-22. Ref 22

CONTENTS: 1. Sea, desert, sand-desert, marsh, swamy etc. words history has been traced in Sanskrit and Pasi works 2. Difference between these words etc. 3. European-Asiatic and Austro-Asiatic : origin of words and views of scholars on these

488 SURYANARAYANAN SS. Mathara and Paramartha. JRAS. 1931, 3 : 623-39. Ref 14

CONTENTS: 1. Difference between Marthara-Vrtli and Paramartha's commentary 2. Mention five subtle elements and their projection (five senses - their works and philosophy) 3. Organs of action in his second difference, Six Kosas (blood, flesh, tendon, semen, brain and bone)

489 BAGCHI PC. On foreign element in the Tantra. IHQ. 1931, 7; (1) : 1-16. Ref 7 KEYWORD: KNOWLEDGE TRANSMISSION; MYSTICISM

CONTENTS: 1. Kubjika - tantra 2. Tantras schools 3. Tibetan dialects

490 NATH P. Scripts on the Indus Valley Seals. IHQ. 1931, 7; suppl. : 1-52. KEYWORD: INDUS SEALS

CONTENTS: 1. Sings classification on a syllabic basis 2. Roman characters on words

491 DAS SR. Short chronology of Indian astronomy. IHQ. 1931, 7; (1) : 137-49. Ref 19 KEYWORD: CHRONOLOGY - ASTRONOMY

CONTENTS: 1. Vedic year 2. Jyotisa vedanga (Metrical treatise) 3. Dark age of Hindu astronomy 4. Jain astronomy 5. Aryabhata and his disciples

492 GANGULI R. Cultivation in ancient India. IHQ. 1931, 7; (1) : 19-27. Ref 10 KEYWORD: AGRICULTURE

CONTENTS: 1. Preparation and application of manure 2. Collection and treatment of seeds 3. Construction of agricultural implements 4. Ploughing the land 5. Sowing, planting, treatments of plants etc.

493 VENKATASUBBIAH A. Two tantrik stories. IHQ. 1931, 7 ; (3) : 515-22. Ref 11 KEYWORD: MYSTICISM

CONTENTS: 1. Pancatantra 2. Floating rock and dancing apes 3. No milk without milking

494 SARKAR AK. Coins and weight in ancient India. IHQ. 1931, 7; (4) : 689-702. Ref 43

CONTENTS: 1. Weight system 2. Manusamhita is weight-metres

495 ROY JC. Fire-arms in ancient India. IHQ. 1931, 7; (4) : 703-08. Ref 3 KEYWORD: WEAPON

CONTENTS: 1. Classification of arms 2. Divine weapons - divya - astras (by godly power) require mantra to be uttered before its use 3. Bow and arrow

496 GHOSE M. Chando-vedanga of Pingala. IHQ. 1931, 7; (4) : 727-34. Ref 24

CONTENTS: 1. Identification of the work 2. Date of the work

497 PISHAROTI KR. Manasara on Nagaradivibheda. IHQ. 1932, 13; (2) : 350-57. KEYWORD: ARCHITECTURE

CONTENTS: 1. 3 Style of architecture, Nagara, Dravida and Verara 2. Measuring and application of styles

498 DATTA B. Relation of Mahavira to Sridhara. ISIS. 1932, 17 : 25-33. Ref 32

CONTENTS: 1. Mahavira's Ganita-Sara-Samgraha 2. Sridhara Trisatika, Pati-ganita, and Bija-ganita 3. Range of topics,-Area of a segment of a circle, -plane firgures 4. Method of treatment - operations of addition and subtraction, - operation of multiplication, divisions 5. Points of difference - arrangement of fundamental arithmetic operation of addition and subtraction, -solution of rational quadrilateral

499 GHOSH E. Studies in Regvedic deities - astronomical and meteorological. JPASB. 1932, 28 : 1-122. KEYWORD: METEOROLOGY; ASTRONOMY; RGVEDA

CONTENTS: 1. Deities physical nature interpreted from astronomical and meteorological points of view 2. 21 papers of A S of Bengal meetings of two years 3. Heaven, firmament

(Antariksa), and earth 4. Visible portion of celestial sphere 5. Celestial equator, Equinoxes and solstices, Sun's shifting in a year 6. Stars, asterisms and constellations 7. Atmospheric objects 8. General and Sanskrit indicesare given

501 HORA SL. Mud-fishing in lower Bengal. JPASB. 1932, 28 : 197-205. KEYWORD: FISHING; FISH TRAP

CONTENTS: 1. Mud-fishing 2. Fish trap used in mud-fishing 3. 8 figures and given

502 COYAJEE JC. Sraosh Yasht: its place in the history of mysticism. JPASB. 1932, 28 : 225-37.

CONTENTS: 1. Abode and locality assigned to Sraosha 2. Numerology 3. Sraosha as worshipper of two producers and preserves 4. Sraosha as religious teacher, protector of saints 5. Trace of esoteric doctrine in Yasht

503 SARBAHI DS. Endoskeleton of Labeo rohita (Ham. Buch). JPASB. 1932, 28 : 295-347. Ref 26 KEYWORD: FISH

CONTENTS: 1. Endoskeleton - Axial skeleton, -skeleton of median fins, -Ribs, -skull 2. Appendicular skeleton -pectoral girdle and fin, -Pelvic girdle and fin 3. Figures are given

504 SOHONI VV. Weather types associated with nor'Westers in Bengal. JPASB. 1932, 28 : 349-57. Calcutta KEYWORD: WEATHER SCIENCE

CONTENTS: 1. A day of thunderstorm phenomena 2. Thunderstorms and duststorms 3. Method of treatment 4. 1914 report on thunderstorm 5. Temperature contrast type

505 BISWAS K. Living conifers of the Indian empire. JPASB. 1932, 28 : 359-77. Calcutta; Herbarium, Royal Botanic Garden

CONTENTS: 1. 106 species their location in Indian empire 2. 12 photographs of conifers are given

506 SANKAR KG. Date of the 11th Paripadal of Nallanduvanar (16th June, 634 AC). JRAS. 1932, 3 : 541-45. KEYWORD: MONSOON - STARS POSITIONS

CONTENTS: 1. Astronomical date given (text indicates actual position of leading planets and stars in Monsoon)

507 MULLER RGF. Zum Alter der fruhen Fachuberlieferungen der indischen medizin, der Samhita des Caraka, Susruta und Vagbhata. JRAS. 1932, 4 : 789-814. KEYWORD: MEDICINE

CONTENTS: 1. Study the time period to speak constants etc., language etc.

508 BAILEY TG. Time taken by the strike of cerebal. JRAS. 1932, 4 : 971-72. KEYWORD: CEREBRAL - TIME TO SPEAK CONSTANTS; SPEECH THERAPHY

CONTENTS: 1. Study time period to speak constants etc., language etc.

509 NATH P. Scripts on the Indus Valley seals. IHQ. 1932, 8; (2) : 1-32 Suppl.

CONTENTS: 1. Elam compared 2. Linear scripts 3. Homo-sings 4. Tantrik texts symbols

510 DEB HK. Macedonian month-Name in a Brahmi Inscription. IHQ. 1932, 8; (1) : 117-20. KEYWORD: DATING

CONTENTS: 1. Modes of dating

511 LAW NN. Mohenjo-daro and the Indus Valley Civilization. IHQ. 1932, 8; (1) : 121-64. Ref 1

CONTENTS: 1. Areas 2. Stratification 3. Building within the city 4. Great bath 5. Hot-air bath or hypocaust 6. Pillared hall etc.

512 RAY JC. Fire-arms in ancient India. IHQ. 1932, 8; (2) : 267-71.

CONTENTS: 1. Secret weapons agni-bana, agni-dharana, Ksepyam agni-yoga (fire missile) 2. Supposed guns

513 DAS SR. Jaina school of astronomy. IHQ. 1932, 8; (1) : 30-42. Ref 32 KEYWORD: SCHOOL - JAINA; JAINA

CONTENTS: 1. Jyotisa Vedanga 2. Earth description 3. Stars and planets 4. Place of astronomy in Jainism 5. Sources. Texts

514 KAPADIA HR. Note on Sripati and his Ganitatilaka. IHQ. 1932, 8; (2) : 351. Ref 2 KEYWORD: MATHEMATICS

CONTENTS: 1. Sripati works and later commentaries on it

515 BANERJEE MN. Note on Iron in the Rg-vedic age. IHQ. 1932, 8; (2) : 364-66. Ref 3 KEYWORD: RGVEDA

CONTENTS: 1. Ayas (Iron or steel) 2. Ksura (razor) 3. Khadi or khadu(iron ring)

516 KAPADIA HR. Note of Candra-prajnapti. IHQ. 1932, 8; (2) : 381-82. Ref 4 KEYWORD: ASTRONOMY

CONTENTS: 1. Bibliographical details of work 2. It is a Jain works of Digambar school

517 DAS SR. Jaina school of astronomy. IHQ. 1932, 8; (3) : 565-70. Ref 10 KEYWORD: SCHOOL - JAINA; JAINA

CONTENTS: 1. Suryaprajnapti (2 suns) 2. Two suns and their motion 3. Motion of the sun 4. Tapaksetra (part of Jambudvipa which on each day is illuminated)

518 RAY JC. Fire-arms in ancient India. IHQ. 1932, 8; (3) : 583-88. Ref 2

CONTENTS: 1. Guns and gun-powder 2. Powder and gun art of Indian invention

519 DAS SR. Jaina Calendar. ISIS. 1933, 19 : 432-37. Ref 3 KEYWORD: CALENDAR - JAINA

CONTENTS: 1. Name of months - modern names and Jaina names 2. Year of Sam Vatsara 3. Days, months, and seasons (Lunar)

520 DATTA B. Algebra of Narayana. ISIS. 1933, 19 : 472-85. Ref 12 KEYWORD: NARAYANA'S ALGEBRA

CONTENTS: 1. Narayana's time and place 2. Narayana's works 3. Narayana's sources, influence 4. Algebra -notation, -operations with zero, -surds ( a number whose square root cannot be extracted exactly), -approximate value of surds, -indeterminate equations of the first degree, square-nature, -cyclic method of integers equation

521 COOMARASWAMY AK. Translation: Maya, Deva, Tapas. ISIS. 1933, 19 : 74-91. Ref 39

522 THOMPSON RC. Assyrian chemist's Vade-mecum. JRAS. 1933 , 4 : 771-85. Ref 1 KEYWORD: CHEMIST - ASSYRIAN; DRUG

CONTENTS: 1. Medicine 130 durgs mentioned, dyes, chemicals etc.

523 VIRA R. Implements and vessels used in vedic sacrifice. JRAS. 1933, 2 : 283-305. Ref 1 KEYWORD: SACRIFICE - VEDIC

CONTENTS: 1. Use of skin, reeds/bamboo, grass, metal clay in making vessels with figures

524 MULLER RFG. Origin of the Caraka and Susruta Samhitas. JRAS. 1933, 2 : 323-37. KEYWORD: SURGERY - OPTHALMIC;

OPTHALIMIC; CARAK SAMHITA; SUSRUTA SAMHITA

CONTENTS: 1. Carak Samhita at Taksasila 2. Susrutas at Kasi in surgery 3. Nemi as author of Opthalimic Surgery

525 MACKAY E. Sumerian representation of an Indian stand. JRAS. 1933, 2 : 335-38. Ref 2 KEYWORD: POTTERY

CONTENTS: 1. Pottery stand 2. Indian modha (Reed wicker products

526 THOMPSON RC. Assyrian words for "Whetstone" and "Corundum". JRAS. 1933, 2 : 343-46. Ref 5 KEYWORD: STONES - MEDICINE

CONTENTS: 1. Powdered stones properties and medicinal uses

527 SEN SN. Meteorological aspects of the recent "rains of fishes" in the Muzaffarpur district. JPASB. 1933, 29 : 111-16. Ref 1

CONTENTS: 1. Monsoon front 2. Occlusion of Bay air 3. Meteorological conditions over Bihar on the 10th July, 1933

528 SARKAR HB. Date of the introduction of the saka year in Java. JPASB. 1933, 29 : 17-21. Ref 6 KEYWORD: YEAR; TIME

CONTENTS: 1. Saka system in Jawa 2. Saka era & sanjaya era 3. History of Jawa

529 CONGER GP. Cosmic persons and human universe in Indian philosophy. JPASB. 1933, 29 : 255-70. Ref 110

CONTENTS: 1. Brahmanas, lepanishads, vedanta & Samkhya systems, Bhagavad Gita, caraka samhita, Puranas; Tantras, other sectarian litrature, medieval mystics, Hinduism religion, Indian philosophy contact with west

530 HORA SL. Note on the biology of the precipitating action of the mucus of Boro fish, Pisoodonophis boro Ham. Buch. JPASB. 1933, 29 : 271-74. Ref 7 KEYWORD: FISH

CONTENTS: 1. Habitat of P.boro 2. Chemical action of mucus Habits

531 RAYACHAUDHURI S; MAJUMDAR B. Note on the chemistry of the precipitating action of slime water obtained from Boro fish, Pisoodonophis boro Ham. Buch. JPASB. 1933, 29 : 275-83. Ref 1 Calcutta; Dept. of chemistry, University. college of Sc. & Technology. KEYWORD: FISH

CONTENTS: 1. Sensitization of colloidal solutions 2. Sensitizing action of slime water obtained from Boro fish 3. Slime prepared with-conductivity water - tap water 4. Electro-osmotic measurement

532 SUBRAMANIAM MK. Temperature as a factor in sex determination in man. JPASB. 1933, 29 : 289-306. Ref 16 KEYWORD: SEX DETERMINATION

CONTENTS: 1. Methods & material 2. Probable errors 3. Temperature & Madras 4. Heat & Cold as affecting human system(i)Measurement of body heat 5. Sex ratio & fertility (ii) Temp. effecting combination of chromosomes 6. Birth & Temperature statistics(iii) Factors effecting combination of chromosomes

533 MAJEED MA. Short note on the occurrence and distribution of diatoms in the Punjab. JPASB. 1933, 29 : 307-09. Ref 6 KEYWORD: ALGAE

CONTENTS: 1. Sub-aerial association 2. Aquatic assiciation 3. Epiphytic diatoms

534 HORA SL. Worship and propitiation of wild animals at uttarbhag, lower Bengal. JPASB. 1933, 29 : 31-37. Ref 12 KEYWORD: ZOOLOGY

CONTENTS: 1. Panch-pir, Bon Bibi (Dakshinder), Manasu worship etc

535 GHOSH E. Experimental study of the asphyxiation of some air-breathing fishes of Bengal. JPASB. 1933, 29 : 327-32. Ref 8 KEYWORD: FISH

CONTENTS: 1. History of surface breathing of fish 2. Experiment to air-breathing in fish 3. Period of 'drowning' in 19 species

536 SIDDIQUI MZ. Science of medicine under the Abbasids. JPASB. 1933, 29 : 39-50. Ref 44

CONTENTS: 1. Interest of caliphs 2. Baghdad as a great literary centre 3. Different groups of medical men 4. Early translators 5. Arabic medical technical terms 6. Pathological terms 7. Development of medical arabic nomenclature 8. Indian Physician in Baghdad

537 CHAKRAVARTI C. Note on the age and authorship of the Tantras. JPASB. 1933, 29 : 71-79. Ref 40

CONTENTS: 1. Trantras compilation 2. Authorship of Tantras

538 RIBEIRO S. Insects found associatd with the bitter-gourd, Momordica charantia Linn(cucurbitaceae) in Calcutta. JPASB. 1933, 29 : 89-93. Ref 13

CONTENTS: 1. Insects from plants of bitter gourd 2. 16 species are described -on foliage -in flower -in fruit

539 HORA SL. Rains of fishes in India. JPASB. 1933, 29 : 95-110. Ref 11

CONTENTS: 1. Particulars of Rains of fishes hitherto recorded from India 2. Recent Rains of fishes in Muzaffarpur district 3. Species of fish known to have pollen with rains in India 4. Explanation of Rains of fishes

540 THOMPSON RC. Assyrian minerals. JRAS. 1933, 4 : 885-95. Ref 25 KEYWORD: MINERAL - MEDICINE

CONTENTS: 1. List minerals, appearance, name and use as medicine 2. Stones

541 FOSTER W. English factories in India. JRAS. 1933, 4 : 911-12. KEYWORD: INDUSTRIES

CONTENTS: 1. Factory at Surat in 1638

542 WINTERNITZ M. Notes on the Guhyasamaja-Tantra and the age of the Tantras. IHQ. 1933, 9; (1) : 1-10. Ref 26 KEYWORD: MYSTICISM

CONTENTS: 1. Bodhicitta 2. Tibetan and Chinese text 3. Prajnaparamita - sadhana

543 DAS SR. Notion of time in Hindu philosophy. IHQ. 1933, 9; (1) : 149-53. Ref 11

CONTENTS: 1. Definition of time 2. Motion and space 3. Jain's finite & infinte time 4. Buddhism time 5. Time-series

544 SCHANZLIN GL. Monetary system of the moghul Times. IHQ. 1933, 9; (1) : 157-60. Ref 2

CONTENTS: 1. Weight measurement of money/coins

545 OBERMILLER E. Study of the twenty aspects of Sunyata. IHQ. 1933, 9; (1) : 170-87. Ref 62 KEYWORD: RELATIVITY

CONTENTS: 1. Sunyata (Relativity) 20 aspects of (Principle of dependent origination (functional interdependence and relativity

546 TRIVEDI HV. Study of ancient geography. IHQ. 1933, 9; (2) : 470-78. Ref 9

CONTENTS: 1. Agni- purana - Topographical information/names

547 CHATTERJI D(C). Sources of Buddhist logic. IHQ. 1933, 9; (2) : 499-502. Ref 8 KEYWORD: LOGIC

CONTENTS: 1. Perception 2. Inference 3. Validity of a conclusion

548 SUR AK. Origin of Indus Valley Script. IHQ. 1933, 9; (2) : 582.

CONTENTS: 1. Brahmi script. Nath P paper in IHB on 'Scripts on the Indus Valley Seals'

549 SARMA BNK. Bhaskara - a forgotten commentator on the Gita. IHQ. 1933, 9; (3) : 663-67. Ref 31

CONTENTS: 1. Gita - Bhasya by Bhaskara 2. Samkara on Bhaskara

550 HAZRA RC. Influence of Tantra on the tattras of Reghunandana. IHQ. 1933, 9; (3) : 678-704. Ref 132 KEYWORD: MYSTICISM

CONTENTS: 1. Deity 2. Rituals 3. Social and religious ideas of Tantras

551 MUKHERJEE SC. Architecture of the Taj and its architect. IHQ. 1933, 9; (4) : 872-79. Ref 23

CONTENTS: 1. Views of different scholars 2. Italian ideas

552 UPADHYE AN. Varahamihira and Buzurjmehr. IHQ. 1933, 9; (4) : 984-86. Ref 3

CONTENTS: 1. V & B are same question verified 2. B (at Persia)who rendered Pancatantra into Pahlvi

553 VENKATASUBBIAH A. Pancatantra of vasubhaga. IHQ. 1934, 10; (1) : 104-11. Ref 5

CONTENTS: 1. Author of books 2. Period of Books 3. In Europe existence of similar works

554 KAPADIA HR. Note on Jaina Hymns and magic squares. IHQ. 1934, 10; (1) : 148-53. Ref 19 KEYWORD: JAINA

CONTENTS: 1. J.H. associated with odd- and even-numbered magic squares 2. 3rd to 9th order magic squares

555 BOSE AK. Agriculture. IHQ. 1934, 10; (2) : 288-300. Ref 132

CONTENTS: 1. Husbandmen 2. Lands for agriculture 3. Rainfall

556 DAS SR. Jaina Calendar. IHQ. 1934, 10; (2) : 332-36. Ref 2 KEYWORD: JAIN - CALENDAR

CONTENTS: 1. Astronomical chronological period 2. Names of months 3. Motion of sun and moon and planets 4. Seasons

557 CHAKRAVARTI C. Ideals of Tantra Rites. IHQ. 1934, 10; (3) : 486-92. Ref 22 KEYWORD: MYSTICISM

CONTENTS: 1. Allegorical and yogic interpretations to tantras 2. Core and precautions 3. Classes ot Tantras

558 TRIVEDI HV. Studies in ancient geography (Markandeya Purana). IHQ. 1934, 10; (4) : 642-64. Ref 15

CONTENTS: 1. Topographical inf. in M.P.

559 KAPADIA HR. Some references pertaining to agriculture in Jaina literature. IHQ. 1934, 10; (4) : 797-800. Ref 20 KEYWORD: JAIN - SOURCE MATERIALS, AGRICULTURE

560 HORA SL. Trade in Live fish (Jiol Machh) in Calcutta. JPASB. 1934, 30 : 1-15. KEYWORD: FISH-LIVE

CONTENTS: 1. Markets of Jiol Machh 2. Description of joil Machh 3. boats & baskets for trade of Joil Machh (Live fish) 4. Quantity of Joil Machh in beating & number of boats from 1918 to 1932 5. 19 figures given

561 MITRA SC. Plant-lore from Bihar. JPASB. 1934, 30 : 25-28. Ref 2 KEYWORD: BOTANY

CONTENTS: 1. Pipal, Bar & Nim tree their religious values 2. Mock-marriage of Mango Groves

562 MITRA SC. Note on the cult of the agricultural deities Sama and Chako in north Bihar. JPASB. 1934, 30 : 29-31. Ref 2

CONTENTS: 1. Plant & land worships for fertility of land & health of plants

563 HORA SL. Ancient Hindu conception of correlation between form and locomotion of fishes. JASBSC. 1935, 1 : 1-7. Ref 15 Zoological Survey of India KEYWORD: FISH

CONTENTS: 1. Ecological classification of fresh waters - animals 2. Form and locomotion in fishes in Susruta

564 SAYEEDUD-DIN M. Common flowering plants of the Hyderabad state; their distribution, economic and medicinal importance. JASBSC. 1935, 1 : 9-92. Ref 26 KEYWORD: DRUG

CONTENTS: 1. Topography and physical features of Hyderabad -chief hill and mountain ranges, rivers, temperature and rainfall, barometric condition (pressure, wind velocity etc.) 2. Warangal district -rivers, agriculture 3. Systematic account giving vern. names, habitat, uses of 302 species of flowering plants 4. Index to vernacular and English names of Hyderabad plants given

565 SINGH S (SARABJIT). Contribution to Angami and Sema Somatology. JASBSC. 1935, 1 : 93-97. Ref 21 Calcutta; Dept. of Anthropology, Calcutta Univ. KEYWORD: ANTHROPOLOGY

CONTENTS: 1. Stature, nasal and cephalic indices of tribes (Angami and Sema) 2. Grouping and analysis of cephalic and nasal indices

566 MANI MS. Notes on some Indian gall forming psyllidae (Homoptera). JASBSC. 1935, 1 : 99-108. Ref 7

CONTENTS: 1. Insects (Homoptera) jumping plantlice 2. Nine species of Homoptera are described for attacking the plants, their structure etc.

567 JAIN JC. Conception of soul in Jainism. IHQ. 1935, 11 ; (1) : 137-41. Ref 16 KEYWORD: JAINA

CONTENTS: 1. Rgveda 2. Soul qualities 3. Classes of souls

568 SARKAR BK. Kautalya, Economics planning and Climatology. IHQ. 1935, 11; (2) : 328-50. Ref 33

CONTENTS: 1. Economic planning relates to climatology 2. Climate as alleged determinant in Hindi Polity 3. Agricultural statistics 4. Comparision with France and German

569 MUKHERJEE GN. Pre-historic copper celt. IHQ. 1935, 11; (3) : 517-28.

CONTENTS: 1. Copper celt measruement, location, age etc. 2. Bronze age in India 3. List of copper age antiquities in northen India

570 GANGULY DC. Historical information in the Prakrita-Paingalam. IHQ. 1935, 11; (3) : 564-69.

CONTENTS: 1. Prakrta - Paingalam (is a work on Prakrta metrical science 14th C.A.D.) 2. Articles describes only the historical incidents narrated in it

571 KARPINSKI LC. Importance of Greek algebraical problems. ISIS. 1935, 22; 63 : 405. KEYWORD: GREEK ALGEBRA; ALGEBRA-GREEK

CONTENTS: 1. System of indeterminate equations 2. Linear indeterminate equations in India (mentioned)

572 SARTON G. George Catlin "Indian-loving Catlin" (1796-1872). ISIS. 1935, 22; 63 : 77-94. KEYWORD: CATLIN G

CONTENTS: 1. Noted painter, artists 2. Social & ethical life of Indian in his work 3. George Catline (American)

573 SARTON G. First explanation of decimal fractions and measures (1585). Together with a history of the decimal idea and a fascimile (no.xvii) of Stevin's Disme. ISIS. 1935, 23; 65 : 153-244. Ref 119 KEYWORD: FRACTION-DECIMAL; TIME

CONTENTS: 1. Stevin's Thiende 2. Progress of decimal numeration 3. Decimal fraction 4. Decimal moneys, weights and measures 5. Decimal division of angles and time 6. Decimal idea 7. Anglo-American weights and measures

574 MULLER RFG. Natur-und medizingeschichtliches aus dem Mahabharata. ISIS. 1935, 23; 65 : 25-53. KEYWORD: KNOWLEDGE TRANSMISSION

CONTENTS:

575 BROWNE CA. Chemical industries of the American Aborigines. ISIS. 1935, 23; 66 : 406-24. Ref 17 KEYWORD: INDUSTRIES-CHEMICAL; AMERICA

CONTENTS: 1. American fertilizer industry 2. Use of fertiliser by Indian tribes in America/American Indian 3. Mentioned use of various chemical by Indians grease and oil, inorganic colours, pottery, copper, gold and silver, medicinal herbs etc.

576 STEIN O. References to alchemy in Buddhist scriptures. ISIS. 1935, 24; 67 : 119. KEYWORD: BUDDHISM; CHEMISTRY

CONTENTS: 1. Indian (Candragupta)(Sanaka)(Canakya) 2. Goldmaking by Sanaka

577 RAGHAVAN V. Historical information the Prakrita-Paingala. IHQ. 1936, 12; (1) : 151-52. KEYWORD: BHANDARKAR RG

CONTENTS: 1. Negate that P.P. has not been noticed in article on Indian History 2. Give reference of Dr RG Bhandarkar "Report on search Sanskrit Manuscript in Bombay presidency 1887-91. . Report description

578 LEVI S. On a tantrik fragment from kucha (Control Asia). IHQ. 1936, 12; (2) : 197-214. KEYWORD: MYSTICISM

CONTENTS: 1. Black magic 2. Method of tantrik 3. Rites 4. Texts of MSS given

579 CHANDRA P. East India Company and its trade monopology. IHQ. 1936, 12; (3) : 420-31. Ref 27

CONTENTS: 1. Rule of E I Company on trade 2. Hampering the Indian trade 3. Co-operative industries (India and Great Britain)

580 UPADHYAYA B. Geographical data in Panini. IHQ. 1936, 12; (3) : 511-17. Ref 5

CONTENTS: 1. Town villages, rivers, mountains in Panini 2. Afghanistan area 3. Gandhara province 4. Kamboja, Taksa sila Hastinapura, Kuruksetra, Kalinga

581 SARKAR JN. Cotton trade of Patna in early seventeenth century. IHQ. 1936, 12; (4) : 638-61. Ref 137 KEYWORD: TRADE - COTTON

CONTENTS: 1. Patana (Commercial town and Industrial towns) 2. Cotton trade 3. Give information about cotton trade and industry in Patna in 1541-1620 4. Varities of calicoes cloth 5. Towelling (Hammamm)

582 DHARMARAJAN M. Anatomy of Otolithus ruber (Bl. and Schn) Part I. The Endoskeleton. JASBSC. 1936, 2 : 1-72. Ref 64 Madras; University Zoological Laboratory

CONTENTS: 1. Axial skeleton -Head skeleton, skull proper, visceral skeleton, vertebral column, ribs, skeleton of the median fins 2. Appendicular skeleton -pectoral girdle. pectoral fin, pelvic girdle & fin

583 GATES GE; KYAW MH. Citellum and sexual maturity in the Megascolecinae. JASBSC. 1936, 2 : 123-25. Ref 1 Rangoon; Judson College KEYWORD: EARTHWORM

CONTENTS: 1. Earthworms and spermathecal spermatozoa description

584 PRADHAN S. Alimentary canal of Epilachna indica (Coccinellidae: Coleopter) with a discussion on the activity of the mid-gut epithelium. JASBSC. 1936, 2 : 127-56. Ref 45 Lucknow; Zoology Department, Univ. of Lucknow CONTENTS: 1. Course of alimentary canal 2. Histology of alimentary canal 3. Activity of mid-gut epithelium 4. Peritrophic membrane 5. Oesophageal and pyloric valves functions 6. Malpighian tubules 7. Salivary glands 8. Material and technique are described

585 MUKHERJI DD; NAIR KK. Abnormalities in fishes. JASBSC. 1936, 2 : 157-64. Ref 2 Zoological Survey of India KEYWORD: FISHES - ABNORMALITIES

CONTENTS: 1. Eversion of stomach into pharyngeal cavity 2. Description and labelled diagrams are given

586 KYAW MH; GATES GE. Earthworm populations and the formation of castings in Rangoon, Burma. JASBSC. 1936, 2 : 165-70. Ref 2 Rangoon; Judson College

587 BISWAS K. Common diatoms of the Loktak Lake, Manipur, Assam. JASBSC. 1936, 2 : 171-75.

CONTENTS: 1. 9 Species for Diagnosis are taken 2. Diatom and desmid flora of the Indian empire. its description

588 SAYEEDUD-DIN M. Common flowering plants of the Hyderabad state, their distribution and economic importance. Monocotyledons- Part 1. JASBSC. 1936, 2 : 73-94. Ref 14

CONTENTS: 1. Systematic account of 111 species describing vern. name, scientific name and habitat and uses etc. 2. Index to vernacular, english and scientific names given

589 MELLO F DE. Further contribution to the study of the blood parasites of the Indian birds, together with a list of the hemoparasites hitherto recorded. JASBSC. 1936, 2 : 95-122. Ref 34 KEYWORD: PARASITES - BLOOD

CONTENTS: 1. Plasmodids' recorded in Indian birds (14) 2. Haemoproteids recorded in Indian birds 3. Their multiplication in the blood 4. Leucoiytozoow recorded in Indian birds 5. Trypanosomids of Indian birds 6. Toxoplasmids recorded in Indian birds 7. Microfilarids recorded in Indian birds 8. Indian birds which did not show any blood parasite at all

590 POGO A. Cylindrical and diagonal calendars. ISIS. 1936 , 24; 68 : 429-30. KEYWORD: CALENDARS- CYLINDRICAL, DIAGONAL

CONTENTS: 1. Indian Sanskrit names of 7 planets, Planetary house 13th, 14th and 15th day of the month are based on Sanskrit purnama (full moon)

591 DURAND DB. First explanation of decimal fractions. ISIS. 1936, 25; 69 : 134-35. KEYWORD: FRACTION-DECIMAL

CONTENTS: 1. Hindu numerals and astronomical tables in the G. Sarton article 2. Roman and Hindu numerals

592 SARTON G. Hindu decimal ruler of the third millennium. ISIS. 1936, 25; 70 : 323-26. Ref 10

CONTENTS: 1. Mohenjo-daro and the Indus civilization excavation on decimal ruler 2. Decimal ideas from the systems of weights

593 STAPLETON H. Further notes on the Arabic alchemical manuscripts in the libraries of India. ISIS. 1936, 26; 71 : 127-31. KEYWORD: ARAB - SOURCE MATERIALS, CHEMISTRY CONTENTS: 1. Khalid (672 A.D.-720 AD) alchemist 2. Adh-dhakhirat al-Iskandariyah by Hermes and discovered by Balinas gave it to Aristotle, who dedicated it to Alexander 3. Risalah on kimiya of Aristotle for Alexander

594 UDADHYAYA K. New Verses of Panini. IHQ. 1937, 13; (1) : 167-71. Ref 14

CONTENTS: 1. Poet Panini with the grammarian Panini 2. Biogrpahy not S & T Information in the article

595 THIEME P; CHATTOPADHYAYA K. Panini and the Rkpratisakhya. IHQ. 1937, 13; (2) : 327-49. Ref 26

CONTENTS: 1. Panini borrowed directly sutra from Rkpratisakhya is under question 2. P, knowledge of veda and subject etc.

596 MITRA K. Maniya Math and Snake-cult. IHQ. 1937, 13; (4) : 690-96. Ref 14

CONTENTS: 1. Naga (Snake) as storer and giver of rain-water 2. Ancient naga belief 3. Mani Naga 4. M. Math gives snake pictures on it

597 MERAS H. Supposed sumero- Babylonian inscription. IHQ. 1937, 13; (4) : 695-703. Ref 4

CONTENTS: 1. Give various symbols reading and understanding and meaning

598 SARTON G. Hindu decimal ruler of the third millennium (Second note). ISIS. 1937, 26; 72 : 304-05.

CONTENTS: 1. Photographs of Indian Linear measure from Mohenjo-daro C. B.C.3000 2. Description of linear measure

599 . Theory of the diurnal rotation of the earth. ISIS. 1937, 26; 72 : 392. KEYWORD: EARTH-DIURNAL ROTATION

CONTENTS: 1. Astronomical conceptions of ptolemaic, Copernican, Tychonic & De Systematealio. 2. Classical commentaries on revolution of earth, Brahamasphuta siddhanta,Aryabhata views and others. (7th Century) 3. Seventeenth Century view on earth rotation

600 AL-HAKIM. Cornelius van Alenvan Dyck (1818-1895). ISIS. 1937, 27; 73 : 20-45. Ref 9 KEYWORD: CORNELIUS VAD; KNOWLEDGE TRANSMISSION

CONTENTS: 1. Transmission of knowledge into Arabic during the Umayyad and Abbasid Dyansties (662-1258 AD) from Greek, Hindu and other sources 2. Article describes in details biography of Cornelius VAD

601 KARPINSKI LC. Is there progress in mathematical discovery and did the Greeks have analytic geometry ?. ISIS. 1937, 27; 3 : 46-52. KEYWORD: GREEK ANALYTICAL GEOMETRY GEOMETRY-ANALTICAL

CONTENTS: 1. Hindu Science in analytic geometry & idea transmission 2. Descarte's

Analytical geometry 3. Trigonometry development 4. Pure geometry 5.Logical geometry etc

602 SHAW GE; SHEBBEARE EO. Fishes of Northern Bengal. JASBSC. 1937, 3 : 1-37. KEYWORD: FISH

CONTENTS: 1. List of 131 fishes is given 2. Identifical table for principal groups 3. Systamatic description Habit & Habitat with text-figure 4. Index of scientific vernacular, Bengali script and Hindi Script names are given

603 SENGUPTA PC. Astronomical references from the Mahabharata and their significance. JASBL. 1937, 3 : 101-19. Ref 40 KEYWORD: MAHABHARATA-ASTRONOMY; ASTRONOMY

CONTENTS: 1. Day of Bhisma's death and the Ekastaka 2. Calculation of Date of Bharata battle by two methods 3. Astronomical examination of the year - 2526 of Saka era or 2449 BC 4. Date of Bharata Battle & the antiquity of vedic literature

604 CHAKRAVARTI C. Tantric family of the Deccan settled at Benares. IHQ. 1938, 14; (1) : 131-36. Ref 20 KEYWORD: TANTRA; MYSTICISM

CONTENTS: 1. Sivarcana- Candrika (work) by the Srinivasa and other works by family 2. Gives family life

605 SHARMA HD. Nirnaya- Kaustubha or Loghu-nirnaya- kaustubha of Visvesvarabhatta. IHQ. 1938, 14; (2) : 345-52.

606 ABBOT N. Arabic numerals. JRAS. 1938, 2 : 277-80. Ref 8 KEYWORD: NUMERALS - ARABIC

CONTENTS: 1. Method of writing numbers 2. Alphabet letters for numerals

607 HAMMETT FS. Ideas of the ancient Hindus concerning man. ISIS. 1938, 28; 76 : 57-72. Ref 3 KEYWORD: MAN

CONTENTS: 1. Biological Knowledge of ancient Hindu 2. Main is five-fold 3. Senses of man 4. Man & offspring 5. Ages of man

608 SARTON G. Geroge Catlin, "Indian-loving Catlin" (1796-1872). ISIS. 1938, 29; 78 : 101-102. KEYWORD: CATLIN G

CONTENTS: 1. Made corrected serious errors in the biography of Catlin

609 CHANDY M. Histology and physiology of the intestine and hepato-pancreas of two Isopods, Ligia exotica roux, and armadillio elevatus verhoeff. JASBSC. 1938, 4 : 1-16. Ref 17 KEYWORD: ISOPODA

CONTENTS: 1. Anatomy of the digestive organs 2. Histology and cytology of the intestine and hepatopancreas 3. Physiological activities with reference to absorption 4. Secretory function of Hepato-pancreas

610 ROONWAL ML. Recent advances in insect embryology, with a complete bibliography of subject. JASBSC. 1938, 4 : 17-105. Ref 760 KEYWORD: EMBRYOLOGY - SOURCE MATERIALS; INSECTS

CONTENTS: 1. Technique 2. Gastrulation and Germ layers 3. Alimentary canal and malpighian tubules 4. Inversion of cell polarity in early germs band 5. Segmentation of head 6. Pleuropodia 7. Abdominal appendages 8. Embryonic blood sinuses 9. Polyembryony and embryonic membranes 10. Provisional dorsal closures of embryo 11. Blastokinesis 12. Viviparity 13. Genital cells 14. Body sclerites 15. Corpora allata 16. Embryological problems 17. Comprehensive bibliography is given (1) alphabetical list (2) classified list

611 SENGUPTA PC. Bharata-Battle traditions. JASBSC. 1938, 4 : 393-413. Ref 60 KEYWORD: DATE CALCULATION

CONTENTS: 1. Aryabhat tradition -astronomical kaliyuga, astronomical kali-reckoning. Aryabhata I, conflict of Aryabhata tradition with Mahabharata evidences, Mahabharata kaliyuga, Krsna's prediction of the day of Bhisma's expiry, evidences of mahabharata kali-reckoning 2. Puranic traditions and evidences -sequence of Puranas, puranic dynastic lists, excellence of Mahabharata astronomical references, Puranic evidences by position of great bear

612 SENGUPTA PC. Solstice days in vedic literature. JASBSC. 1938, 4 : 415-34. Ref 24 KEYWORD: DAYS

CONTENTS: 1. Method of finding the solstice days in vedic literature 2. Meaning of month of Magha for vedic cycles 3. Vedic standard month of Magha in present times 4. Meaning of the 'first night of the year' 5. Not beginning of spring but of winter

613 SENGUPTA PC. Madhu-vidya (or the science of Spring). JASBSC. 1938, 4 : 435-43. Ref 11

CONTENTS: 1. Determination of beginning of spring (mathematical calculation) 2. Religious description of spring in Indian literature. Rg-veda 3. Start alpha arietis

614 SENGUPTA PC. When Indra became Maghavan. JASBSC. 1938, 4 : 445-53. Ref 8

CONTENTS: 1. Rg-veda description of maghavan 2. Positions and calculation of stars to maghavan

615 CHATLEY. Date of the Hsia calendar Hsia Hsiao Cheng. JRAS. 1938, 4 : 523-33. Ref 1 KEYWORD: CHINA CALENDAR

CONTENTS: 1. Calendar includes comments on weather, stars, animals and vegetable life all in 12 moons of the year 2. Motion of moon and stars 3. Translated texts for 1st-12th moon 4. Riddle of Yao Tien calendar

616 MALIK GM. Food of the mountain Barbel Qreinus McClelland, and its probable bearing on the introduction of Brown trout in Himalayan streams. JASBSC. 1938, 5 : 1-6. Ref 11 KEYWORD: HIMALAYAS; FISH; CARNIVOROUS-FISH

CONTENTS: 1. Gut contents analysis of oreinus, feeding materials and habits

617 MORELAND WH. Ships of the arabian sea about AD 1500. JRAS. 1939, 1 : 63-74. Ref 6 KEYWORD: SHIP BUILDING; SHIP

CONTENTS: 1. Indian-ships, building, structure given in chronology 13th and 14th C

618 MITRA K. Magic and miracle in Jaina literature. IHQ. 1939, 15; (2) : 175-82. Ref 30 KEYWORD: MYSTICISM

CONTENTS: 1. Use of thread on arms and waist to protect from evil eyes 2. Magic sleep 3. Art of opening locks 4. Magic shawl 5. God & goddess in Jain literature

619 CHAUDHURI N. Rudra-siva- as an agricultural Deity. IHQ. 1939, 15; (2) : 183-196. Ref 59

CONTENTS: 1. Brahmanical Society 2. In Rgveda 3. In the Later Vedas and vedic literature 4. In Epic 5/ in Puranas

620 RAY JC. Soma plant. IHQ. 1939, 15; (2) : 197-207. Ref 2

CONTENTS: 1. Bhanga - Botanical feature/pharmaceutical 2. Sana, Soma 3. In vedic literature 4. Habit and habitat

621 SARKAR JN. Silk trade of Patna in early seventeenth century. IHQ. 1939, 15; (2) : 208-27. Ref 45 KEYWORD: TRADE - SILK

CONTENTS: 1. Silk trade activities in Patna 2. Influence of English traders 3. Economic conditions (price etc) in silk trade 4. Types of silk

622 SANKALIA HD. Two rare images of a buddhist tantrick deity Padmanarttesvara. IHQ. 1939, 15; (2) : 278-80. KEYWORD: MYSTICISM

CONTENTS: 1. Description of deity images, size, figure, etc.

623 GHOSH DP. Kalinga style of architecture. IHQ. 1939, 15 ; (3) : 432-34. Ref 2

CONTENTS: 1. Orissan style 2. Temples at Khajuraho and Bhuvanesvara 3. History of temple

624 SIRCAR DC. Date of Patanjali's Mahabhasya. IHQ. 1939, 15; (4) : 631-38. Ref 7

CONTENTS: 1. Observations by different scholars 2. Description of MSS which indicates some date

625 MORELAND WH. Ships of the Arabian sea about AD 1500. JRAS. 1939, 2 : 173-92. Ref 4 KEYWORD: SHIP BUILDING; SHIP

CONTENTS: 1. Chinese, India, Malabar, Konkan, Gujarat ports etc.

626 PRASAD G. Age of Baudhayana srauta sutra. JRAS. 1939, 2 : 417-21. Ref 12

CONTENTS: 1. Srona, citra and Krittika a position, measurements and compared with Barth or Winternitz

627 CHATTERJEE D. Studies on the endemic flora of India and Burma. JASBSC. 1939, 5 : 19-67. Ref 105

CONTENTS: 1. India and the surrounding countries 2. Plant geographical regions of India 3. Inter-relationship of the Indian Regions 4. Endemism 5. Non-endemic elements in Indian flora 6. Commentary on endemism, relationship and other special features of some Indian families 7. List of endemic genera of Indian and Burma 8. Details of numbers showing endemic species, etc. in each family

628 MALIK GM. Cases of mortality of Brown trout, salmo trutta fairo linn., in the hatcheries in the Hazara district, N.W.F. Province. JASBSC. 1939, 5 : 7-18. Ref 10

CONTENTS: 1. Mortality of brown trout in Jabori Hatchery -growth of algae on gills, thyroid tumour or goitre 2. Morality of brown trout in Shinu Hatchery -malstripping, conditions favourbale for stripping, malnutrition, visceral abnormalities 3. Plates and illustrations are given

629 SARMA SR. Schorer's account of the coromandel coast. IHQ. 1940, 16; (4) : -837. KEYWORD: SCHORER

CONTENTS: 1. Trade metric system at (coast) 2. Ceylon area (Coast) 3. Trade items mentioned food, metals, cloth etc.

630 SASTRI SSSN. Advaita, causality and human freedom. IHQ. 1940, 16; (2) : 331-369. Ref 26 KEYWORD: SCIENTIFIC METHOD

CONTENTS: 1. Causes and effects, reason for these in modern science and acceptance of them by human freedom/responsibility free nature 2. Scientific determinism 3. Determination of physical laws 4. Principle of indeterminacy 5. Scientific determinism 6. Space and time relationship

631 SASTRI SS. Proto- Indian ceramics. IHQ. 1940, 16; (3) : 511-23. Ref 33 KEYWORD: HITTKARI; CERAMICS - PROTO-INDIAN

CONTENTS: 1. Evolution of the art of ceramics 2. Colour and shape of ceramics wares 3 Poly chrome pottery 4. Hittites people (Hittkari) 5. Sumerian pottery etc.

632 RENOU L. On the identity of the two Patanjalis. IHQ. 1940, 16; (3) : 586-91.

CONTENTS: 1. Grammatical theory 2. Terms in Grammar in Mahabhasya and yogasutra

633 THOMAS EJ. Interpretation of the indus seals. IHQ. 1940, 16; (4) : 683-688.

CONTENTS: 1. Language development 2. Interpretation of symbols

634 GORDON ME; GODON DH. Survivals of the Indus culture. JASBL. 1940, 3 : 61-71. Ref 11

CONTENTS: 1. Pottery & paints in indus culture 2. Human structure in indus culture

635 CHATTERJEE NC. Psychological study of arithmetical ability with reference to the students of secondary schools. JASBSC. 1940, 6 : 113-29. KEYWORD: PSYCHOLOGY

CONTENTS: 1. Standardization of text 2. Arithmetical Reasoning 3. Questions & students response 4. Comparative study of arithmetical abilities of students of secondary schools in Bihar & Bengal

636 MENON CPS. Magic Cakras and rectilinear orbits in ancient astronomy. JASBSC. 1940, 6 : 51-67. Ref 34

CONTENTS: 1. Mural picture of universe 2. Orbit of mercury 3. Rasi Cakra 4. Rahu Cakra 5. Shadow-measurement

637 CHAKRAVARTY M. Observations on two Myxosporidians Zschokkella lissemysi N. Sp. from the gall-bladder of the tortoise, Lissemys punctata and Zschokkella auerbachi (weill) from the gall bladder of Bufo melanostictus, with a note on the Genus. JASBSC. 1940, 6 : 69-76. Ref 15

CONTENTS: 1. Description of Zschokkella Habit, Habitat, Locality & vegetative form, dimensions etc.

638 HORA SL; GUPTA JC. Collection of Fish from kalimpong Duars and siliguri Terai, Northern Bengal. JASBSC. 1940, 6 : 77-83. Ref 9

CONTENTS: 1. Physical condition of Kalimpong Duars & Siliguri for fish fauna 2. Taxonomic notes are given on Lepidocephalus guntea, semiplotus semiplotus & Barbus (Puntius) titius Hamilton

639 MURRAY EFO. Ancient workers of western Dhalbhum. JASBL. 1940, 6 : 79-104. Ref 20

CONTENTS: 1. Santals Mundas & Kols's cephalic & Nasal & stature Index 2. Copper workers 3. Stone beads workers, pottery Roamgarh

640 PASRICHA CL. Note on the History of Bacterioloy and some of the early workers in India. JASBSC. 1940, 6 : 85-91. Ref 1

CONTENTS: 1. Early workers in Bacteriology who come to India 2. Nationality of 330 early workers in Bacteriology & number of workers in each decade in nineteenth century

641 HORA SL. Life-history & wanderings of Hilsa in Bengal waters. JASBSC. 1940, 6 : 93-112. Ref 41

CONTENTS: 1. Hilsa fishery of Bengal 2. Earlier Hilsa investigations in Bengal 3. Hilsa invesitigation at Pulta waterworks & in Hooghly river, Eastern Bengal 4. Marine life of Hilsa

642 MACFARLANE EWE. Tibetan and Bhotia blood groups distributions. JASBSC. 1941, 7 : 1-5. Ref 8 USA Michigan; Collaborator, Asiatic Research at Univ. of Michigan, USA KEYWORD: ANTHROPOLOGY

CONTENTS: 1. Percentages in different groups and frequencies were calculated 2. Type M & N anti-fluids were examined

643 MACFARLANE EWE. Blood groups among Balahis (weavers), Bhilo, Korkuus, and Mundas,

with a note on Pardhis and Aboriginal blood types. JASBSC. 1941, 7 : 15-24. Ref 16 USA Michigan; Collaborator in Asiatic Research, Univ. of Michigan, USA KEYWORD: ANTHROPOLOGY

CONTENTS: 1. Munda blood groups 2. Blood types in Mundas and Bhils 3. Distribution of blood groups, percentages and frequencies etc. 4. M & N type and serologic value in blood groups

644 SENGUPTA PC. Time indications in the Baudhayana Srauta Sutra. JASBL. 1941, 7 : 207-14. Ref 13

CONTENTS: 1. Rule for performing Naksatresi sacrific 2. Baudhayana rule for Pancasaradiya 3. Division of ecliptic 4. Calculation for 887-886 B.C, 884 B.C. showing the beginning of seasons and Luni-solar-stellar phenomena

645 MISRA PL. Observations on an intestinal flagellate, Tetratrichomastix hegneri sp. nov., from the 'Skipping frog' Rana Limnocharis Meig. JASBSC. 1941, 7 : 25-33. Ref 23

CONTENTS: 1. Material and methods 2. Observations on textratrichomastix hegneri sp. nov 3. Systematic position, previous work on the subject

646 MAJUMDAR DN. Racial affiliation of the Gonds of the central provinces. JASBSC. 1941, 7 : 35-56. KEYWORD: ANTHROPOLOGY

CONTENTS: 1. Statistical analysis of anthropometric data 2. Nasal index and cephalic index in different groups 3. Orbitonasal index and total facial index etc.

647 SEN P. Observations on the method of carp culture in the so-called salt lakes near Calcutta, with a note on the fish fauna of the lakes. JASBSC. 1941, 7 : 7-13. Ref 13 KEYWORD: FISH

CONTENTS: 1. Salinity in salt lakes 2. List of fishes and crustacea available in salt lakes with scientific and local names, 37 & 6 respectively 3. Culture of carps in salt lakes area 4. Natural enemies of the fry (eggs)

648 SENGUPTA PC. Solar eclipse in the Rgveda and the date of Atri. JASBL. 1941, 7 : 91-113. Ref 20 KEYWORD: ECLIPSES

CONTENTS: 1. Text and translations of Rgveda related to solar eclipse 2. Solar eclipse on 21st July, 3146 B.C. calculations 4. Description of solar eclipse on July 26th 3928 B.C. and others 5. Appendix for moon's equation, longitude of sun and moon, latitude of moon, parallax etc. 6. Method of finding a central eclipse near a past date

649 HENNING WB. Astronomical chapter of the Bundahishn. JRAS. 1942; 3 : 229-48. Ref 81 KEYWORD: STAR

CONTENTS: 1. Creation of lights (fixed stars. the not-fixed stars) 2. Notes on astronomical terms.measures,degree & minute of a circle, spheres,polaris,lunar mansions table Satavaesa,vanant,Tistryaeni

650 JOHNSTON EH. Ctesias on Indian Manna. JRAS. 1942; 1 : 29-35. Ref 14

CONTENTS: 1. Describes names of trees & animals 2. Zetacora name of the tree in Indian language conifers 3. Pine-Sugars 4.Persian names of animals (Little descriptions)

651 GANGULY KK. Early Indian Jewellery. IHQ. 1942, 18; (2) : 110-27. Ref 80

CONTENTS: 1. Precious ornaments in temples 2. Head-ornaments, neck-ornaments .3. Arm-ornaments. etc.

652 DE JC. Earliest phases of the company's Indigo trade. IHQ. 1942, 18; (2) : 137-55. Ref 4 KEYWORD: TRADE - INDIGO; DYE

CONTENTS: 1. 16th C to 17th C accounts of Indigo trade 2. Herbs descriptions and method to extract indigo 3. Place of grow and trade 4. Agra, Ahmadabad and Surat area of trade Lahore

653 MANKAD DR. Manvantara. IHQ. 1942, 18; (3) : 208-30. Ref 4 KEYWORD: ASTRONOMY; TIME

CONTENTS: 1. Manvantaras in Puranas. names. Text from Purana 2. Number of years allotted to a manvantara 3. Unit of astronomical time-measure 4. List of Manus 5. Manvantara computation 6. Vayu's method of Samasa and Vyasa

654 FALK M. Amavasya in mythical and philosophical thought. IHQ. 1942, 18; (1) : 26-45. Ref 12 KEYWORD: PHILOSOPHY

CONTENTS: 1. Surya and Asvins as process of Amavasya in Rg-veda 2. Moon motion in Suryasukta 3. Psycho- cosmic reality

655 GANGULI KK. Early Indian Jewellery. IHQ. 1942, 18; (1) : 46-59. Ref 45

CONTENTS: 1. Mohenjo-daro in Sind and Harappa in Punjab Jewellery 2. Chalcolithic jewellery. Stones and metals 3. Shapes and decription of jewellery 4. Anklets, neck-ornament, rings, girdles

656 HOPKINS LC. Sunlight and Moonshine. JRAS. 1942, 2 : 102-110. Ref 2 KEYWORD: ECLIPSES

CONTENTS: 1. Kuo Mo-jo (Chinese scholar in ancient history of his century) 2. Lunar eclipse 3. Solar eclipse

657 BISWAS K. Notes on the occurrence of a cosmopolitan blue-green algae in the hotsprings at Wairaki in the thermal region of the north Island of New Zeland. JASBSC. 1942, 8 : 1-4. Calcutta; Royal Botanic Garden

CONTENTS: 1. Species of Phormidium are described 2. Occurrence and noting of species

658 MAJUMDAR DN. Tharus and their blood groups. JASBSC. 1942, 8 : 25-37. Ref 14 KEYWORD: ANTHROPOLOGY

CONTENTS: 1. Dominance of women in Tharu Society, Garhwal Kumaun area 2. Incidence of malaria among the tharus 3. Blood groups 4. Features of men and women

659 HORA SL; LAW NC (NIRMAL CHANDRA). Respiratory adaptations of the South Indian Homalopterid fishes. JASBSC. 1942, 8 : 39-46. Ref 10 KEYWORD: FISH - RESPIRATION

CONTENTS: 1. Respiratory movements of Balitora, Hemimyzon and Bhavania are discussed and collected with habitat & form of fishes 2. Functions of various structures of fish are explained.

660 SENGUPTA PC. Gupta era. JASBL. 1942, 8 : 41-56. Ref 19 KEYWORD: ERA GUPTA

CONTENTS: 1. Gupta year 165 calculations by -modern method, siddhantic method 2. Mentioned dates by earlier works on Gupta Era

661 SAKSENA RD. Anatomical studies on Indian plant Galls-Part I. JASBSC. 1942, 8 : 5-23. Ref 70

CONTENTS: 1. Material and methods 2. Acarina galls 3. Diptera galls -stem galls, long galls, bud galls, flower galls 4. Longitudinal and transverse sections of plants part are given 5. 5 families are discussed

662 COOMARASWAMY AK. Prana-citi. JRAS. 1943; 1 : 105-109. Ref 15 KEYWORD: PRECEPTION

CONTENTS: 1. Human senses & their philosophical explanations

663 KINCAID CA. Romance of the Indian Calendar. JRAS. 1943 ; 2 : 255-59. Ref 4

CONTENTS: 1. Equinoxes,lunar months & solar year 2. Greek names of week days 3. India's week & Egypt months divisions 3. Two musulman year systems in India 4. Solar year

664 DROWER ES. Mandaean book of Black Magic. JRAS. 1943, 1 : 149-81. KEYWORD: PREGNANCY; MAGIC - BLACK

CONTENTS: 1. Curative and against evil eye 2. To induce pregnancy or protect embryo against enemies etc. 3. Text Harsia Bisia and its translation

665 RAJWADE VK. Metrical and Grammatical lapses due to the imposition of rules of Modern Grammar and prosody on Rks. IHQ. 1943, 19; (2) : 147-55.

CONTENTS: 1. Syllabically and rhythmically metres/rules in writing, change of meaning etc.

666 BHATTACHARYYA DC. Purusottamadeva's commentary on the Mahabhasya. IHQ. 1943, 19; (3) : 201-13. Ref 21

CONTENTS: 1. First page completely published in 2. Commentaries on the work 3. Purusottama's data and works

667 GANGULY DC. Yadavaprakasa on the Ancient Geography of India. IHQ. 1943, 19; (3) : 214-24. Ref 136

CONTENTS: 1. Vaijayanti (Sanskrit vocabulary indicates name of countries in its Saravati (South east) and Udicya (North-west) of it 4. Gives description of countries their location and relative countries (different names)

668 CHAUDHURI N. Pre-historic tree cult. IHQ. 1943, 19; (4) : 318-29. Ref 73

CONTENTS: 1. Fig tree varieties 2. Pipal 3. Worship forms 4. Description in ancient literature (in vedas civilization) 5. Banyan

669 CHATURVEDI SP. Note on the Alleged Metrical defect in the legend-verse on the seal of the Pipardula copper-plate inscription. IHQ. 1943, 19; (4) : 358-501. Ref 1

670 SARKAR BK. Study of Meyer's Trilogy of Hindu vegetation powers and festivals. IHQ. 1943, 19; (4) : 373-84. CONTENTS: 1. Use of vegetation in festivals in Eruope, India, Nepal, German, France 2. Social behaviour of people in festivals 4. Act on festivals towards their ancestors 5. Use of animals in festivals

671 GODE PK. Notes on the history of the Fig-Does the word 'phalgu' used by Caraka and Susruta mean 'Anjira'. IHQ. 1943, 19; (1) : 62-65. Ref 15 KEYWORD: FIG

CONTENTS: 1. Word 'fig' geneology, in Sanskrit and Arabic literature 2. Materia medica (1374 AD) 3. Phalgu (Carak Samhita) in Phalavarga of the Susrutasamhita

672 COOMARASWAMY AK. Eastern wisdom and Western knowledge. ISIS. 1943, 34; 96 : 359-63. KEYWORD: KNOWLEDGE TRANSMISSION

CONTENTS: 1. Books and biography of Rene Guenon-his views 2. Reactions of Roman catholics to Rene Guenon 3. Issues of apartness between East and West

673 HORA SL; MISRA KS. Small collection of fish from Iraq. JASBSC. 1943, 9 : 1-15. Ref 15

CONTENTS: 1. Thirteen species are studied arabic names, scientific names and description of fish with diagram

674 ROONWAL ML. Notes on the colour of the iris, of the bare orbital skin around the eyes and of the edges of the eye-lids in the Indian Ring-Dove. JASBSC. 1943, 9 : 17. Ref 1 KEYWORD: IRIS - COLOUR

675 LINDBERG K. Notes sur quelques cystocercoides de Crustace's de 1'Inde. JASBSC. 1943, 9 : 19-29. Ref 17 KEYWORD: CRUSTACEAE

676 LINDBERG K. Un nouveau Cyclopide de 1'Inde. JASBSC. 1943, 9 : 31-32. KEYWORD: CRUSTACEAE

677 LINDBERG K. Cyclopoides (Crustace's Copepodes) de 1'IndeII.. JASBSC. 1943, 9 : 33-36. KEYWORD: CRUSTACEAE

CONTENTS: 1. Eucyclops gibsoni (Brady) description, Habitats etc. & mesocyclops splendidus sp. nov.

678 BISWAS K. Systematic position of a common tree fern of the eastern Himalaya with a note on the genus Cyathea. JASBSC. 1943, 9 : 43-47. Ref 6 KEYWORD: HIMALAYAS

CONTENTS: 1. Distribution in India & Burma 2. Description of plants

679 CHAKRAVARTY M; KAR AB (AMIYA BHUSAN). Observations on two Coccidia, Eimeria trionyxae n.sp. and Eimeria triangularis n.sp., from the Intestine of the Turtle Trionyx gangeticus Cuv.. JASBSC. 1943, 9 : 49-54. Ref 9

CONTENTS: 1. Description & Diagnosis 2. Text figures are given 3. Parasite, Oocysts & sporocyst, locality & host are given in table indicating, form, size, residue etc.

680 BHADURI JL; GHOSE KC. Notes on the dorso-lumbar veins in the Common Indian toad Bufo melanostictus Schneider. JASBSC. 1943, 9 : 55-62. Ref 11 CONTENTS: 1. Renal-portal veins are shown & explained 2. Statistical analysis of dorso-lumbar veins in both sexes of toad

681 RAY HC. Report on a collection of Molluscs from Santal Parganas, Bihar. JASBSC. 1943, 9 : 63-80. Ref 3

CONTENTS: 1. Description names of different numerous species and the reference of a document in which detail studies are given

682 MAJUMDAR DN. Blood groups of Tribes and Castes of the U.P with special reference to the Korwas. JASBSC. 1943 , 9 : 81-94. Ref 21 KEYWORD: UTTAR PRADESH

CONTENTS: 1. Distribution of the tribes 2. Blood groups of the tribal population 3. Blood groups of Korwas 4. Negroid affiliation of Korwas

683 ELWIN V. Conception, Pregnancy and birth among the tribesmen of the Maikal Hills [Satpura Mountains]. JASBC. 1943, 9 : 99-148. Ref 56 KEYWORD: STERILITY

CONTENTS: 1. Desire for Children 2. Cure for sterility - Prescriptions 3. Attitude of contraception 4. Mechanism of conception 5. Function of teste 6. Pregnancy & birth in tribal mythology 7. Pregnant woman 8. General precautions during pregnancy 9. Cravings of pregnancy 10. Sex of unborn child 11. Onset of labour (false pains) 12. Place of birth 12.1 Midwives (assistants) 13. Course of labour 14. Placenta 15. Puerperium 16. Abnormal births 17. Lucky and unlucky children 18. Stillborn child 19. Caul (Child born in foetal membranes) 20. Multiple pregnancy

684 RANDLE HN. Saurashtrans of South India. JRAS. 1944; 2 : 151-64. Ref 38

CONTENTS: 1. Silk weavers & Industry in Madura (Madras) (Not in detail) 2. Society description etc

685 SZCZESNIAK B. Penetration of copernican theory into Feudal Japan. JRAS. 1944, 1 : 52-61. Ref 9

CONTENTS: 1. Spread of science alongwith catholicism in china & far East 2. Copernican theory & Japanese Studies

686 MELLO AFB. Further additions to the list of Ciliates living in the intestine of Rana cyanophlyctis form Nova-Goa. JASBSC. 1944, 10 : 1-7. Ref 6

CONTENTS: 1. 10 ciliates & their description, length & width

687 SAKSENA RD. Anatomical Studies on Indian plant Galls-Part II. JASBSC. 1944, 10 : 119-24. Ref 5

CONTENTS: 1. Six species diagrams are given alongwith their description

688 CHATTOPADHYAY KP. Two Indian Rafts. JASBSC. 1944, 10 : 13-14.

CONTENTS: 1. Sola Raft of North Bengal 2. Gourd raft photos given

689 BENTHALL AP. Note on the Corypha palms of Bengal. JASBSC. 1944, 10 : 15-25. Ref 14

CONTENTS: 1. Corpha umbraculifera - Corypha Talie - Corypha elata Palms 2. Structure & description of tree photo

690 KUNDU BC. Anatomy of Jute stem with special reference to cambial activity and distribution of fibres in relation to leaf-trace system. JASBSC. 1944, 10 : 27-52. Ref 20

CONTENTS: 1. Anatomy of internode 2. Extra-stelar secondary growth 3. Mucilage cells & canals 4. Crystals of calcium oxalate 5. Tannin - containing cells 6. Phyllotaxis and leaf-trace system 7. Cambial activity & fibre distribution in relation to leaf-trace

691 BHADURI JL. Two salientian tadpoles,Rana blanfordii Boulenger and Bufo himalayanus Gunther, from the Ha Valley, Bhutan Eastern Himalayas. JASBSC. 1944, 10 : 53-57. Ref 5 Calcutta; Zoology Department, university Calcutta

CONTENTS: 1. Distribution Head & body, eye & nostrils, mouth disc, Tail, sense organs etc

692 MOONEY HF. List of Plants recorded from the pats of Ranchi and palamau districts and the states of Jashpur and Surguja. JASBSC. 1944, 10 : 59-118.

CONTENTS: 1. Geology, climate 2. Biotic factors, forest 3. 630 species, location & height etc.

693 MISRA KS. Description of Day's Specimens of Dentex nufar (val.) from Sind. JASBSC. 1944, 10 : 9-11. Ref 3 Zoological survey of India

CONTENTS: 1. Measurement of body structures 2. Description of cheimerius nufar

694 BOSE AN. Early Industrial organisation in North India. IHQ. 1944, 20; (2) : 167-78. Ref 18 KEYWORD: ORGANISATION- INDUSTRIAL

CONTENTS: 1. Emphasis on organisation, combination and laws regulating business instead, tools and medical power 2. Origin of combination 3. Urge to union 4. Organisational structure/localisation of industries: in theory: in practice 5. Heredity of occupation 6. Apprenticeship (no heredity) in Pali canon 7. Terms of apprenticeship 8. Guild-laws: evolution 9. Apportionment of shares and liabilites 10. Guild-laws: -Judicial authority, -funds, -Relation with the state etc.

695 COOMARASWAMY AK. Gradation and evolution. ISIS. 1944, 35; 99 : 15-16. Ref 1

CONTENTS: 1. Existence of living creatures of (1) Gradation and (2) evolution 2. Causes of existence

696 BOYER CB. Fundamental steps in the development of numeration. ISIS. 1944, 35; 100 : 153-68. Ref 99 KEYWORD: NUMERALS

CONTENTS: 1. Hindu-arabic system of notation 2. Hindu numerals - zero concept 3. Babylonian and Mayas numeration development 4. Pentenary symbols and multiplicative and subtractive principles 5. Ionian numerals 6. Egyptian - cipherization 7. Greek fractional notation 8. Kharosthi script (Brahmi Characters)

697 SCHMIDT OH. Computation of the length of daylight in Hindu astronomy. ISIS. 1944, 35; 101 : 205-10. Ref 28 KEYWORD: DAYLIGHT-LENGTH

CONTENTS: 1. Numerical tables of Hindu astronomy 2. Types of source of Hindu astronomy 3. Mathematics used in length of daylight 4. Trigonometric tables

698 SARTON G. Grassmann - 1844. ISIS. 1944, 35; 102 : 326-30. Ref 9 KEYWORD: GRASSMANN HG; RGVEDA - DICTIONARY; SANSKRIT

CONTENTS: 1. Grassmann HG - Great mathematician 2. Grassmann love sanskrit and translated Dictionary of the Rgveda (1872-75) and a german translation of it (1876-77) 2. Biography and his works 3. His days with Hamilton WR (1805-65)

699 GOTTHEIL R. Fragment on pharmacy from the Cairo Genizah. JRAS. 1945, 1 : 123-44. Ref 60

CONTENTS: 1. Original text and its translation 2. Laxative, perfume for treatment, strengthening heart, brain, digestive organs

700 SZCZESNIAK B. Notes on the penetration of the Copernican theory into China (17th to 18th C). JRAS. 1945, 1 : 30-38. Ref 31

CONTENTS: 1. Conflict of traditional astronomies 2. Religious disagreement 3. Copernican solar system

701 AHRENDT LWA. Some new and little known Berberis from India. JASBSC. 1945, 11 : 1-5. KEYWORD: INFLORESCENCE

CONTENTS: 1. 14 species given, describing frutex, inflorescentiae, location, folia etc.

702 AHMAD N. Small collection of vertebrates from high altitudes in Kashmir state. JASBSC. 1945, 11 : 119-21.

CONTENTS: 1. Six species, description of specimen

703 CHAKRAVORTI S. Preservation of Botanical specimens with cellulose Acetate foil. JASBSC. 1945, 11 : 123-25. Ref 3

CONTENTS: 1. Method of preserving botanical specimens

704 NAIR KK. Descriptions of some new and little-known midge galls from Travancore. JASBSC. 1945, 11 : 17-20. Ref 4 KEYWORD: GALL

CONTENTS: 1. Stem gall, leaf gall

705 SUN YC. Silurian stratigraphy of Yunnan. JASBSC. 1945, 11 : 20-23. China; Geological Society of China KEYWORD: STRATIGRAPHY; CHINA

CONTENTS: 1. Silurian sedimentation of Yunnan 2. Stratigraphical summary

706 SUN YC. Devonian Subdivisions of Yunnan. JASBSC. 1945, 11 : 23-26. Ref 5 China; Director, Geological Dept. National Peking University

CONTENTS: 1. Devinian fossils 2. Boundaries of Devinion of Yannan 3. Subdivisions of Devonian of Yunnan

707 WANG HC. Silurian Rugose Corals of northern & eastern Yunnan. JASBSC. 1945, 11 : 26-27. Ref 9

708 WANG HC. Middle Devonian Rugose Corals of Eastern Yunnan. JASBSC. 1945, 11 : 27-30. CONTENTS: 1. Stratigraphy, faunas

709 WANG HC. Stratigraphical position of the Devonian fish - bearing series of eastern Yunnan. JASBSC. 1945, 11 : 31-32. Ref 8 KEYWORD: STRATIGRAPHY

710 NAIR KK. Systematics of certain Eel-like fishes of the Ganges. JASBSC. 1945, 11 : 32-36. Ref 14 KEYWORD: FISH

CONTENTS: 1. Pisoodonophis hijala 2. Moringua (Raitaboura)

711 CHAKRAVARTY M, KAR AB. Studies on Haemosporidia from Indian Birds series I. JASBSC. 1945, 11 : 36-39. Ref 6

CONTENTS: 1. 3 species Haemoproteus

712 RAY HC. Descriptions of some unnamed shells of Cypraeacea. JASBSC. 1945, 11 : 39-53. Ref 131

CONTENTS: 1. 32 species describe locality & Distribution 2. Publications references along with species description chronologically

713 GATES GE. Indian EarthwormsII. JASBSC. 1945, 11 : 54-91. Ref 4

CONTENTS: 1. Systematics - Moniligastridae - Drawida - Megascolecidae family - putellus genus - perionyx - octochaetoides - Hoplochaetella - Tonoscolex, Describes External & Internal anatomy Characteristics regeneration, abnormality etc

714 DAS SM. Collection of Monascidians from Madras. JASBSC. 1945, 11 : 6-17. Ref 20

CONTENTS: 1. Historical review of literature 2. Description of species -Ascidiidae - Styelidae - pyuridae describing Diagnosis (External characters, distribution, Test, mantle, Branchial sac, tentacles Dorsal lamina, Dorsal Tubercle, Alimentary canal, and Gonads) etc.

715 NATARAJ S. Two new species of solenocera (crustacea Decapoda : Penaeidae) with notes on solenocera pectinata (spence Bate). JASBSC. 1945, 11 : 91-98. Ref 21 Zoological survey of India

CONTENTS: 1. Description, locality etc

716 HORA SL. Fish farms : objectives and requirements. JASBSC. 1945, 11 : 99-117. Ref 11 Bengal; Director of fisheries

CONTENTS: 1. Objectives & classification of farms 2. New methods of Demonstration 3. Fish farms in Bengal, Punjab & Madras 4. Requirements of a fish farm 5. Fish farm Engineering problems 6. Commercial fish farms

717 HENNING WB. Manichaean Fasts. JRAS. 1945, 2 : 146-64. KEYWORD: CALENDAR

CONTENTS: 1. Sogdian manichaean calendar 2. Text and its translation given 3. Dates, years etc.

718 NARAHARI HG. Nirukta and the theory of tansmigration. IHQ. 1945, 21; (2) : 118-23. Ref 27 KEYWORD: SOUL - TRANSMIGRATION

CONTENTS: 1. Five-fires (Pancagni) 2. Soul to heaven and back to body (Path of soul) as Soma, rain, food, seed and embryo 3. School of opinion on fate of man after death 4. Three classes of opinion- Soul stays for ever in brahmaloka; No permanent destiny; know nothing 5. Return of soul depends upon nature of conduct in previous existence

719 HAZRA RC. Samvatsara- pradipa. IHQ. 1945, 21; (1) : 49-55. Ref 15

CONTENTS: 1. Location of manuscripts. It tells works deals in months of the year etc. 2. Solar and lunar eclipes

720 HORNELL J. Primitive types of water transport in Asia: distribution and origins. JRAS. 1946; 2 : 124-41.

CONTENTS: 1. Use of float with net 2. Design of swimming float-log of wood,tigari(coracle-like basin of eathern ware) 3.Inflated bullock skin, tub-boat 4. Viking boats.

721 AKHTAR. SA. New species of Nematode Parasite of a Lizard. JASBSC. 1946, 12 : 1-2. Ref 5

CONTENTS: 1. Alaeuris rifkii sp.nov description, male & female characteristics are described

722 DASEN DD. Fundamentals in fishery research. JASBSC. 1946, 12 : 13-44. Ref 23

CONTENTS: 1. Training in fisheries Research 2. History of fishery research in Europe 3. Scope & nature of fisheries research 4. Hydrographical investigation 5. Biological investigation and fisheries 6. Statistical studies & their significance 7. Some fishery problems

723 CHAKRAVARTYM; KAR AB. Observations of two reptilian Coccidia. JASBSC. 1946, 12 : 3-5. Ref 7 KEYWORD: REPTILIA

CONTENTS: 1. Caryospora species (six) studies its systematic position Description & Habitat

724 HARDENBERG JDF. Development of Marine Fisheries. JASBSC. 1946, 12 : 45-51. Batavia; Head of the Laboratory for Investigation of the Sea KEYWORD: FISHERIES - MARINE

CONTENTS: 1. Practical fishery methods - fisheries with hook & line, fisheries with nets, - with drift nets, seines, trawl, throw nets, cross nets, purse seines- fish culture, fisheries with traps 2. Fishery organization

725 GUDGER EW. Pomacentrid fishes symbiotic with giant sea Anemones in Indo-pacific waters. JASBSC. 1946, 12 : 53-76. Ref 23 USA, American Museum of Natural History KEYWORD: FISH - SYMBIOTIC; SYMBIOSIS

CONTENTS: 1. Historic records of observations (1868-1936) 2. Relationship between fish & anemone - a symbiosis, - Do anemones have nematocysts?, - Why are damselfishes immune to nematocysts of anemones?, - Do damselfishes & anemones recognize each other by chemical substances emanating from each other?

726 PURI GS. Further observations on the systematic position of Ficus Krishnae C. de Candolle. JASBSC. 1946, 12 : 7-12. Ref 13

CONTENTS: 1. Cuticular structure of Ficus bengalensis & F.Krishnae (leaf stomato) - lower surface, - upper surface

727 BISWAS K. Prospect of cultivation and systematic study of the TUNG OIL YIELDING TREES of India. JASBSC. 1946, 12 : 77-114. Ref 45 KEYWORD: TREES; OIL

CONTENTS: 1. Early history and industrial uses of Tung Oil 2. Cultivation of Tung trees in USA & in parts of world other than India 3. Cultivation in India & Burma 4. Methods of vegetative propagation 5. Biology of flowers 6. Systematic enumeration of Species

728 WINTER HJJ; ARAFAT W. Ibn al-haitham on the paraboloidal focussing mirror. JASBSC. 1946, 15 : 25-40. Ref 44 UK, Exeter; University College of the South-West of England

CONTENTS: 1. Rays of Sun proceed from sun to the surfaces of all types of mirrors & to all objects on straight lines. - proofs are given

729 PALMER JAB. Identification of Ptolemy's Dounga. JRAS. 1946, 2 : 165-73.

CONTENTS: 1. Western Ghats, Nanaghat 2. Bor Ghat etc., Thana, Kalyan etc.

730 PHISHAROTI KR. Studies in Indian Architecture: Ayadisadvarga. IHQ. 1946, 22; (4) : 291-99. KEYWORD: SPACE AND TIME; TIME; BUILDING CONSTRUCTION

CONTENTS: 1. Gives formulae/rules of measurement for good building (six formulae) 2. Space and time formulae 3. Space (size and orientation materials 4. Naksatra relation in construction and their measurement with human life

731 BISWAS DK. Maga Ancestry of varahamihira. IHQ. 1946, 25; (3) : 175-83. Ref 26 KEYWORD: ASTRONOMY

CONTENTS: 1. Devotee of the sun (quotes paragraph/lines indicating the sun relation 2. Magas sect 3. Iranian sun cult. 4. Avanti a native place of varahamihira

732 FARMER HG. Ghosts : an excursus on Arabic muscial bibliographies. ISIS. 1946, 36; 104 : 123-30. Ref 22 KEYWORD: PARASYCHOLOGY; ARAB

CONTENTS: 1. Belief in Ghosts 2. Bibliographies of Mss and their location 3. Idea of natural sound

733 SCHLESINGER, AM. American historian looks at science and technology. ISIS. 1946, 36; 105-106 : 162-66. KEYWORD: AMERICA; HISTORY OF SCIENCE - INTERNATIONAL

CONTENTS: 1. Historian view of dynasties striven for the mastery of nations 2. Note of science and technology ideas.

734 ADAMS CW. Age at which scientists do their best work. ISIS. 1946, 36; 105-106 : 166-69. KEYWORD: SCIENTOMETRICS

CONTENTS: 1. Quoted many names (Scientists) their best work and the age at which it was done

735 TUCCI G. Minor sanskrit texts on the Prajna-paramita 1.The Prajna-paramita-pindartha of Dinnaga. JRAS. 1947 , 1 : 53-66. Ref 30

CONTENTS: 1. Sanskrit text with translation given 2. Ten descriptions have been given 2. D.of non-existence, existence 3. D.based upon positive assumption 4. Negative assumption 5. Negative assumption of identity, assumption of diversity,assumption of on essence, Dist.based upon the assumption that names correspond to thing, that thing correspond to name etc.

736 BOSE U (UMA) MS. Psychological approach to the origin of religion and the development of the concepts of God and ghost in children. JASBSC. 1947, 13 : 1-30. Ref 16 Benaras; lecturer in Psychology, Basant College for women KEYWORD: PSYCHOLOGY

CONTENTS: 1. Analysis of date 2. Idea of ghost - appearance of Ghost (in different religions - characters of Ghost - child's reaction towards Ghost - source of the knowledge of Ghost 3. Idea of God - characters of God - child's attitude to God - Sources - after life 4. Form of questionnaire is attached

737 AKHTAR SA. Fisheries of Afghanistan. JASBSC. 1947, 13 : 31-34. Ref 13 Kabul; Professor of biology, faculty of science

CONTENTS: 1. fishing & fish-eating

738 SAPRE SN. Balantidium macacae N. Sp., from the Caecum of the Indian monkey, Macaca mulatta (zimm.) . JASBSC. 1947, 13 : 35-41. Ref 15 Poona; Biological Products Station, Animal Husbandary Dept.

CONTENTS: 1. Body, length, Breadth 2. Nucleus length, Breadth

739 BASU B. Problems of snow survey in the Eastern Himalayas . JASBSC. 1947, 13 : 43-56. Calcutta; Dept. of Geography, Calcutta University

CONTENTS: 1. Application of Nevada percentage method of snow survay 2. Precipitation etc 3. Water content of snowcover & stream flow

740 MITRA E (MISS) EVA. Morphological studies on two species of Indian Bnalanophora. JASBSC. 1947, 13 : 57-68. Ref 14

CONTENTS: 1. Vegetative organs 2. Anatomical relation of host & parasite 3. Anatomy of rootstock 4. Reproductive organs - inflorescence - male & female spikes - Anatomy of male & female flower 5. Summary of results 6. Review of previous works 7. Material and methods 8. Observations

741 DASGUPTA SR. Carplarvae and their predators in the river padma at Rajshahi Bengal. JASBSC. 1947, 13 : 69-74. Ref 5 West Bengal; District fishery officer

CONTENTS: 1. 3 species described 2. Notopterus chitala (Ham) - Rita rita (Ham) Glyptothorax sp 3. Analysis of Gut-contents

742 BISWAS K. Lichen flora of India. JASBSC. 1947, 13 : 75-114. Ref 52 Calcutta; Superintendent, Royal Botanic Garden

CONTENTS: 1. Ecological distribution of Lichen flora 2. Economic uses of lichens 3. Classification of lichens (listing 678 Species) with their location

743 RAWLINSON HG. Indian influences on Western culture. JRAS. 1947, 2 : 142-50. Ref 10 KEYWORD: INDIAN - INFLUENCE; TRANSLATION

CONTENTS:

744 RICHESON AW. First arithmetic printed in English. ISIS. 1947, 37; 107-108 : 47-50. Ref 40 KEYWORD: ARITHMETIC - FIRST IN ENGLISH

CONTENTS: 1. 1537, Book printed in English devoted entirely to arithmetic and dealing with the Hindu-Arabic system of numbers titled "An Introduction for to Lerne to Recken with the pen, or with the counter 2. Lists others books in Arithmetic

745 SARTON G. Qualifications of teachers of the History of Science. ISIS. 1947, 37; 107,108 : 5-7. USA, Cambridge, Mass KEYWORD: EDUCATION - TEACHING

CONTENTS: 1. Stress on Modern Science 2. Traits of teachers are mentioned

746 RUDNYKH SP. Teaching of the History of Science in Australia, in university of Sydney, and in Russia. ISIS. 1947, 37; 107-108 : 76-78. KEYWORD: EDUCATION - TEACHING

CONTENTS: 1. Graduate students - course 2. Knowledge of subject etc.

747 SARTON G. History of Science in university of London, and in Brown University, Providence, RI. ISIS. 1947, 37; 107-108 : 78-79. KEYWORD: EDUCATION - TEACHING

CONTENTS: 1. Graduate students - course 2. Knowledge of subject etc.

748 KENNEDY ES. Al-kashi's "Plates of Conjunction". ISIS. 1947, 38; 111-112 : 56-59. Ref 8 KEYWORD: INSTRUMENT-ASTRONOMY

CONTENTS: 1. Description of manuscript, biography 2. Description of the instrument 3. Operation of the instrument

749 RICHMON D. Williams college. ISIS. 1947, 38; 111-112 : 98. KEYWORD: EDUCATION

CONTENTS: 1. Semesters course 2 semesters 2. Details topics in 2 semesters

750 . History of Science at Cornell Unversity. ISIS. 1947, 38; 111-112 : 99. KEYWORD: EDUCATION

CONTENTS: 1. Publications and programmes at university

751 WALES HGQ. Culture change in Greater India. JRAS. 1948 : 2-32. Ref 34

CONTENTS: 1. Western zone - Gupta, Pallava, Pala 2. Eastern zone - Java (and Bali), Champs, Cambodia 3. Indo-Javanese art 4. Postscript : some implication for political history (word 'Sailendra' was given

752 CONZE E. Text Source and bibliography of the Prajnaparamita-hrdaya. JRAS. 1948, 1 : 33-51. Ref 53

CONTENTS: 1. Negation of the negation of knowledge 2. Text & its translation given 4 Bodhisattva 3. Chinese commentaries 4. Manuscripts translations,& commentaries in foreign language a list is given

753 HORA SL. Sanskrit names of fish and their significance. JASBSC. 1948, 14 : 1-6. Ref 5 Banaras; Director, Zoological survey of India

CONTENTS: 1. Food and feeding habits of fish in ancient Hindus knowledge 2. Trygon marginatus(Indian Ray) Balitora Gray, Bhavani Hora 3. Travancoria Hora

754 MOOKERJEE HK; MOOKERJEE S. Two cases of atypical development in Chick. JASBSC. 1948, 14 : 11-16. Ref 13 Calcutta; Zoological Laboratory, University of Calcutta.

CONTENTS: 1. Anatomy of two atypical cases of chick describes external morphology & internal morphology 2. conjoint development 3. Atypicality in head, neck region etc, illustrations & photos given

755 ALIKUNHI KH. Hermaphrodite species of microphthalmus (Polychaeta-Hesionidae) from the sandy beach, Madras. JASBSC. 1948, 14 : 17-25. Ref 13 Madras,; Freshwater Biological Research Station, Government Fisheries

CONTENTS: 1. Hesionid polychaete genus Microphthalmus-hermaphroditism 2. Descriptive account - External characters - Reproductive organs - Diagnostic features

756 MANI MS. Cecidooza and Zoocecidia from India (Paragitic beings). JASBSC. 1948, 14 : 27-195. KEYWORD: PARASITE

CONTENTS: 1. Collection, preservation and preparation for study 2. General part - cecidozoa - Biology - Ethology Zoocecidia - morphology & Taxonomy - physiology - Biology - Actiology and Teleslogy 3. Systematic part - Cecidowa - Nematoda - Acarina - Hexapoda - Zoocecidia - Helminthocecidia - Acarococidia - entomocecidia

757 HORA SL. Knowledge of the ancient Hindus concerning fish and fisheries of India. 1. References to fish in Arthasastra (ca 300 BC). JASBSC. 1948, 14 : 7-10. Ref 3 KEYWORD: ARTHASASTRA-FISH

CONTENTS: 1. Role of fisheries during shortage of food 2. Fishery management 3. Rendering fishes poisonous 4. Miscellaneous uses of fish 5. Fish metaphors

758 LANE A; SERJEANT RB. Pottery and Glass fragments from the Aden Littoral, with historical

notes. JRAS. 1948, 2 : 108-33. Ref 114 KEYWORD: PAINTED WARES

CONTENTS: 1. Sites, Chinese procelain - literary sources 2. Import of China-ware to South Arabia 3. Arab glass in far east 4. Stone-ware, painted ware

759 GHOSH A. Prehistoric exploration in India. IHQ. 1948, 24; (1) : 1-18. Ref 51 KEYWORD: STONE AGE

CONTENTS: 1. Stone age (i) Palaeolithic age (exploration sites), (ii) Microlithic industries (findings sites studies), (iii) Neolithic age

760 CHOWDHURY JN. Commerce and industry in the pre-mughal period. IHQ. 1948, 24; (2) : 123-33. KEYWORD: MUGHAL

761 GHOSHAL SN. Importance of the 'Prakrita-Paingala' and its place in Indian literature. IHQ. 1948, 24; (3) : 195-200.

CONTENTS: 1. Grammar sutras, laws 2. Metric laws for writing laws of metre

762 MAJUMDER PC. Karanda vyuha: Its metrical version. IHQ. 1948, 24; (4) : 293-99. Ref 15

763 RAJAGOPAL CT; VENKATARAMAN A. Sine and Cosine power-series in Hindu Mathematics. JASBSC. 1949, 15 : 1-13. Ref 7 Tambaram, Madras; Madras Christian College, S. India

CONTENTS: 1. Hindu terminology 2. Enunciations of sine & cosine series 3. Lemma from yukti-Bhasa 4. Appendum: note on the style of the yukti-Bhasa by George KM, Madras Christian College

764 FOSTER AR. Soil fertility and balanced manuring. JASBSC. 1949, 15 : 105-19. Ref 14

CONTENTS: 1. Soil composition 2. Minerals in fertilizers etc. 3. Principles of plant nutrients 4. Indian conditions & suggestions

765 BAGCHI H. Novel properties of cyclides and hyper-cyclides. JASBSC. 1949, 15 : 121-30.

CONTENTS: 1. Ordinary cyclides (of a 3-space) 2. n-Cyclides in an (n+1)-space

766 BAGCHI H. Notes on the sextactic points of a cubic and its hessian. JASBSC. 1949, 15 : 131-38. Ref 3 Calcutta; Head of the Dept. of Pure mathematics, Calcutta University KEYWORD: GEOMETRY - CONFIGURATION

CONTENTS: 1. Bicursal cubic & its Hessian have all nine harmonic polars in common 2. Geometrical configuration of two trials - sextactic points which lie upon any of nine harmonic polars (Cognate)

767 ROY AC. Lecithin and Venom haemolysis. JASBSC. 1949, 15 : 139-47. Ref 22

CONTENTS: 1. Cobra and Russel's Viper venom analysis 2. Experiments with intact corpuscles 3. Experiments with etherial extracts of r.b.c 4. Cobra venom-Lecithin Russell's viper venom - lecithin for time for complete haemolysis, and surface tension

768 ROY AC. Haemolysis by bile salts. JASBSC. 1949, 15 : 15-24. Ref 12

CONTENTS: 1. Technique employed for haemolysis of bile salts 2. Four bile salts their abnormal haemolytic behaviour 3. Action of sodium taurocholate & Sodium glycocholate

769 CHATTERJEE B (MRS). Geometrical interpretation of the motion of the sun, moon and the five planets as found in the mathematical syntaxis of Ptolemy and in the Hindu astronomical work. JASBSC. 1949, 15 : 41-89. Ref 81 KEYWORD: MOTION - SUN, MOON, PLANETS

CONTENTS: 1. Green trigonometry - Motion of sun according to ptolemy, Motion of moon according to ptolemy, Motion of five planets according to ptolemy, Planetary system of Hindus, Ancient Hindu Planetary theory 2. Hindu trigonometry - Hindu epicyclic theory to find the sun's motion, Hindu excentric theory to find the sun's motion, Hindu theory of moon, Hindu planetary theory, Hindu epicyclic method to detect a planet's motion, Hindu excentric method to detect a planet's motion, Hindu values of Apogees & circumferences of 1st & 2nd epicycles

770 NANJUDIAH TS. Contributions to affine differential geometry-I. JASBSC. 1949, 15 : 91-104. Ref 2 KEYWORD: GEOMETRY

CONTENTS: 1. Osculating conics 2. Systems of Conics C 3. Affine invariants

771 SPEAR P. Bentick and the Taj. JRAS. 1949, 2 : 180-87. Ref 33

CONTENTS: 1. Sale of Taj,its marble architect 2. Different statements by various British persons 3. A Hindu wishes to buy Taj to carryaway marble & erect a temple to his own idols at Bindrabund 4. Beautiful piece of architecture Moti Masjid at Agra for Rs.125,000 & to sell its marble

772 DISKALKAR DB. Excavations at kasrawad. IHQ. 1949, 25; (1) : 1-18. Ref 20 KEYWORD: POTTERY CONTENTS: 1. Pottery cones, perforted comparision with Mohenjodero and Harappa and Indus valley excavation 2. Buddhist stupas and residential blocks etc. 3. Household pots and clay objects, jewellery/ornaments 4. Metal objects (gold, silver, copper and coins) 5. Stone objects

773 YASHPAL. Surgery and Medicine in the days of Gautama. IHQ. 1949, 25; (2) : 102-9. Ref 81

CONTENTS: 1. Scriptures of a certain religion account on the subject not the treatise on the subject 2. Constitutents of the body 3. First four stages in the formation of foetus 4. Physician Jivaka komarabhacca at court of king Bimbisara and Gautama as well Jivaka komarabhacca learned at univ. of Taksasila (for 7 yrs) 5. Performed a intestine operation and head operation (and taken out 2 worms) 6. Extracting foreign matter especially arrow-heads etc. from the body 7. Digestion was believed to be related to internal heat 8. Skin diseases 9. Fruits and vegetable in medicines

774 LALOU M. Eleven years of works on Buddhism in Europe. IHQ. 1949, 25; (4) : 229-60. KEYWORD: KNOWLEDGE TRANSMISSION; TIME

CONTENTS: 1. Science of religions 2. Notion of time in Indian philosophy (time concept in Indian works) 3. Myths, cults, legends, faiths etc. 4. Artistian works

775 GHOSHAL SN. Date of the Prakrita-Paingala. IHQ. 1949, 25; (1) : 52-57. Ref 24

CONTENTS: 1. Remarks on C M Ghosh work 2. Comparision of others work on PP and finding exact date

776 . Chemistry in Ancient India. ISIS. 1949, 40; 120 : 119. Ref 1

CONTENTS: 1. Announcement of the compilation of History of chemistry in ancient and medieval India

777 SARTON G. Qualifications of teachers of the Hisotry of Science.(II). ISIS. 1949, 40; 122 : 311-13. KEYWORD: EDUCATION

CONTENTS: 1. Three dimensions of History of Science - time (chronology), -area (geography), -subject (Scientific Knowledge) 2. Teachers qualifications

778 DRIVER GR. Plague of the Philistines (1 Samuel V,6-vi,16). JRAS. 1950; 1 : 50-52. Ref 17

CONTENTS: 1. Disease symptons 2. Cause of Disease

779 HILL EG. Colouring principle of the flowers of Nyctanthes Arbortristis. JPASB. 1950, 1 : 102-05. KEYWORD: FLOWERS - COLOURING

CONTENTS: 1. "Harsinghar" in Urdu 2. Aqueous extract, alcoholic extract 3. Preparation of colouring principle 4. Properties of colouring matter

780 ANNANDALE N. Contribution to oriental Herpetology III-notes. Part 2. JPASB. 1950, 1 : 139-51. Ref 7

CONTENTS: 1. Dibamus, Lacertidae, Scincidae - Mabuia, Lygosoma 2. List 156-242 3. Dimensions of type, diagnosis

781 SILBERRAD CA. Notes on a decomposition product of a peculiar variety of Bundelkhand Gneiss. JPASB. 1950, 1 : 168-71. KEYWORD: CLAY; GNEISS

CONTENTS: 1. Report on clay from Bundelkhand gneiss

782 BURR M. Earwigs of the Indian museum, with descriptions of new species. JPASB. 1950, 1 : 27-31.

CONTENTS: 1. 4 new species, description, larvae etc. location and other species, measurement of size

783 VREDENBURG E. Occurrence of the genus Apus in Baluchistan. JPASB. 1950, 1 : 33-34. Geological Survey of India

CONTENTS: 1. Thalonk (Place) environment description 2. Description of Crustacean - Phyllopods

784 ANNANDALE N. Hydra of the Calcutta tanks. JPASB. 1950, 1 : 72-73. Ref 3 Deputy Superintendent of the Indian Museum

CONTENTS: 1. Description, coloration of the -Hydra

785 HILL EG. Composition of the oil from Bir Bahoti or the "Rains insect"(Trombidium grandissimum). JPASB. 1950, 1 : 74-80.

CONTENTS: 1. Determination of saponification value 2. Unsaponifiable matter 3. Fatty acids etc.

786 ANNANDALE N. Contributions to oriented Herpetology II-Notes on the oriental Lizards in the Indian Museum, with a list of the species recorded from British India and Ceylon. Part I. JPASB. 1950, 1 : 81-93. Deputy Superintendent of the Indian Museum

CONTENTS: 1. Photos given 2. Gymnodactylus, calotes, Agama etc. measurement of body, description coloration, diagnosis

787 SENGUPTA PC. Researches in ancient Indian Chronology. JASBL. 1950, 16 : 1-12. Ref 31

CONTENTS: 1. Researches in astronomical chronology 2. Transmission through millennia of years of the Mahabharata & vedic literature 3. Vedic data for vedic antiquity 4. Max Muller & Winternitz on vedic antiquity 5. Note on variation of lunar months in vedic period

788 WINTER HJJ; ARAFAT W. Discourse on the concave spherical mirror by Ibn al-haitham. JASBSC. 1950, 16 : 1-16. Ref 38 UK, Exeter, Univ. college of the South-West of England KEYWORD: MIRROR

CONTENTS: 1.Burning mirrors of circular shape-proofs are given

789 CHAUDHURI S (SIBADAS). Bibliographical survey of the contributions of the Royal Asiatic

Society of Bengal towards the study of Buddhism: 1788-1949. JASBL. 1950, 16 : i-xxvi Suppl. Bengal; Deputy Librarian, Royal Asiatic Society of Bengal KEYWORD: INFORMATION SCIENCE

CONTENTS: 1. Introduction - Society, Publication, Proceedings, Oriental Publications, Bibliotheca Indica 2. Bibliography (245 items) 3. Analytical subject-Index to the bibliography (numerals indicate the serial number in the bibliography) - Antiquity:, Architecture:, Biography:

790 PACHORI GN. Occurrence of an accessory hind-limb in Rana Tigrina Daud. JASBSC. 1950, 16 : 115-22. Ref 7 Agra; Agra College

CONTENTS: 1. External features 2. Lymphatic system 3. Musculature 4. Skeleton

791 MENON PMG. Regeneration of the caudal fin in Coilia reynaldi Cuv. and val.. JASBSC. 1950, 16 : 123-26. Ref 11 KEYWORD: REGENERATION - FIN, VAL

CONTENTS: 1. Skeletal structure of normal tail and regeneration tail 2. Description of fins

792 GEORGE JC. Case of conjoined twins in the Indian domestic goat with a reference to the general problem of domestication . JASBSC. 1950, 16 : 127-32. Ref 29 Calcutta; Vertebrate Zoologist, Dept. of Anthropology Calcutta, Indian Museum KEYWORD: TWINS - CONJOINED

CONTENTS: 1. External features 2. Internal features 3. Specimen was of cephalo-thoracopagus type (showing duplicity of limbs, ears and hind parts of body), internally there were 2 trachae and lungs, 2 hearts, 2 spleen, 2 livers and single pancreas 4. Illustrations (photos) are given

793 BAGCHI HD; CHATTERJI PC. Note on hermite's function Hn(z) and its associated equations (functional and differential). JASBSC. 1950, 16 : 17-26. Ref 1 Calcutta; Head of the Dept. of Pure Mathematics, Calcutta Univ., KEYWORD: EQUATION-DIFFERENTIAL, FUNCTIONAL

CONTENTS: 1

794 WINTER HJJ; ARAFAT W. Algebra of 'Umar Khayyam'. JASBSC. 1950, 16 : 27-78. Ref 210

CONTENTS: 1. Classification of equations 2. Equations having two terms 3. Quadratic equations 4. Cubic equation -with 3 terms, with 4 terms 5. Equations involing fractional terms 6. Index given in last

795 HORA SL. Knowledge of the ancient hindus concerning fish and fisheries of India. 2. Fishery legislation in Asoka's Pillar Edict v (246BC). JASBL. 1950, 16 : 43-56. Ref 17 Calcutta; Director, Zoological Survey of India, Indian Museum

CONTENTS: 1. Asoka & his inscriptions 2. Identification of fishes 3. Significance of protection 4. Asoka's fishery legislation 5. Lunar periodicity in spawning 6. Scientific analysis of Asoka's law related to fishery 7. Evolution of fishery laws in ancient India 8. Present-day of fishery laws in India

796 BISWAS K; SHARMA AK. Sargassa of Indian seal (Seaweed). JASBSC. 1950, 16 : 79-97. Ref 24 Calcutta; Superintendent, Indian Botanic Garden

CONTENTS: 1. Table showing distribution of seaweed 2. Description of family and genus 3. Key to identification of Indian species 4. Description of Indian species 5. Illustrations are given

797 LAL BB. Painted grey ware of the upper gangetic basin. JASBL. 1950, 16 : 89-102. Ref 23 KEYWORD: PORCELAIN; POTTERY

CONTENTS: 1. From Hastinapura (31 items) 2. From ahichchhatra (17 items) 3. Pottery-types from Hastinapura (3) & Ahichchhatra (7) which do not appear in half-tone blocks

798 BISWAS K; KUMARAN MAS. Indian dioscoreas and their role in dietary. JASBSC. 1950, 16 : 99-114. Ref 20

CONTENTS: 1. Dioscoreas (food plant), history, distribution, habitat and as food 2. Systematic description of 25 species, methods of cultivation

799 MITRA KP. Dramaturgy in the Mahapurana of Puspadanta. IHQ. 1950, 27; (3) : 232-40.

800 SARTON G. Notes on the reviewing of learned books. ISIS. 1950, 41; 124 : 149-58. Ref 14 KEYWORD: INFORMATION SCIENCE; REVIEW

CONTENTS: 1. How one should read a book? 2. Main points of a good review 3. Editor view about reviewer

801 SARTON G. Critical Bibliographies of ISIS . ISIS. 1950 , 41; 125-126 : 291-98. KEYWORD: INFORMATION SCIENCE

CONTENTS: 1. Objective of CB 2. Method of compilation of CB 3. Method of classification used in CB

802 TRITTON AS. Catalogue of oriental manuscripts in the library of the Royal college of physicians. JRAS. 1951; 2 : 182-92.

CONTENTS: 1. Given Arabic,persian & Turkish mss 2. List of 63 mss given 3. Describe size, Author,Title subject etc.

803 HORA SL. Knowledge of the ancient hindus concerning fish and fisheries of India. 3. Matsyavinoda or a chapter on angling in the Manasollasa by king Somesvara (1127 AD). JASBL. 1951, 17 : 145-69. Ref 38 Calcutta; Director, Zoological survey of India KEYWORD: CLASSIFICATION -FISH; FISH-CLASSIFICATION

CONTENTS: 1. Somesvara's classification of fishes - Marine & Estuarine forms - freshwater Reverine forms 2. Fishing Tackle - line - rod - hock 3. Ground Baiting 4. Somesvara's advice on fish

804 HORA SL . Maintenance of irrigation tanks through fishery revenue in ancient India. JASBL. 1951, 17 : 41-50. Ref 9 KEYWORD: FISHERIES

CONTENTS: 1. List (Table) of tanks discussed by Venkayya history, maintenance etc 2. List (Table) of Tamil inscriptions of 14th-16th C.A.D. indicating fishery in irrigation tanks 3. Development of fishery in irrigation tanks 4. Irrigation tanks in Bengal 5. Irrigation Tanks & fish-farming

805 AGRAWALA VS. Some chronological consideration about Panini's date. IHQ. 1951, 27; (4) : 269-86. Ref 3 KEYWORD: PANINI

CONTENTS: 1. Ten naksatras in Sravistha and other stars. views of scholars 2. Panini association with different people to get date his mention in other literary works

806 GHOSHAL SN. Was Sumati - the author of the Prakrita-Paingala. IHQ. 1951, 27; (4) : 333-36. Ref 5

CONTENTS: 1. Pingala - saravikasini and Bhabisattakha description of PP 2. Metrical system in sanskrit literature

807 FOOTE GA. Place of Science in the British Reform Movement 1830-50. ISIS. 1951, 42; 129 : 192-208. Ref 84

CONTENTS: 1. Criticism of existing British Science 2. Attacks and views on the Royal Society of London and the British Association for the Advancement of Science

808 SARTON G. History of Science in India. ISIS. 1951, 42; 127 : 44. KEYWORD: HISTORY OF SCIENCE - SYMPOSIUM INDIA

CONTENTS: 1. Symposium on History of Science in South Asia, Nov. 1950 Delhi University by

National Institute of Sciences in India & UNESCO 2. History of Indian medicine, -Zoological knowledge, -Social factors in the History of Science in astronomy, -Indian sciences in the far East

809 WALES HGQ. Pre-India basis of Khmer Culture. JRAS. 1952; 2 : 117-23. KEYWORD: INDO-CHINA; IRRIGATION

CONTENTS: 1. Megalithic culture (Indo-china) describes irrigation work 2. Khmer culture and art

810 SEED G. Lord William Bentick and the reform of education. JRAS. 1952; 1 : 66-77. Ref 35

CONTENTS: 1. Policy of East India Company in Education(1823) 2. Committee of public instruction (emphasied)English teaching 3. Sanskrit & Arabic were languages of religions 4. No govt. funds for printing in sanskrit & Arabic books 4. Institutions were founded to teach English language/western ideas about 1835

811 WINTER HJJ; MIRZA A. Concerning the persian version of Lilavati. JASBSC. 1952, 18 : 1-10. Ref 44 Exeter; England University college of the south-west of England KEYWORD: LILAVATI-PERSIAN VERSION

CONTENTS: 1. Commentaries translations & editions on Lilavati 2. Persian MS - Investigation of mixture - Rule of three - Rule of Nine & Eleven

812 MAJUMDAR B. Rivers in the Bengal delta: river problems in West Bengal and their solution. JASBSC. 1952, 18 : 103-21.

CONTENTS: 1. Maps showing river system and rivers in Bengal 2. Irrigation and waterways schemes 3. Paths of rivers: Ganga, and is tributaries, etc. 4. Drainage of city and surrounding area 5. River problems -flooding, dry, does not received enough water, water logging, water does not remain long enough, soil erosion, construction of embankments, bridges, canals, etc. their solutions

813 NAG AC. Construction of fertility tables for India and studies therefrom. JASBSC. 1952, 18 : 11-22. Ref 6 Calcutta; Lectures, Vital Statistics, Calcutta Univ. KEYWORD: POPULATION; BIRTHCONTROL

CONTENTS: 1. Bengal exhibits higher fertility rate 2. Age and age-gaps between parents 3. Frequency of birth 4.Fertility in different income levels of society

814 CHAKRAVARTI C. Application of vedic mantras in tantric rites. JASBL. 1952, 18 : 113-15. Ref 6

CONTENTS: 1. Mantra of surya 2. Tantric worship

815 HORA SL. Parallel evolution of Pseudecheneis blyth and similar fishes of south-east Asia. JASBSC. 1952, 18 : 123-28. Ref 7 KEYWORD: ASIA - SOUTH-EAST; FISH

CONTENTS: 1. Distribution of Pseudecheneis and allied genera 2. Catfishes and there structures development 3. Three genera pseudecheneis, parapseude cheneis and propseudecheneis origin

816 JONES S; PANTULU VR. Metamorphosing stages of the talabon eel, Muraenesox talabon (Cantor), with descriptions of some leptocephali from the estuaries of Bengal and Orissa. JASBSC. 1952, 18 : 129-40. Ref 9 Barrackpore; Central Inland Fisheries Research Station KEYWORD: METAMORPHOSIS

CONTENTS: 1. Metamorphosis of muraenesox talabon (Cantor) 2. Leptocephali from the Burhabalang estuary 2. Leptocephalus (? Muraena tile) from the Hooghly Estuary 3. Juvenile of Muraena tile (Hamilton)

817 HORA SL . Lac and the lac-insect in the atharvaveda. JASBL. 1952, 18 : 13-15.

CONTENTS: 1. Habits & Habitats of lac-insect 2. Work of Dave KN on lac & the lac-insect in Atharva-veda

818 RUDRA MN. Toxic factor of lathyrus sativus and experimental lathyrism in animals. JASBSC. 1952, 18 : 141-46. Ref 38 Bihar; Professor of Medical Chemistry, Darbhanga Medical College, Laheriasarai KEYWORD: TOXICITY

CONTENTS: 1. Toxic factor of L. Sativus -selenium content of pulses 2. Experimental Lathyrism in monkey -position of hallux before and after stimulation

819 JOB TJ; DAVID A; MOTWANI MP. Observations on the fish and fisheries of the Damodar basin with reference to the multipurpose projects of the valley. JASBSC. 1952, 18 : 147-72. Ref 9 Barrackpore (Calcutta); Central Inland Fisheries Research Station

CONTENTS: 1. Topography : general features of Damodar drainage 2. Collection of data and methods employed 3. Occurrence of fishes with reference to seasonal physical and hydrological -changes of rivers in Valley, Barakar, Konar, Damodar 4. List of fishes (89 species)

820 SILAS EG. Respiratory adaptations of torrential fishes, with special reference to the Glyptosternoid fishes of the family sisoridae: (order siluroidea). JASBSC. 1952, 18 : 173-80. Ref 11 Laboratories of the Zoological Survey of India KEYWORD: FISH - RESPIRATION

CONTENTS: 1. Glyptosternoid fishes and their adaptations 2. Mode of respiration in Glyptothorax Blyth 3. Role of dentition and lips in respiration 4. Gill-openings and pharyngecal chambers 5. Branchial structure 6. Respiratory convergence and conclusion

821 CHATTOPADHYAY KP. Korku physical type and racial affinities. JASBSC. 1952, 18 : 23-34. Ref 7 KEYWORD: ANTHROPOLOGY

CONTENTS: 1. Stature, caphalic index, Nasal index 2. Calculations of racial affinities

822 ALIKUNHI KH; CHAUDHARI H; RAMACHANDRAN V. Response to transplantation of fishes in India, with special reference to conditions of existence of Carp Fry. JASBSC. 1952, 18 : 35-53. Ref 47 Barrackpore; Central inland fisheries Research station KEYWORD: TRANSPLANTATIONS-FISH; FISH

CONTENTS: 1. Exotic food fishes in Indian waters 2. Salt water species acclimatized to fresh water 3. Fluviatile carps transplanted in ponds - physico-chemical conditions of environment - nature of pond - Growth in stagnant water - Predatory organisms - intensity of stocking - Availability & quality of food

823 SEN N. Astronomical basis for fixing the age of kalidasa(criticism). JASBL. 1952, 18 : 5-6. Tripura, Agartala; Professor of Sanskrit, Maharaja Bir Bikram college KEYWORD: TIME

CONTENTS: 1. Criticism on Prof. Sengupta's paper Time indications in Kalidasa JRASB Lets 9(3)1945

824 CHATTERJI BK; KUMAR GD. Somatic characters and racial affinities of the Kannikars of travancore state. JASBSC. 1952, 18 : 55-70. Ref 14 KEYWORD: ANTHROPOLOGY

CONTENTS: 1. Stature, cephalic index and nasal index 2. Facial index table

825 HORA SL. I..n in the Ramayana. JASBL. 1952, 18 : 63-69. Ref 15 Calcutta; Director, Zoological Survey of India, Indian Museum KEYWORD: RAMAYANA-FISH

CONTENTS: 1. Species of fish referred to in the Ramayana 2. Description of the Pampa lake in Ramayana 3. Present Day geographical features of the Pampa sarovara 4. Changes in fish fauna of Pumpa saravara 5. Methods of catching fish referred to in Ramayana 6. Fish preparations referred to in Ramayana 7. Shooting fish with Bow & arrow 8. Illustrations given

826 SENGUPTA PC. Note on Dr. Sen's criticism of a chapter of the work, Ancient Indian Chronology. JASBL. 1952, 18 : 7.

827 CHATTERJI BK; KUMAR GD. Somatic characters and their racial affinities of the Santals of the Santal Parganas. JASBSC. 1952, 18 : 71-82. Ref 15 KEYWORD: ANTHROPOLOGY

CONTENTS: 1. Variation and distribution of the Santals - male and female 2. Mean with standard errors, standard deviation etc. of stature, colour, facial index, cephalic index, nasal index

828 HORA SL. Major problems of the fisheries of India, with suggestions for their solution. JASBSC. 1952, 18 : 83-101. Calcutta; Director, Zoological Survey of India, Indian Museum

CONTENTS: 1. Fishing industry in India -geographical basis of Indian fisheries, fresh water fisheries, saltwater fisheries, fishery technology and distribution, socio-economic problems, fishery research, fishery information services 3. Major problems with suggestions for their solution -some aspects of socio-economic, productive methdos of fishing, fishery exploration work, co-ordination and integration of fishery work, fishery extension work, periodic assessment of development works

829 CHAUDHURI S. Concordance of the fauna in the Ramayana (to be continued). IHQ. 1952, 28; (2) : 135-141. Ref 22

CONTENTS: 1. List of 1-22 animals, their location in Ramayana. Reference position

830 BHATNAGAR OP. Smallpox vaccine - its introduction under Wellesley. IHQ. 1952, 28; (2) : 186-89. KEYWORD: VACCINE - SMALLPOX

831 LAL KS. Food and drink in medieval India. IHQ. 1952, 28; (2) : 192-96. KEYWORD: DRINKS

832 AGRAWALA VS. Ancient Indian geography. IHQ. 1952, 28; (3) : 205-14.

CONTENTS: 1. Vedic geography (north-west in Rgveda, Ganga, Sindhu, Sarasvati etc.) 2. Epics: Ramayana and Mahabharata 3. Puranas (Sthala - mahatmyas (topography in details) 4. Panini in Astadhyayi 5. Buddhist literature (Pali Trpitaka) 6. Chines Tripitakas (fan-fan-Yu : a geographical dictionary) 7. Ptolemy, periplus (scholars) on geography 8. Jain sources Sanskrit literature (Kalidasa, Bana, Rajasekhara, Varahamihira, Kalhana) 9. Arab geographers 10. Scheme of an Indian Geographical Dictionary (vol. 1-5)

833 CHAUDHURI S. Concordance of the fauna in the Ramayana. IHQ. 1952, 28; (3) : 249-56. Ref 56

CONTENTS: 1. List of 25-56 animals their locations in Ramayana (Reference to AK, Adi and Ayo) their English name also given

834 SIRCAR DC. Astrologer at the village and the court. IHQ. 1952, 28; (4) : 342-49. Ref 5

CONTENTS: 1. Almanacs called (Panji, Panjika or Pancanga) 2. Astrologer calculating positions of stars (heavenly bodies) and deciding the course of action of people 3. Astrologers at royal court

835 CHAUDHURI S. Concordance of the fauna in the Ramayana. IHQ. 1952, 28; (4) : 350-59. Ref 102

CONTENTS: 1. List of 57-102

836 . International Congress on Medical Librarianship in London in July 1953. ISIS. 1952,

43; 132 : 126. KEYWORD: INFORMATION SCIENCE

837 SAHA MN. different methods of date-recording in ancient and medieval India and the origin of the Saka era. JASBL. 1953, 19 : 1-24. Ref 20 KEYWORD: ERA-SAKA

CONTENTS: 1.Stages in India's Calendar development 2. Date-recording of Ashoka of Satavahanas and their early Indian Dynasts 3. Introduction of Era 4. Introduction of Era in India 5. method of date recording 6. Omission of figures for hundreds in the counting of the years of Era 7. Old saka Era 8. Date-recording by western sakas 9. Saka and satavahana Chronology 10. Astronomical notes

838 GAUSDAL J. Ancestral and sacrificial clans among the santals. JASBSC. 1953, 19 : 1-97. Ref 258 KEYWORD: ANTHROPOLOGY

CONTENTS: 1. Pans or Ancestral clans with notes 2. Khuts or sacrificial clans with notes 3. Chart of the Santal clans 4. Social picture of santal villages

839 NAG AC. Estimation of the rate of natural growth of population for west Bengal. JASBSC. 1953, 19 : 103-14. Ref 5

CONTENTS: 1. Determination of fertility rates 2. Suitablity of applying the sample fertility rates to the population 3. Determination of an Index for population growth

840 CHATTERJI BK; KUMAR GD. Short note on the proportion of the upper and lower extremities of the body of the hill tribes of Travancore and Cochin. JASBSC. 1953, 19 : 115-16. Ref 3 KEYWORD: ANTHROPOLOGY

CONTENTS: 1. Brachial index & Tibio radiale index

841 SARKAR HL. Collection of fish from Rajgir, Patna district, Bihar. JASBSC. 1953, 19 : 117-19. Ref 6

CONTENTS: 1. 13 species described locations etc.

842 MITRON JN; DATTA RM. Synadenium grantii Hook F. Fam: (Euphorbiaceae)- its occurrence in Bengal and some morphological and anatomical observations . JASBSC. 1953, 19 : 121-31. Ref 38 Calcutta; Department of Botany,presidency college, 34 paraganas, Barrackpore, Jute Agricultural Research Institute

CONTENTS: 1. Inflorescence, Ovule & polen grains structures 2. fibres & description of flower

843 SARKAR H. Fish-hooks from the Indus valley. JASBSC. 1953, 19 : 133-39. Ref 20

CONTENTS: 1. Fish-hooks of the Indus valley 2. Relative sizes of the Indus valley fish-hooks 3. Evolution of Barbs at chanhudaro (fish-hooks) 4. Fish-hooks at other archaeological sites 5. Photos given

844 PACLT J. Genera of the Hepialidae (Insecta, Lepidoptera). JASBSC. 1953, 19 : 141-48. Ref 51

CONTENTS: 1. Geographical distribution of the Hepialidae 2. New Synonymy in Genera Hepialidae

845 GANAPATI SV, CHACKO PI, SRINIVASAN R. Hydrobiological conditions of the Gangadhareswarar temple tank, Madras. JASBSC. 1953, 19 : 149-59. Ref 28 KEYWORD: HYDROBIOLOGY

CONTENTS: 1. Methods 2. Physical & Chemical conditions 3. Biological conditions

846 KACHROO P. Ferns of Assam. JASBSC. 1953, 19 : 161-74. Ref 6

CONTENTS: 1. Description of Genera (fern) 2. Enumeration & distribution of species (69)

847 JOB TJ; PANTULU VR. Fish trapping in India. JASBSC. 1953, 19 : 175-96. Ref 44 KEYWORD: FISH TRAP

CONTENTS: 1. Methods of Trapping - Groping - stranding - Bunding - Screening - Screen wall - Sieve trap - Waterfall Basket - Platform-cum-Basket - Leap traps-traps Boat-Traps Raft-Platform-cum-bush - Bush traps etc 3. Photos of traps are given

848 DAVID A. Notes on the Bionomics & some early stages of the Mahandi Mahseer. JASBSC. 1953, 19 : 197-209. Ref 18 KEYWORD: FISH

CONTENTS: 1. Topography & Hydrological features 2. Seasonal variations in Condition of sex-organs 3. Post-larvae 4. Growth of scale 5. Cultural possibilities

849 HORA SL. Knowledge of the ancient hinuds concerning fish and fisheries of India. 4. fish in the Sutra nd Smrti literature. JASBL . 1953, 19 : 63-77. Ref 37 Calcutta; Director, Zoological Survey of India, Indian Museum KEYWORD: SUTRA - SOURCE MATERIALS; SMRTI - SOURCE MATERIALS

CONTENTS: 1. Reference to fish as food 2. Reference to fish as forbidden food 3. Religious injunctions pertaining to fish and ceremonial fishing 4. Sin accuing from fishing 5. Fish protection 6. Identification of fishes -chronology, true fishes, aquatic animals other than fishes

850 CHATTERJI BK; KUMAR GD. Short note on the relation between the stature and sitting height of the different hill tribes of Travancore. JASBSC. 1953, 19 : 99-102. Ref 4 KEYWORD: ANTHROPOLOGY

CONTENTS: 1. Brachycormic, Metriocormic & Macrocormic (cormic indices) 2. Sitting height, stuture index

851 AGRAWALA VS. Geographical Data in Panini. IHQ. 1953, 29; (1) : 1-34. Ref 6

CONTENTS: 1. Astadhyay (geographic inf. in it) 2. Country (Geographical horizon), division of country 3. Mountains, rivers and forest 4. Janapadas (both a state and a cultural unit) 5. Town and villages (nomenclature)

852 CHAUDHURI S. Concordance of the fauna in the Ramayana (to be continued). IHQ. 1953, 29; (2) : 121-28. Ref 162

CONTENTS: 1. List of 130-168

853 CHAUDHURI S. Concordance of the fauna in the Ramayana (to be continued). IHQ. 1953, 29; (3) : 276-285. Ref 189

CONTENTS: 1. List 169-192

854 CHAUDHURI S. Concordance of the Fauna in the Ramayana (to be continued. IHQ. 1953, 29; (4) : 378-86. Ref 230

CONTENTS: 1. List 193-230

855 CHAUDHURI S. Concordance of the Fauma in the Ramayana (to be continued). IHQ. 1953, 29; (1) : 56-63. Ref 128

CONTENTS: 1. List 103-128

856 GUERLAC H. Proposed revision of the ISIS critical bibliography. ISIS. 1953, 44; 137 : 226-31. USA; Institute of Advanced Study, Princeton KEYWORD: INFORMATION SCIENCE; BIBLIOGRAPHY

CONTENTS: 1. Opinion about continuation of CB 2. Proposed system of classification for ISIS CB. A- History of Science, General references and tools, 1-5 (Subdivisions), B- Science and its History from special points of view (includes works involving two or more subjects or chronological periods), 10-15, C- Histories of the special sciences (works involving two or more chronological periods)

20-29, D- Chronological classification 30-38 3. Basis of classification in CB

857 SARTON G. Why ISIS?. ISIS. 1953, 44; 137 : 232-42. Ref 29 KEYWORD: ISIS - WHY

CONTENTS: 1. Homage to Lorenz Oken 2. Why ISIS? 3. Oriental and international tendencies of ISIS 4. Integral humanism 5. Growth of ISIS 6.Earlier ISIS occurrences

858 BOWEN J. East India Company's education of its own servants. JRAS. 1954; 2 : 105-23. Ref 86

CONTENTS: 1. 1800 onwards EI Company plans 2. College of fort William, Calcutta 3. Civil servants in India rules

859 BAILEY HW. Buddhist Sanskrit. JRAS. 1954; 1 : 13-24. Ref 66

CONTENTS: 1. Describe various works in sanskrit,Buddhist Sanskrit,Iranian form langauge etc 2. Given scientific names (Colour acid,salt etc)

860 TRIPATHI YR. Some observations on Parasities of Hilsa. JASBSC. 1954, 20 : 75. Barrackpore; Central Inland Fisheries Research Station

861 HORA SL. Symposium on Hilsa and its fisheries - a review. JASBSC. 1954, 20 : 1-5. Chairman Hilsa Sub-Committee, Indo-Pacific Fisheries Council

CONTENTS: 1. Present status (1954) 2. Recent researches 3. Races of Hilsa 4. Effect of dams & artificial propagation

862 HORA SL. Biology of the Hilsa. JASBSC. 1954, 20 : 15-18. Ref 4 Calcutta,; Chairman, Hilsa-sub-committee, Indo-pacific fisheries council KEYWORD: FISH

CONTENTS: 1. Certain new Trends in Hilsa fisheries - Winter fishery for the Hilsa - Probable changes in the Hilsa stocks - Intensification of fishing - Hilsa and public Health

863 HORA SL. Proverbs and popular sayings concerning the Hilsa fish current in Bengal. JASBSC. 1954, 20 : 19-27. Calcutta; Zoological Survey of India KEYWORD: FISH

CONTENTS: 1. Hilsa, a fish of high quality & taste 2. Ancient close season for Hilsa 3. Prediction for successful Hilsafishing 4. Hilsa fish in recent Bengali literature 5. Popular sanskrit & Bengali sayings concerning Hilsa fish

864 BISWAS JC. Trends of Hilsa in Calcutta markets. JASBSC. 1954, 20 : 29-31. Calcutta; President, fish Retailers' Association

CONTENTS: 1. Source of Hilsa supplies and their relative value 2. Trends of Hilsa prices 3. Causes of high prices of Hilsa

865 MITRA GN; DEVASUNDRAM P. Hilsa of the Chilka lake with a note on the occurrence of Hilsa in Orissa. JASBSC. 1954, 20 : 33-40. Ref 10 Cuttack; Dy. Director of Fisheries Orissa,; Balugaon, Senior Research Asst.

CONTENTS: 1. Fishing methods, seasons & area 2. Fish export from Chilka lake 3. Present condition of fishery & its trend 4. Migration, price of Hilsa & occurrence of Hilsa in Orissa

866 VARMA CP. Hilsa fishery in Bihar. JASBSC. 1954, 20 : 41-43. KEYWORD: FISHERIES

CONTENTS: 1. Hilsa fishery is of secondary importance to Bihar 2. Types of Nets

867 KAUSHIVA BS. Preliminary note on Hilsa fisheries of Uttar Pradesh. JASBSC. 1954, 20 : 45-46. Lucknow; Ex-deputy Director of Fisheries, Uttar Pradesh

CONTENTS: 1. Hilsa at Allahabad, Benaras (in Ganga, Yamuna & tons)

868 KULKARNI CV. Hilsa Ilisha (Hom) on the West Coast of India. JASBSC. 1954, 20 : 47-53. Ref 10 Bombay; Assistant Director of fisheries

CONTENTS: 1. Up-stream migration of Hilsa in Narbada 2. Upper limit of migration of Hilsa in Narbada 3. Production of Hilsa 4. Depletion of fishery

869 CHACKO PI. Past, present and future of the Hilsa fisheries in the Madras state. JASBSC. 1954, 20 : 55-58. Ref 9 Madras; Directorate of fisheries

CONTENTS: 1. Madras state dams, anicuts across the Godavari, Krishna & Cauvery rivers giving Hilsa fisheries

870 QUARESHI MR. Palla of Sind. JASBSC. 1954, 20 : 59-60. Karachi; Dy. fisheries Development Adviser to the Govt. of Pakistan

CONTENTS: 1. River indus, sukkur Barrage & fish-ladder

871 PANIKKAR NK. Progress of Hilsa investigations in India from 1938 to 1950 : a review. JASBSC. 1954, 20 : 61-63. Ref 14 Mandapam; Chief Research Officer, Central Marine Fisheries Research Station

CONTENTS: 1. Hilsa investigations in Bengal 2. Hilsa investigations in Madras & Bombay

872 JONES S. Hilsa investigations at the central inland fisheries research station aims and achievements. JASBSC. 1954, 20 : 65-67. Borrackpore; Research Officer (Estuarine) Central inland fisheries Research Station

CONTENTS: 1. Embryonic studies 2. Morphological & parasites on Hilsa 3. Contributions on Hilsa from the Central inland fisheries Research station -8 articles

873 PILLAY TVR . Morphological and serological characters of the Hilsa, Hilsa ilisa (ham)with reference to racial investigations . JASBSC. 1954, 20 : 69-74. Ref 14 Calcutta; Hilsa fish Enquiry, Indian Council Medical Research

CONTENTS: 1. Morphological characters 2. Serological characters

874 RAY NB. Trade,Agriculture, and industry in West Bengal 1200 to 1608 as found in the persian and arabic sources. JASBL. 1954, 20 : 7-13. Ref 33 KEYWORD: PERSIAN SOURCES; BENGAL

CONTENTS: 1. Tabaqat-i-Nasiri 2. Kitab Mirat-ul maani fil Idrakat alam ul Insani 3. Architecture, Agriculture produce etc

875 AHMAD N. Hilsa fishery of East Bengal. JASBSC. 1954, 20 : 7-14. Pakistan; Comilla, Dy. Director of Fisherman. East Pakistan KEYWORD: FISHERIES; BENGAL

CONTENTS: 1.Capture operations - Seine & drag nets _ Purse nets - Dip nets 2. Artificial propagation of Hilsa 3. Migration, spawning, production etc 4. Export of Hilsa from East Bengal (Statistics)

876 NAIR KK. Calcutta, Directorate of fisheries, West Bengal. JASBSC. 1954, 20 : 77-79. Ref 6

CONTENTS: 1. Methods for conservation of Hilsa fishery 2. Breeding grounds on migratory fish Hilsa

877 UKIL AC. Science and the Health of India. JASBSC. 1954 , 20 : 81-132. Ref 25

CONTENTS: 1. Historical perspective - public health progress in British India - Medical Education & Research - Environmental Hygiene - Mortality rates 2. Present status of health in India (1954) - Mortality rates - sickness rates - vocational education - sanitary & medical legislation - Drugs & intruments in Indian Union 3. Probing into future - integrated rural health scheme - comments on Bhore committee's proposals - Health section of India's first 5-year Rehabilitations & development plan

878 CHAUDHURI S. Concordance of the fauna in the Ramayana. IHQ. 1954, 30; (2) : 148-153. Ref 37

CONTENTS: 1. List 231-237

879 ROY UN. Fortifications of cities in ancient India. IHQ. 1954, 30; (3) : 237-44. Ref 74

CONTENTS: 1. Categories of fortifications (i) Natural (ii) Artificial 2. Description of king fortifications 3. Kind of natural fortification 4. Three methods of artificial fortifications (i) Wall (ii) trenches (iii) plantation of forests. i.e. in India, Greek and Roman town 5. Ancient India description, (Ayodhya, Lanks, ....)

880 BANERJI SC. Ancient Indian geography (as revealed in Dharma-Sutras). IHQ. 1954, 30; (4) : 332-38.

CONTENTS: 1. Geographical names in following heads in alphabetical order. River, Mountains and Hills, Places

881 AGRAWALA VS. Janapada and the Greek City State. IHQ. 1954, 30; (10) : 38-49.

CONTENTS: 1. List janapada (state) in India in Panini work 2. Greece (city-states) description 3. Evolution of J & G 4. Cultural and social life there in

882 SEN RK. Susruta, Carak and Bharata. IHQ. 1954, 30; (10) : 68-80. Ref 7

CONTENTS: 1. Link between S.C. & B 2. Bharata positions to know Carak and Susruta 3. Mental symptoms and diseases 4. Concept of vyabhicaribhava

883 HORA SL. Fish in the Jataka sculptures. JASBSC. 1955, 21 : 1-13. Ref 8

CONTENTS: 1. Note on Indian sculpture 2. Bharhut its location and antiquity 3. Fish in Bharahut Reliefs - Naga Jataka - Uda Jataka - Mahakapi Jataka - Kuranga mrga Jataka - Makara medallion - Timingila Jataka 4. Photos given

884 HORA SL; SARASWATI SK. Fish in the Jataka Tales. JASBSC. 1955, 21 : 15-30. Ref 44

CONTENTS: 1. Varieties of fish 2. habits & instincts of fish 3. Enemies of fish 4. Methods of catching fish 5. Fish as food 6. Fish in religious ceremonies 7. Fish as medicine 8. Fish omens 9. Fish & human settlements 10. Fishing village 11. Fish & law 12. Sale of fish 13. Beauty contest among fish 14. Fish metaphor

885 BHATTACHRYA RS. Characteristics of the ancient Vrttis on the Astadhyayi. IHQ. 1955, 31; (2) : 168-74.

CONTENTS: 1. Functions of vrttis 2. Patanjali in mahabhasya and relation of ancient vrttis 3. Variation of vrttis use

886 KANJILAL DK. Picture of ancient India as revealed in Patanjali's Mahabhasya. IHQ. 1955, 31; (4) : 375-82. Ref 5 KEYWORD: PATANJALI

CONTENTS: 1. Maha, gives information about then society, methods of agriculture, system of irrigation... etc. 2. Agricultural aspects discussed 3. Trade system 4. Geographical description 5. Women's position of society life

887 KOPE L. Bird ANUQ (a lexicological study concerning arabic zoology). JRAS. 1956; 2 : 157-64. Ref 18 KEYWORD: BIRDS

CONTENTS: 1. The use of ANUQ in animals names & characteristics 2. Origin of word ANUQ

888 STERN SM. Ibn al-Samh. JRAS. 1956; 1 : 31-44. KEYWORD: ARISTOTLE

CONTENTS: 1. His Aristotelian interpretation & philosophy,logician 2. His works 3. Aristotle's physics as interpreted by ibnal-Samh & his disciple Abu'l-Husayn (or Abu'l-Hasan) AL-Basari 4. Aristotle's Rhetoric,from the copy of al-samh

889 LEVEY M, KREK M, HADDAD H. Notes on the Chemical Technology in an Eleventh Century Arabic work on Bookbinding. ISIS. 1956, 149 : 239-43. Ref 34 KEYWORD: LIBRARY SCIENCE - BOOKBINDING; ARAB

CONTENTS: 1. Tanning 2. Glues 3. Dyes 4. Paints and inks 5. Arabic and ancient mesopotamian technology

890 BUDDHAPRAKASH. Geographical and cultural aspects of the northern itinerary of Raghu as described by Kalidasa. JASBSC. 1956, 21 : 245-60. Ref 88 KEYWORD: GEOGRAPHY

CONTENTS:

891 CHAUDHURI S. Studies in indology in 1953 : A bibliographical survey of Periodical literature. JASBSC. 1956, 22 : 1-43. KEYWORD: INFORMATION SCIENCE

CONTENTS: 1. Author index 2. Analytical index of titles & subjects 3. Supplementary author index - English - Sanskrit - Bengali - Hindi - Telugu 4. List of abbreviations

892 SENGUPTA S. Food prohibition in smrti texts. JASBSC. 1956, 22 : 163-209. Ref 125

CONTENTS: 1. smrtis (14) 2. Appendix : Asoka's pillar Edict

893 SENGUPTA PC. Historicity of the mahabharata on the basis of astronomical data. JASBL. 1956, 22 : 75-84. Ref 12

CONTENTS: 1. Krisna's Birthday (21 July 2501 BC) 2. Yudhisthira's Birthday 3. Arjuna's Birthday 4. Tropical longitude Calculations

894 . Indian Journal of the History of Medicine. ISIS. 1956, 47; 150 : 423. KEYWORD: NEW JOURNAL - INDIAN JOURNAL OF THE HISTORY OF MEDICINE

CONTENTS: 1. From June 1956 - twice a year founded by Indian Association of the History of Medicine 2. Address 187 Poonamallee Road, Madras - 7, India

895 CHAUDHURI S (SIBADAS) COMPILED BY. Index to the publications of the Asiatic Society 1788-1953 Vol.I part I and Vol I part II. JAS. 1956,1959, I&II Calcutta; The Asiatic Society, 1 Park Street KEYWORD: INFORMATION SCIENCE

CONTENTS: Volume I Part I Author-Index to Asiatick Researches, Journal (three series upto 1953) Memoirs, and Miscellaneous Publications Part 2 Index of Bibliotheca.Indica and Monographs(Publications arranged alphabetically) Author-Index to Notices in the Proceedings, list of plates, Analytical classified Index. Volume II Subject-Index Religion and philosophy Language and literature, History and social sciences, Art and Archaeology, Anthropology Science, Miscellaneous.

896 SANKALIA HD; DEO SB. Saddle querns and stratigraphy. JRAS. 1957, 2 : 209-12. Ref 12 KEYWORD: POTTERY; DYE

CONTENTS: 1. Black-on-red pottery, Red&Black Pottery

897 GLASENAPP HV. Influence of Indian thought on German science, philosophy and literature. JASBL. 1957, 23; (2) : 1-10. KEYWORD: KNOWLEDGE TRANSMISSION

CONTENTS: 1. How German got the influence about India 2. German scholar who knew

Sanskrit-Friedrich Schlegel 3. Use of Sanskrit and English by German 4. Development of Indian studies in Germany

898 CHATTERJI BK; KUMAR GD. Somatometric and Somatoscopic observations and their affinities of the mannans of Travancore - Cochin. JASBSC. 1957, 23 : 1-18. Ref 36 KEYWORD: SOMATOSCOPY; ANTHROPOLOGY

CONTENTS: 1. Stature, cephalic index, facial index, orbito-nosal index etc.

899 CHATTERJI BK; KUMAR GD. Comparative study of the somatometric and somatoscopic observations and their racial affinities of the polyans and the malapantarams of Travancore-Cochin. JASBSC. 1957, 23 : 19-42. Ref 38 Calcutta; Dept. of Anthropology, Govt. of India KEYWORD: SOMATOSCOPY; ANTHROPOLOGY

CONTENTS: 1. Cephalic index, length and breadth index, skin colour, standard errors and standard deviations 2. Stature, nasal index

900 VRAT S. Conception of space (Dik) in the Vakyapadiya. JASBL. 1957, 23; (2) : 21-26. Ref 20 KEYWORD: TIME

CONTENTS: 1. Space to be a power or force along with time 2. Space infinite and finite things 3. Bhartrhari idea of space 4. Advaitin's view of space

901 CHAUDHURI AKR. Does memory yield true and valid knowledge. JASBL. 1957, 23; (2) : 27-34. Ref 5

CONTENTS: 1. True knowledge 2. Truth and validity of memory 3. Jayanta, view of memory. Udayana and Hamilton, Loird etc. Sridhara, Prabhachandra etc.

902 SENGUPTA M. Porcelain in prehistoric India. JASBL. 1957, 23; (2) : 35-42.

CONTENTS: 1. Indus valley culture 2. Mohenjo-daro and Harappa 3. Collection of Indian Museum

903 SWAMY BGL; GOVINDARAJALU E. Anatomical studies in polyploid strains of Parthenium argentatum. JASBSC. 1957, 23 : 43-54. Ref 3 KEYWORD: BOTANY

CONTENTS: 1. Methods and materials 2. Gross anatomical features, minute anatomical features 3. Chromosome study

904 VARMA KM. Animal Anatomy. IHQ. 1957, 33; (4) : 342-48. Ref 6 KEYWORD: ANIMAL

CONTENTS: 1. As a branch of study in Indian art 2. Angula (as a basic unit of measurement) not a standard but a relative 2. Shape and features 3. Methods of measurement

905 GHOSHAL SN. Note on the Eastern and Western Manuscripts of the prakrit-paingala. IHQ. 1957, 33; (1) : 50-61. Ref 7

CONTENTS: 1. List of conjunct-words and sounds in manuscripts

906 THAKUR W. Development and growth of Tantrick religion in Mithila. IHQ. 1957, 34; (1) : 193-204. Ref 32 KEYWORD: MYSTICISM

CONTENTS: 1. Medieval period 2. Theory of Tantra 3. Historical aspects 4. Worship and Tantras

907 SARTON G. Publication of ISIS. ISIS. 1957, 48; 153 : 338. KEYWORD: INFORMATION SCIENCE

908 RICE DS. Early signed Islamic glass (8-9thC.). JRAS. 1958, 1 : 8-16. Ref 23 KEYWORD: GLASS-ISLAMIC

CONTENTS: 1. Glass work 2. Coloured glass, blue,green,yellow 3. Glass window

909 JAYASUNDARA A. Sinhala tree press. JASBL. 1958, 24 : 53-54. KEYWORD: PRESS

CONTENTS: 1. Trees press for extracting oil 2. Parts of tree press description 3. Method of setting tree press

910 NAVLAKHA S. Oil press among Bhils of Rajasthan. JASBL. 1958, 24 : 55-56. KEYWORD: PRESS

CONTENTS: 1. Chatawa-Bhil oil press 2. Use of oil press and its description

911 CHATTOPADHYAY KP. Bondo tree press and its links. JASBL. 1958, 24 : 7-12. Ref 12 KEYWORD: PRESS

CONTENTS: 1. Oil press, plank press etc. 2. Tree press in Santal villages in Muruda Pargana, Mayurbhanj 3. Bastor tree press 4. Distribution of tree press and plank press in India used for extracting oil from various kinds of seeds, roots, bark and herbs of medicinal value etc.

912 KANJILAL DK. Ancient Indian Geography in Banabhatta. IHQ. 1958, 34; (2) : 123-38. Ref 110

CONTENTS: 1. Rivers Ajiravati and metropolis Sthanvisvara 2. Punjab as (Aratta in Arthasastra, Takka, cheh-ka by Hiuen tsang 3. Kamboj, Sindhu, and Sauvira Sravasti, Mandara, Cakora etc.

913 GUERLAC H. Award of the George Sarton Medal to Lynn Thorndike. ISIS. 1958, 49; 156 : 107-08. KEYWORD: HISTORY OF SCIENCE - INTERNATIONAL; THORNDIKE L

CONTENTS: 1. Work of Thorndike L 2. G Sarton Medal History

914 . George Sarton Memorial Foundation, Inc.. ISIS. 1958, 49; 157 : 342. KEYWORD: SARTON G

915 . Indian Society for the History of Science. ISIS. 1958, 49; 155 : 76. KEYWORD: HISTORY OF SCIENCE - INDIAN SOCIETY

916 . History of Science Society: The idea and Henery Schuman Prize in the History of Science. ISIS. 1958, 49; 156 : Last Page. KEYWORD: HISTORY OF SCIENCE - INTERNATIONAL

917 BASU SK. Finger dermatoglyphic study of the riangs of Tripura. JAS. 1959, 1; 2 : 161-68. Ref 6 KEYWORD: DERMATOGLYPHY - FINGER; FINGER

CONTENTS: 1.Distribution of papillary patterns 2.Whorls and Loops (R+U) in different digits 3.Genotype, zygote and V, R, U factors

918 BANERJEE NR. Technique of the manufacture of stone beads in ancient Ujjain. JAS. 1959, 1; 2 : 189-96. Ref 25 KEYWORD: BEADS

CONTENTS: 1.Literature citing stone beads works 2.Shapes in stone beads, photographs given 3.Stages of stone beads formations

919 CLAUSON G. Turkish numerals. JRAS. 1959, 1 : 19-31. KEYWORD: NUMERALS-TURKISH

CONTENTS: 1. Numerical series in Turkish. a. Cardinals b.Ordinals c.Distributives e.Collectives 2. Inter-vocalic consonants in numerals 3. Units and corresponding tens-relationship 4. Quinqual on a decimal system 5. Mongolian words of numerals

920 PAL N. Floral organogeny and development of male and female gametophytes in two species of

Scrophulariaceae. JAS. 1959, 1; 3 : 245-49. Ref 28 Calcutta; Biology Department, Maulana Azad College KEYWORD: ORGANOGENY - FLORA

CONTENTS: 1.Materials and methods - buds and mature flowers 2.Organogeny of flower 3.Microsporogenesis and development of male gametophyte

921 KACHROO P. Aquatic vegetation of Damodar valley.I.Phanerogamic flora of freshwater ponds and marshy lands with particular reference to Damodar-Eden canal area of west Bengal. JAS. 1959, 1; 4 : 271-98. Ref 9 New Delhi,; Indian Council of Agricultural Research KEYWORD: PHANEROGAMS

CONTENTS: 1.Damador valley: description,vegetation-acquatic & semi-aquatic habitats 2.Depth of water nature of substratum in relation to aquatic vegetation 3.Pond contour in relation to marginal flora 4.Plant associations and changes in local pond flora 5.List of aquatic and semi-aquatic plants of lower Damodar valley

922 KACHROO P. Aquatic vegetation of Damodar Valley II. further observations on association of aquatic vegetation with anopheline breeding within Damodar-Eden canal area of West Bengal. JAS. 1959, 1; 4 : 299-324. Ref 19 New Delhi; IARI.

CONTENTS: 1. Classification of ponds in relation to anopheline breeding 2. Algal flora in relation to anopheline breeding 3. Anopheline breeding within Damodar- Eden canal area 4. Temperature and pH tolerance 5. List of algae collected from districts Burdwan & Hoogly

923 KOROLKOV D; SOBOLEVA N. New Data on the radio waves form sunspots. JAS. 1959, 1 : 3-7.

CONTENTS: 1. Polarization 2. Measurements of fields in sun spots

924 BHARGAVA KN; MALVIYA G. Variations in the third metatarsal bone of man. JAS. 1959, 1; 4 : 325-29. Ref 9

CONTENTS: 1. Variations in basal articular surface for lateral cuneiform bone,medial side of base, lateral side of base, site & number of nutrient foramina,surface & borders of shaft, configuration of head.

925 DAVID A. Effects of Calcutta sewage upon the fisheries of the kulti estuary and the connected cultivable fisheries. JAS. 1959, 1; 4 : 339-63. Ref 32 Allahabad; Central Inland fisheries Research Sub-station

CONTENTS: 1. Hydrographical features of Kulti Estuary 2. Mode of sewage dischange & dissipation 3. Methods & materials 4. Physico-chemical nature of affected water 5. Biotal studies, Plankton, Lirroral and burrowing forms 6. Effects of sewage pollution, fish-life & fisheries, Availability of fish 'seed' & cultivated fisheries

926 MAJUMDAR RC. Eras in Nepal. JAS. 1959, 1 : 47-49. Ref 13

CONTENTS: 1. Era No. 1 386 of Manadeva 2. Era No. II AD 879 current Newan Samvat

927 MITRA E (EVA). Practical methods for controlling the water weeds in fishery tanks. JAS. 1959, 1 : 51-71. Ref 27 KEYWORD: WEED - WATER; FISH

CONTENTS: 1. Recent work on the problem 2. Use of poisonous chemical 3. Observations on plants, plankton, quality of water fish - By means of manual labour, mechanical means, drying the tanks, biological

928 SINHA BM. Endoskeleton of Wallago attu (Bl. & Schn.) Part I- the skull. JAS. 1959, 1 : 9-22. Ref 25

CONTENTS: 1. Skull - cranium, occipital region, auditory region, sphenoidal, orbital, ethmoidal ; Visceral skeletal, mandibular arch, hyoid arch, branchial arches

929 STERN SM. Notes on Al-kindi's treatire on definitions. JRAS. 1959, 1 : 32-43. Ref 64

CONTENTS: 1. Gives definitions & description of things Mss(FT Hududal-Ashya'wa-Rustimiba) 2. Matter,Element,action,instance,Space,quantity & quality 3. Text & Translations gives

930 DUNLOP DM. Translations of Al-Bitriq and Yahya (Yunanna) b.AL-Bitriq. JRAS. 1959, 2 : 140-5.

CONTENTS: 1. Metioned original mss. in science & medicine 2. Aristotle translation & commentaries are mentioned

931 MUKHERJEE BN. Critical examination of Alberuni's charges against the copyists of the Puranas. IHQ. 1959 , 35; (2) : 132-38. Ref 29 KEYWORD: AL-BIRUNI

CONTENTS: 1. Puranas have come to us people in their manuscripts forms (these manuscripts were written by copyists or scribes who were man of little or of no learning thus misunderstood & omitted many things 2. Difference between the mss. of puranas concerning the same subject .3 Error in geographical section given various examples of river, places, people etc.

932 MISHRA V. Ancient Mithila, its boundary, names and area. IHQ. 1959, 35; (2) : 151-65. Ref 58

CONTENTS: 1. Studies through texts. Buddhist Puranas etc.

933 BHATTACHARYYA VC. Magical Kamya-rites in the Samavidhana-brahmana. IHQ. 1959, 35; (4) : 312-26. Ref 32 KEYWORD: MYSTICISM

CONTENTS: 1. Ayusya-section or the section on longevity 2. Section on fortune 3. Section on material wealth (fame, prosperity heavenly enjoyment etc. 4. Section on prescience (rites to know before hand life, wars, disputes)

934 . Pfizer award and publication program. ISIS. 1959, 50 ; 160 : 151. KEYWORD: HISTORY OF SCIENCE - PRIZE

935 JARCHO S. Origin of the American Indian as suggested by Frag Joseph de Acosta (1589). ISIS. 1959, 50; 162 : 430. KEYWORD: KNOWLEDGE TRANSMISSION; AMERICA

CONTENTS: 1. How Indian travelled to America : land, sea descriptions/classifications 2. Peruvian Indians 3. Old world to New world

936 WILCZYNSKI JZ. Presumed Darwinisim of Alberuni eight hundered years before Darwin. ISIS. 1959, 50; 162 : 459-66. Ref 22

CONTENTS: 1. Darwin and his consulted work in "Origin of Species" 2. Alberuni's works entitled "India" discussed for Darwinisim 3. Hindu tradition - history of mankind 4. Hindu classification of living beings

937 WRIGHT HRC. James Augustus Grant and the Gorakhpus Opium,1789-1796. JRAS. 1960, 1 : 1-16. KEYWORD: DYE

CONTENTS: 1. Trade & EI Company policy to do trading in opium 2. Price of opium & Social relations 3. Indigo etc

938 LAW NN. Astronomical evidence on the age of the Rgveda. IHQ. 1960, 36; (suppl) : 16-72. Ref 246 KEYWORD: ASTRONOMY

CONTENTS: 1. Seasons 2. Naksatra epoch in vedic times 3. Pole star

939 WHOL R. Buffon and his project for a New Science. ISIS. 1960, 51; 164 : 186-99. Ref 48 KEYWORD: SCIENTIFIC METHOD

CONTENTS: 1. Criticism of science and scientific method 2. Buffon's distinguished two kinds of truth -mathematical, -physical 3. Senses perceptions 4. True order in world 5. Buffon's idea of species

940 . Classification scheme sought. ISIS. 1960, 51; 164 : 206-07. KEYWORD: INFORMATION SCIENCE - CLASSIFICATION SCHEME

CONTENTS: 1. American Library Association to develop classification scheme for special materials

941 BOCKSTAELE P. Note on the first arithmetics printed in Dutch and English. ISIS. 1960, 51; 165 : 315-21. Ref 10 KEYWORD: ARITHMETICS - DUTCH AND ENGLISH; KNOWLEDGE TRANSMISSION

CONTENTS: 1. First Arithmetic in English (1537) 2. Dutch (1508)- "Die maniere om te leerne cyffren na die reahte consten Algorismi"

942 HOPWOOD D. Islamic arabic manuscripts in the Mingana collection. JRAS. 1961, 2 : 100-05 .

CONTENTS: 1. Alphonse Mingana collected 2700 items now preserved in migana collection of oriental mss.in selly oak colleges Lib.44 Brimingham 2. Work of Albumasar (Abu Ma'shar (died 272/886) on Astronomy in Min.176 3. And other words on grammar, history,music etc.

943 KURIYAN GH; NAYAR SG. Common Indian fishing gear preservatives - a general account. JAS. 1961, 3; 1 : 21-37. Ref 19 KEYWORD: FISH

CONTENTS: 1. Indigenous fishing gear preservatives and their methods of their application 2. Chemical preservations 3. Coal-tar treatment of fishing gear 4. Tanning fixation process by oxidising agents 5. Tables showing different preservatives and under continuous immersion

944 MAJUMDAR B. Cinchona cultivation in India - its past, present and future: symposium on (Held 18th Sept. 1958 at Asiatic Society, Calcutta). JAS. 1961, 3; 2 : 63-80. KEYWORD: DRUG

CONTENTS: 1. Vegetable drugs - Quinine 2. Economic aspects of Cinchona, its cultivation, harvesting, drying and manufacture of quinine 3. Grafting and hybridizing in Cinchona 4. Syntheic drugs and vegetable durgs 5. History of plant, mosquito control 6. Species of cinchona

945 SINHA BM. Endoskeleton of Wallago Attu (Bl. & Schn.) Part II- the vertebral Column, median fins and appendicular skeleton. JAS. 1961, 3; 1 : 7-13. Ref 25 Meerut; Departement of Zoology, Meerut College KEYWORD: FISH

CONTENTS: 1. Vertebral column, trunk region, caudal region, ribs 2. Median fins, dorsal fin, anal fin, caudal fin 3. Appendicular skeleton, pectoral gridle, pectoral fin, pelvic girdle, pelvic fin 4. Illustration fo vertebrae, fins and girdle given

946 RAJAN KVS. Architectural affiliations and early Saurashtra Temples. IHQ. 1961, 37; (1) : 1-7. KEYWORD: TEMPLES - SAURASHTRA; ARCHITECTURE

CONTENTS: 1. Arch. description of temple 2. Description base, walls roof and pillar support etc.

947 SINGH MM. India's oversea trade as known from the Buddhist Canon. IHQ. 1961, 37; (2) : 177-. KEYWORD: TRADE - OVERSEAS

948 DIKSHIT MG. Two prakrit verses on weighing scales. IHQ. 1961, 37; (2) : 189-91. KEYWORD: BALANCES

CONTENTS: 1. Jeweller balance description 2. Naraji (Iron balance) one pan and beam graduated of 7-8th C. AD. De-ription of weights methods

949 CHAUDHURI S. Place-name in the Garuda Purana. IHQ. 1961, 37; (1) : 53-77. Ref 5

CONTENTS: 1. list of place names (alphabetically) 1-343

950 . Symposium on the History of Sciences in India. ISIS. 1961, 52; 170 : 587. KEYWORD: HISTORY OF SCIENCE - SYMPOSIUM

CONTENTS: A symposium on the History of Science in ancient and medieval India was held on August 4th and 5th, 1961 of the Bose Institute, Under the auspicious of the National Institute of Sciences of India. Inaugurated by Dr DM Bose, Director of the Institute, the symposium opened with remarks by Dr. AC Ukil about the importance recently assumed by the History of Science in India and about problems confronting workers in the field, particularly in connection with source materials.

951 . Hyderabad: Indian Institute of History of Medicine. ISIS. 1962

952 BASSETT DK. Trade of the English East India Company in Cambodia, 1651-1656 . JRAS. 1962, 1 : 35-61. Ref 107 CONTENTS: 1. Items in trade 2. Study of Cambodia items 3. Social aspects & history of English men to establish trade 4. Measures of items in EIC & Camboja

953 LARWOOD HJC. Western Science in India before 1850. JRAS. 1962, 1 : 62-76. Ref 25

CONTENTS: 1. Activities in India up to about 1850 2. Man who pursued the science in India up to about 1850 3. Observation & descriptions in the part were as scientific work 3. European writers in India giving articles in philosophical transaction before 1780 4. Indian science reveals to the west 5. Detailed account of English men & their works in India is given 6. Publications Journals/& other started publishing in India giving S&T

954 TRIPATHI MP. Geomorphological knowledge of ancient Indians. IHQ. 1962, 38; (2-3) : 168-82. Ref 113

CONTENTS: 1. Rgvedic aryans 2. Ramayana-mountains, plateau topography, text quoted. plains, and swamps and deserts 3. Underground water 4. Arthasastra and nitisastra 5. Puranas, Grammarians (Panini) 6. Manasara (earliest circa 25 B.C. work of architectural treatise) works on architecture; types of soils; smell of soils 7. Jaina works 8. Buddhist literature

955 SHASTRI AM. Bhattotpala, his date, life and writings. IHQ. 1962, 38; (4) : 247-59.

956 MAJUMDAR RC. Date of the periplus of the Erythraean sea. IHQ. 1962, 38; (2-3) : 89-97. Ref 9 KEYWORD: GEOGRAPHY

CONTENTS: 1. Geographical description of India in details 2. Comparison by Ptolemy work of geography

957 LAL RK. Place-names in the Kalacuri Records. IHQ. 1962 , 38; (2-3) : 98-110. Ref 94

CONTENTS: 1. Analysis and classification of place-names 2. Classified semantic and morphology

958 GHOSH AK,CHAUDHURI DNR. Preliminary account of bionomicsand taxonomy of Aphids from Assam-II. JAS. 1962, 4; (3-4) : 101-13. Ref 5

CONTENTS: 1. 10 species described in the text,morphological characters, locality & host plants, measurement of specimen

959 GHOSH AK; CHAUDHURI DNR. Schoutedenia Bougainvilleae (Theobald): Redescription and its synonyms. JAS. 1962, 4; 1 : 1-4. Ref 8 KEYWORD: BOUGAINVILLEAE

CONTENTS: 1. Measurement of specimen 2. Host plant, location etc., Apterous viviparious female specimen colonizing Bougainivillia in East Africa

960 GORAI AK; CHAUDHURI DNR. Food and feeding habits of Anisops bouvieri Kirk (Heteroptera: notonectidae). JAS. 1962, 4; 3-4 : 135-39. Ref 17

CONTENTS: 1. Copepods & Cladocerans an used as food by insect

961 PAL S (MRS) (SUNANDA). Prothallus of Phymatodes nigrescens (BL.) J.SM. JAS. 1962, 4; 3-4 : 151-56. Ref 17

CONTENTS: 1. Spore germination 2. Development & Structure of gametophyte 3. Prothallus,rhizoids descriptions etc.

962 SEN SN. Joseph Tieffenthaler and his geography of Hindustan. JAS. 1962, 4; 3-4 : 75-99. Ref 181

CONTENTS: 1. Tieffenthalar's Travels in India 2. Ticffenthaler's works in India 3. Bernoulli's contribution 4. Geography of Hindustan into 2 sections,Dissertation & preliminary Researches 5. Tieffenthalar's sources

963 REINGOLD N. Subject analysis and description of manuscript collections. ISIS. 1962, 53; 171 : 106-12. KEYWORD: SCIENTOMETRICS

CONTENTS: 1. Personal papers in National Union Catalog of Manuscripts 2. Description of contents of manuscript in precise/condensed form on library cards 3. Describing the items in the manuscripts

964 WOOLF H. Manuscripts and the History of Science conference on science manuscripts. ISIS. 1962, 53; 171 : 3-4. KEYWORD: INFORMATION SCIENCE

CONTENTS: 1. Held on 5-6 May, 1960. Powell Auditorium of Cosmos Club. Washington

965 BALLABH R. Research in ancient Indian Mathematics and Astronomy in India. ISIS. 1962, 53; 174 : 502-04. KEYWORD: UNIVERSITY LUCKNOW

CONTENTS: 1. 70 manuscripts in Deptt. of Mathematics, Lucknow University on mathematics and astronomy 2. Description of persons and their work on mathematics and astronomy 3. "History of Hindu Mathematics" by B Dutta and AN Singh Published in Parts I & II in 1935 and 1938 respectively 4. Research in maths and astronomy at university - publication

966 GROVER WG. Role of the archivist in the preservation of scientific records. ISIS. 1962, 53; 171 : 55-62. KEYWORD: INFORMATION SCIENCE

CONTENTS: 1. Notes of scientific works which never documented by scientist 2. Documentation - of archival records 3. National archives description

967 SHRYOCK RH. Viewpoint of an historian and a manuscript Librarian. ISIS. 1962, 53; 171 : 9-13. KEYWORD: INFORMATION SCIENCE; MANUSCRIPT LIBRARIAN

CONTENTS: 1. Historian looks for contents in record 2. Manuscript Librarian looks for preservation of record

968 AGRAWALA VS. Symbolism of the two boilers : a study of Rgveda X.114-1-10. VIJ. 1963, 1; 2 : 275-86. Ref 7 Banaras; Hindu University KEYWORD: MAN

CONTENTS: 1. Soma Ritual 2. Planes of Mind, life and Matter 3. Three lights 4. Mortal human principal 5. Immortality in Death 6. Rest and Motion

969 GUPTA PS. Sankuka : a Buddhist logician. VIJ. 1963, 1; 2 : 315-19. Ref 8 Aligarh; Muslim University KEYWORD: SANKUKA-LOGICIAN; BUDDHIST; LOGIC

CONTENTS: 1. Rasanumeyatavada or inference theory of Rasa 2. Mental state through poetical characters/actors 3. Five reasons of Sankuka to prove him as Buddhist logician

970 SARMA KV. Drgganita of Paramesvara : Critically edited with introduction. VIJ. 1963, 1; 2 : as suppt-32pp. Panjab; V.V.R.I.

CONTENTS: 1.System of astronomical computation prevalent in Kerala (parachita system of Astronomy) 2. Advent of Drggnita 4. Pramesvara's works, dates & biography present with sanskrit text

971 BUTTER BCM (OXFORD). Nephrite Jade in West Pakistan. JRAS. 1963, 2 : 130-39. Ref 20 KEYWORD: GEM STONE

CONTENTS: 1. Description of the pebbles 2. Location of the nephirite pebbles 3. Geological occurrence of nephrites 4. Origin of the nephrite pebbles 5. Occurrences of jade minerals in Asia

972 DOE DB. Pottery sites near Aden. JRAS. 1963, 2 : 150-62. Ref 9

CONTENTS: 1. Style and manufactured methods in pottery in pre-islamic period 2. Sites to the east of the Bana 3. Sites on the west of the wadi Bana

973 IMAM A. Sir Alexander Cunningham (1814-1893) The first phase of Indian archaeology. JRAS. 1963, 2 : 194-207. Ref 19

CONTENTS: 1. Development of the Science of archaeology in world & in India Comparison for its delay development in concepts & techniques 2. Burgess methods in architecture & archaeology 3. Excavations at different places enumerated

974 HALDAR DP; CHAKRAVARTY MM. Localization of alkaline phosphate in three speceis of termite flagellates. JAS. 1963, 5; 3-4 : 113-15. Ref 12

CONTENTS: 1. Intestinal flagellates of wood-eating termites 2. Enzyme and alkaline phosphate in holomastigotides and trichonymphid flagellates

975 BASU BD. Description of two spiders of the family Thomisidae (Arachnida Araneae) from India. JAS. 1963, 5; 1-2 : 23-26. Ref 14

CONTENTS: 1. Xysticus hindusthanicus (spider) description measurement, abdomen etc. 2. Thomisus beautifularis (spider) description

976 . India's New Institute of History of Medicine and Medical Research. ISIS. 1963, 54; 175 : 139-40.

CONTENTS: 1. November 15, 1962 marked the day Prime Minister Nehru of India laid the foundation stone for the Institute of History of Medicine and Medical Research

977 PINGREE D. Astronomy and astrology in India and Iran. ISIS. 1963, 54; 176 : 229-46. Ref 142 KEYWORD: NEUGEBAUER O

CONTENTS: 1. Interrelationship of Babylonian, Greek and Indian astronomy and astrology 2. Professor O Negebauer cuneiform materials/sources 3. Naksatras 4. Candragupta Maurya, Bindusara and Asoka period 5. Uni-solar calendar of Jyotisavedanga of Lagadha 6. Historical development (nvasion in India) 7. Ptolemy's time (130-160) 8. Teaching of Yavanesvara (150 AD) 9. Kalpa, Yugas etc. 10. Iranian astronomy and astrology 11. Ibn Hibinta astrologer

978 . New History of Science unit at New Delhi. ISIS. 1963 , 54; 176 : 279. KEYWORD: CSIR - HISTORY OF SCIENCE UNIT

979 MUKHERJEE SN. Sir William Jones and the British attitudes towards India. JRAS. 1964, 1 : 37-47. Ref 56

CONTENTS: 1. James mill & Jones points of ideas 2. Jones Botanical & Hindu Science ideas are given 3. Modern India moving with Western ideas & alongwith their glory part in the world etc.

980 RAMACHANDRAN KS; KRISHNAMURTI C. Hermaphroditism and early ardhanarisvara figures in India. VIJ. 1964, 2; 1 : 123-25. Ref 6 Nagupr; Archaeological Survey of India, Excavation Branch Madras; Lecturer in Archaeology, University of Madras.

CONTENTS: 1.Conception of man-woman 2. Statue of Hermaphroditos at Istanbul museum (Greek) 3. Taxila, Mathura Museum Siva etc. 4. Description of figures only

981 SRIVASTAVA B. Balances in ancient India: from the earliest times upto circa 300 AD. VIJ. 1964, 2; 1 : 131-34. Ref 22 Varanasi; Asst. Curator, Sanskrit University

CONTENTS: 1. History of the word tuta (Balance) 2. Use of balance 3. Description of length, weight of balances 4. Heavy balances six types

982 BHAN S . Late Phase of Harappa civilization. VIJ. 1964 , 2; 2 : 344-52. Ref 32

CONTENTS: 1. Metal, pottery, Terracotta bangles etc in Harappa civil 2. Pre-Harappan elements sites mentioned 3. Post Harappan elements sites mentioned

983 KESSLER C. Above the Ceiling of the outer ambulatory in the dome of the rock in Jerusalem. JRAS. 1964, 2 : 83-94. Ref 53

CONTENTS: 1. Architecture of ceiling 2. Readings on the wood inscripts on it 3. Replacement of roof given with old plank pictures

984 BORK AM. Fourth dimension in nineteenth - century physics. ISIS. 1964, 55; 181 : 326-38. Ref 49 KEYWORD: TIME

CONTENTS: 1. Time as fourth dimension of real body 2. "The Time Machine" by HC Well's (Novel) in 1895 3. Hermann Minkowski's papers on a space-time

985 MITRA E (EVA). Contributions to our knowledge of Indian freshwater plants. 4. On some aspects of the morphological and anatomical studies of turions of Hydrilla verticillata (Linn. F.) Royal. JAS. 1964, 6; 1 : 17-27. Ref 14 Barrackpore; Central Inland Fisheries Research Institute KEYWORD: PLANTS - FRESHWATER

CONTENTS: 1. Formation and types of turions, floating, underground 2. Structure and development of the turions 3. Period of maturity and sprouting 4. Germination of turions in the presence of sunlight 5. 10 plates are given

986 ROONWAL ML; BHANOTAR RK. Phase-significance of elytron-width and the length/width ratio in the desert locust. JAS. 1964, 6; 1 : 33-41. Ref 11 KEYWORD: LOCUST - DESERT)

CONTENTS: 1. Elytron-width (WI) 2. Ratio length/width (E/W) of elytron

987 BOSWORTH CE, CLAUSON G. Al-xwarazmi on the peoples of central Asia. JRAS. 1965, 1 : 2-12. KEYWORD: AL-KHWARIZMI

CONTENTS: 1. (Bosworth, "A pioneer Arabic encyclopedia of sciences at khwarizmi's keys of sciences".ISIS,LIV,1963,97-111) 2. ( Mafatihat-'ulum 977 shortly after) 3. Knowledge of peoples of central Asia 4. Translation of Al-xwarajmi's text 5. Commentary

988 SARMA KV (ED & TRANSLATOR). Grahanamandana of Paramesvara(of Kerala) (V.L. series 34) . VIJ. 1965, 3 ; 2 : 1-44. KEYWORD: KERALA; ASTRONOMY

CONTENTS: 1. Biography & works of Paramesvara 2. Grahana - Mandana - Eclipses - latitude (moon) - Half duration in eclipse -

graphical representation - deviation 3. Calculations of orbital diameters of Sun, Moon & Earth 4. Text & index, translation given

989 SOGANI KC. Fundamentals of Jaina Mysticism. VIJ. 1965, 3; 2 : 255-72. Ref 108 Udaipur,; Reader in philosophy, Rajasthan University KEYWORD: JAIN-MYSTICISM

CONTENTS: 1. Mystic way 2. Metaphysician & the Mystic 3. Dark period of self: Mithyatva Gunasthana 4. Conversion, Moral, intellectual & spiritual 5. Awakening of self 6. Mystic's, journey after spiritual coversion 7. Purgation 8. Process before higher Ascession, yogas 9. Dark night of soul after illuminations 10. Transcendental life 11. Characteristics of Arahantas, siddha state, Arhat

990 BURMAN S. Essay on man: an introduction to the philosophy of scientific humanism. JAS. 1965, 7; (3-4) : 181-203. Ref 32 KEYWORD: SCIENCE

CONTENTS: 1. Definitions of science 2. Impact of science on society 3. Evolution of man

991 SRINIVASAN KS. Indian botany in retrospect with particular reference to algal Systematics. JAS. 1965, 7; 1-2 : 49-78. Ref 71

CONTENTS: 1. Botany of 'Ancient India' (1500 BC to 200 BC) vedic and classical literature 2. Plants in Kalidasa's works -Ritusamhara, Mehaduta, Sakuntalam 3. Descriptive botany of ancient authors 4. Lower plants 5. Indian botany under western influence 6. Exploratory voyages and Indian algology (algae) 7. Koeign and algal work in 1768 South India 8. 19th and 20th centuries progress: algal groups and region 10. Description of various families of algae

992 RAYCHAUDHURI SP. Land classification in Ancient India (2500 B.C.-A.D. 600). IJHS. 1966, 1; (2) : 107-11. Ref 10 Delhi; Senior specialist (Land Resources), Planning Commission, Yojana Bhavan. 110 001 KEYWORD: SOIL

CONTENTS: 1. Based on fertility of soil in Arthasastra 2. Based on nature of the soil and climate by Caraka 3. Classification suggested by Shri Misra Cakrapani in Visva-Vallabha 4. Revenue system of land classification

993 SEN SN. Character of the introduction of Western Science in India during the eighteenth and the nineteenth centuries. IJHS. 1966, 1; 2 : 112-22. Ref 22 Calcutta.Indian Association of the Cultivation of Science

CONTENTS: 1. Social and psychological conditions of the people discussed 2. Prohibition order to instruct to any Indian of the art of any kind of surveying by East India Company. Britishers were free 3. Education dispatch of 1854 expected non-teaching characters of the universities in the field of scientific education and research 4. Efforts by Indian themselves

994 TANAKA M. Characteristics of the History of Science and Technology of Modern Japan. IJHS. 1966, 1; (2) : 123-32. Ref 2 Japan; Tokyo Kogyo Daigaku (Tokyo Institute of Technology), O-Okayama, Meguro-ku, Tokyo, Japan KEYWORD: JAPAN - HISTORY OF SCIENCE AND TECHNOLOGY

CONTENTS: 1. Characteristics and problems 2. Division of Periods of Scientific Development in Japan. Germination, Transition, Foundation, Self-standing, present time 3. Scientific activities during years until 1876 4. Science and technology during the period of foundation 5. Influences of social conditions 6. Response of scientists

995 GHORI SAK; RAHMAN A. Paper Technology in medieval India. IJHS. 1966, 1; (2) : 133-49. Ref 40 Delhi; History of Science, Medieval India Unit, National Institute of Sciences of India. 110 001

CONTENTS: 1. Development of paper industry 2. Method of manufacturing paper 3. Specialities of different types of papers made in different parts of the country 4. Maulana Azad Library,

Aligarh-Habibganj Section preserved varities of paper

996 GUPTA RK. Botanical Explorations of Victor Jacquemont (1801-32). IJHS. 1966, 1; (2) : 150-57. Ref 6 Pondichery; French Institute/Present Add: Ecologist, Central Arid Zone Research Institute, Jodhpur KEYWORD: JACQUEMONT V

CONTENTS: 1. Early years of Victor Jacquemount 2. Explorations in India ( Botanical) 3. Botanical results of exploration

997 RAHMAN A; SUBBARAYAPPA BV. Note on the Native Method of Bar Iron Production in South India (Salem Region). IJHS. 1966, 1; (2) : 158-61. Ref 2 Delhi; Research Survey and Planning Organization, CSIR, 110 001 KEYWORD: METALLURGY

CONTENTS: 1. Construction of the furnace 2. Operation 3. Yield and quality 4. Cost

998 SEN S N. Impetus theory of the Vaisesikas. IJHS. 1966, 1; 1 : 34-45. Ref 30 Calcutta. Indian Association for the cultivation of Science. 32 KEYWORD: MOTION

CONTENTS: 1. Impetus theory of motion 2. John Philoponus (6th Century A D) 3. Vaisesika theory & Prasastapada's Bhasya 4. Conception of momentum.

999 RONCHI V. New Hisory of the optical Microscope. IJHS. 1966, 1; (1) : 46-59. Ref 5 Italy Union Internationale d' Historie et de Philosophie des Sciences, Firenze-Arcetri, S.Leonardo 63.

CONTENTS: 1. History of the compound microscope - Mounfiy 1.1 Achromatic microscopes 2. Galileo, G.B. Amici Prof. Mossotti 3. Ernst Abbe's interferential theory of the formation of images in the microscope 4. Energetic Theory 5. Work of Antony was Leeuwenhoek - using only simple microscopes

1000 SUBBARAYAPPA B V. Indian Doctrine of Five Elements. IJHS. 1966, 1; (1) : 60-70. Ref 26 Delhi Survey and Planning of Scientific Research Unit. CSIR. KEYWORD: MATTER

CONTENTS: 1. Stressed that Indian Doctrine seems to be a forerunner of the theory of elements of the Greeks 2. Origins 3. Vaisesika System, Ayurveda - Atomic concept of matter 4. Comparison with Greeks.

1001 BAG A K. Binomial Theorem in Ancient India. IJHS. 1966 , 1; (1) : 68-74. Ref 13 Calcutta History of Sciences, Ancient Period Unit-II, 1 Park Street-16

CONTENTS: 1. Brief of European Binomial theorem: Pascal's Triangle 2. Discovery of the theorem in India occurs several 3. Centuries ealier than that of China, Metrical consideration - Pingala's (200 B.C.) technique in - Pingala - chandahsutra, - meru - prastara as explained by Halayudha.

1002 RENSCH B. Problems of Biological Philosophy with Regard to the Philosophy of the Upanisads. IJHS. 1966, 1; (1) : 75-81. Ref 6 Germany West. University of Munchen. KEYWORD: EPISTEMOLOGY

CONTENTS: 1. Psychistic epistemological theory and parallelism to matter. 2. Psychistic theory and upanisads 3. Process of reduction 4. Epistemological, biological and physical facts

1003 PETRI W. Uigur and Tibetan Lists of the Indian Lunar Mansions. IJHS. 1966, 1; (2) : 83-90. Institut for Geschichte der Naturwissenschaften der Universitat Iinchess. 8 Munchess 22, Deutsches Museam. West Germany KEYWORD: MONSIONS LUNAR

CONTENTS: 1. List with 27 naksatras 2. List with 28 naksatras Astronomical teachinqs in medieval control Asia 3. Uigur with sidereal month and an asterism of seven stars 4. synoptic table of the lunar Mansions

1004 ROY M. Methods of sterilization and sex determination in the Atharvaveda and in the Brhad-Aranyakopanisad. IJHS. 1966, 1; 2 : 91-97. Ref 27 KEYWORD: BIRTH CONTROL

CONTENTS: 1. Sterilization - male, - female (operation, medicinal herbs) 2. For the birth of male child

1005 BAG AK. Trigonometrical series in the Karanapaddhati and the probable date of the text. IJHS. 1966, 1; (2) : 98-106. Ref 17 Delhi; INSA KEYWORD: PIE SERIES

CONTENTS: 1. Sine and cosine series 2. Tan series 3. Date of Karanapaddhati 4. Pie series

1006 RAY P. Theory of Chemical combination in Ancient Indian Philosophies. IJHS. 1966, 1; (1) : 1-14. Ref 12 Calcutta, History of Sciences in India, Ancient period (Unit I) 1, Park Street-16 KEYWORD: CHEMICAL COMBINATION - ANCIENT

CONTENTS: 1. Evolution of matter -Atoms 2. Atomic Theory of - Kanada of Nyaya-Vaisesika system, - Buddhists, - Jains 3. Vedantic View 4. Chemical Action and Heat 5. Arrangement of Atoms in space (Molecular Configuration) 6. Indian and the Greek Atomism-Question of Priority 7. Decline of the Scientific Spirit in India.

1007 RAVETZ JR . What was 'The Scientific Revolution'?. IJHS. 1966, 1; (1) : 15-21. U K, Department of Philosophy, University of Leeds, Leeds 2

CONTENTS: 1. What it was and how it come to be 2. Space & Time of Scientific Revolution Italy, France, England. -1620-1640, Holland, Francis Bacon, Galileo,-Descartes. - Progress in sciences - Mathematics - Chemistry - Life Sciences and medicine - Oppositions to scientific ideas.

1008 HOSKIN M. Stellar distances: Galileo's method and its subsequent history. IJHS. 1966, 1; (1) : 22-29. Ref 47 UK, Cambridge University, Cambridge KEYWORD: GALILEO STELLAR DISTANCES

CONTENTS: 1. Two methods to measure distances of stars 2. Galileo's 'Dialogo' 3. History upto 1838 of the methods & techniques.

1009 PRICE D J DE S; POSPISIL L. Survival of Babylonian Arithmetic in New Guinea?. IJHS. 1966, 1; (1), : 30-33. Ref 11 USA. Department of History of Science and Medicine, Yale University, New Haven, connecticut USA. Department of Anthropology. KEYWORD: ARITHMETIC BABYLONIAN

CONTENTS: 1. Sexagesimal arithmetic 2. Counting practices / South Indian & Others.

1010 SARMA KV (ED & TRANSLATOR). Grahananyayadipika of Paramesvara (V I Series - 35). VIJ. 1966, 4; 1 : 1-40. KEYWORD: ASTRONOMY

CONTENTS: 1. Distances of Sun,Moon from centre of Earth 2. Diameters of Sun, Moon & Earth 3. Eclipes, Sine Drggati 4. Astronomical constants for computations 5. Astronomical concepts in computation of eclipses 6. Text and Translation, index given 7. Deviation, graphical representation of eclipses

1011 LAHIRI A. Indian and Babylonian Demonology :a comparative study. VIJ. 1966, 4; 2 : 186-97. Ref 126 KEYWORD: MYSTICISM

CONTENTS: 1. Works on the demonology 2. fire & figures in demonology 3. National calamities, & diseases as demonology

1012 SOMPURA KF. Buddhist Monument and sculptures in Gujarat a historical survey . VIJ. 1966, 4; 2 : 230-63. Ref 203 Ahmedabad Deptt. of Ancient Indian Culture, H.K. Arts College 9

CONTENTS: 1. Rock-cut (Viharas) 2. Caves(Caitya-Grhas) 3. Pillars, Plasters & arches in Rock

1013 YUN-HUA J. Korean record on Varanasi and Sarnath. VIJ. 1966, 4; 2 : 264-72. Ref 33 Santiniketan; Visva-Bharati University KEYWORD: KOREA

CONTENTS: 1. Travellers note and names 2. Infunction pillars & coming the enrichment of faith

1014 HEILBORN JL. GM Bose: The prime mover in the invention of the Leyden Jar?. ISIS. 1966, 57; 188 : 264-67. Ref 13 KEYWORD: BOSE GM

CONTENTS: 1. Force of electricity

1015 . Teaching the History of Science. ISIS. 1966, 57; 188 : 272. KEYWORD: EDUCATION

CONTENTS: Harvard Univeristy announces the establishment of an independent Department of the History of Science as of 1 July 1966. This is a graduate department only, offering degrees of M.A. and Ph. D. in Hisotry of Science

1016 MITRA E. Contributions to our knowledge of Indian freshwater plants. 5. On the morphology, reproduction and autecology of Pistia Stratiotes Linn. JAS. 1966, 8 ; 2 : 115-35. Ref 19 Barrackpore; Central Inland Fisheries Research

CONTENTS: 1. Habit, morphology - tuber, roots, leaves, anatomy 2. Reproduction - Vegetative and sexual 3. Seeds and their germination, role of roots 4. Physico-chemical conditions of existence, table and description 5. Soil conditions of the natural ponds

1017 TIKADER BK. Description of two new spiders of the genus Xysticus (family: Thomisidae ) from India. JAS. 1966, 8; 4 : 249-52. Ref 19

CONTENTS: 1.Xysticus pynurus sp. nov. and Xysticus joyantius sp. nov. Cephalothorax, abdomen, localities etc

1018 HALDAR A. Doctrine of sarvastivada in the light of modern philosophy and psychology. JAS. 1966, 8; 1 : 51-63. Ref 85

CONTENTS: 1.Traditional view of Sarvastivada 2.Realistic trend in sarvastivada 3.Limitation and Idealism in sarvastivada 4.Spatio-temporal nature of things 5.Pluralism and Neo-realism and sarvastivada 6.Buddhist conception of reality and its psychological significance 7.Different form of applications of Dharma/reality 8.Matter-formation(Rupa-skandha)

1019 NORMAN KR. Notes on Asoka's fifth pillar edict. JRAS. 1967, 1 : 26-32. Ref 26 KEYWORD: ZOOLOGY; ASOKA

CONTENTS: 1. Mentioned animals list 2. Categories of animals 3. Comparision with Jain lists, different copies versions of Asoka edict

1020 RAY P. Origin and tradition of alchemy. IJHS. 1967, 2; (1) : 1-21. Ref 28 Calcutta; 50/1 Hindustan Park KEYWORD: CHEMISTRY

CONTENTS: 1. Aims, ideas characteristic features of alchemy 2. Origin of the names (chemistry and alchemy) 3. Alchemy in Greece 4. Alchemy in China 5. Alchemy in Europe 6. Islamic alchemy 7. Alchemy in India (in detial)

1021 JHALA HI. W M W Haffkine, Bacteriologist- a great saviour of mankind. IJHS. 1967, 2; (1) : 105-20. Ref 42 Bombay; Haffkine Institute. 12 KEYWORD: HAFFKINE WMW; PLAGUE VACCINE; VACCINE PLAGUE CONTENTS: 1. Early life 2. Work of Pasteur Institute, Paris 3. Arrival in India (Calcutta) 4. Discovery of Plague Vaccine 5. Hurdels and handicaps 6. Tragic episode at Malkowal in Punjab (Tetanus death by Haffkine bottle) 7. Institute named after Haffkine 8. After-life of Haffkine

1022 GUPTA RC. Bhaskara I's approximation to sine. IJHS. 1967, 2; (2) : 121-36. Ref 23 Ranch; Birla

Institute of Technology KEYWORD: MAHABHASKARIYA

CONTENTS: 1. Rule 2. Equivalent forms of the rule from subsequent works 3. Discussion of the rule in five approaches

1023 ROY M. Rasarnavakalpa of Rudrayamala Tantra. IJHS. 1967, 2; (2) : 137-42. Ref 6 Calcutta; History of Sciences in India, Ancient Period, 1 Park Street. 16 KEYWORD: MERCURY; ALCHEMY

CONTENTS: 1. Chapters and subdivisions in the manuscript are mentioned 2. It deals with alchemical recipes and mercurial preparations 3. Manuscripts can be divided into three parts -Introduction; -Processes; -mineral substances, plants and the character of soils of different regions for alchemical purposes

1024 SUBBARAYAPPA BV. Estimate of the Vaisesika Sutra in the History of Science. IJHS. 1967, 2; (1) : 22-34. Ref 30 Delhi; NCCHCI. INSA KEYWORD: GREEK SCIENCE

CONTENTS: 1. Origins and probable period of the Vaisesika Sutra 2. Important layers of thought in the Vaisesika Sutra 3. Greek Science and the Vaisesika System - Physical concepts, matter, motion, atomism 4. Internalized limitation of the system

1025 ROY M. Anatomy in the vedic literature. IJHS. 1967, 2; (1) : 35-46. Ref 64 Calcutta; History of Science in India, 1 Park Street. 16 KEYWORD: HUMAN ANATOMY; VEDIC LITERATURE-HUMAN ANATOMY

CONTENTS: 1. Structure of the human body 2. Important bones and bodily parts 3. Samhitas, Brahmanas (general structure, chest, neck, back portion of the body, ribs, abdominal portion,hand) 4. Aranyakas and Upanisads - Heart, nervouş system, sense-organs 5. Human body and the related cosmic body

1026 MOHAN B. History of plus and minus sings. IJHS. 1967, 2; (1) : 47-51. Ref 11 Varanasi; Central Hindu College. Banaras. 5 KEYWORD: PLUS SING; MINUS SING

CONTENTS: 1. Egyptions (1550 BC) work plus and minus sings (Earliest) 2. Diophantus (C. 275) symbols 3. Bokhshali manuscript 4. European symbols for plus and minus

1027 CHOUDHURY M. Embryonic Development and the Human Body in the Yojnavalkya Smrti. IJHS. 1967, 2; (1) : 52-60. Ref 24 Calcutta, National Institute of Science of India, History of Sciences, Ancient Period (Unit II), 1 Park Street. 16 KEYWORD: EMBRYOLOGY

CONTENTS: 1. Date of work 2. Embryology 3. Dhatus 4. Osteology - Table in Yajnavalka Smrti 5. Vital parts of the human body - Table in Yajnavalka Smrti 6. Rasas (fluid) in the human body - Table in Yajnavalka Smrti 7. Veins, sinews, arteries, muscles and nerves - Table in Yajnavalka Smrti

1028 MARZA VD; TARNAVSCHI IT. Problem of the Fertilization and Evolution of Phanerogams in Darwin's work: - a critical study. IJHS. 1967, 2; (1) : 71-104. Ref 61 Iasi; Laboratory of Histology, Institute of Medicine and Pharmacy. Rumania. Laboratory of Morphology of Plants, University of Bucharest KEYWORD: DARWIN; FERTILIZATION - PHANEROGAMS; EVOLUTION - PHANEROGAMS

CONTENTS: 1. Biology of the sexual process in the higher plants in Darwin's work : History of the problem -Directions followed by Darwin in the study of the problem 2. Methodology Employed -Interrelationships between function and structure; -Detailed study of the entire group of plants; -Statistico- mathematical methods; -Interpretation of results; -Generalization of results 3. Evolution of sexual process in plants, reflected in their classification, according to data taken from the three works of Darwin - Hypotheses regarding sex differentiation; -Premises of Hypotheses regarding phylogenetic evolution of the sexual process in plants; -Stage of plylogenetic evolution of the sexual process in plants; -Directions of fertilization development : self-and cross fertilization

1029 MEHRA KL. History of Sesame in India and its cultural significance. VIJ. 1967, 5; 1 : 93-107. Ref 110 Simla; Central Potato Research Institute KEYWORD: PLANT; INDUS VALLEY

CONTENTS: 1.Archaeological records- Indus civilization (3600 to 1750 BC) - khokhrakot mound 2. Historical Records - vedic period (before 800 BC) - sutra period (800-300 BC) - Early Buddhist & Jain period (500-300 BC) - Maurya and surya period (300 BC-75 AD) - Epics and Manusmrti(4th cent. BC-4 cent. AD) - Kusana and saka satavahana period (75 AD - 300 AD) - caraka samhita (1st,2nd cent. AD) - Gupta period (C 300 -750 AD) - Post Gupta period (after 750 AD) 3. Sesame cultivation, oil, geographical movements etc.

1030 POTTAGE J. Vitruvian value of pie. ISIS. 1967, 59; 197 : 190-97. Ref 25 KEYWORD: PIE

CONTENTS: 1. Singer and Hero value of pie, measurement of a circle 2. Vitruvius's Hoelometer

1031 MUKHERJEE BN. Hermaphrodite figure on a silver plate. JAS. 1967, 9; 2 : 180-82.

CONTENTS: 1. North-western Turkey, Lampsacus (now in Istanbul Museum) 2. Description of plate 3. Indian view of hermaphrodite and West view

1032 RISHI K. Head skeleton of tetrodon patoca Ham. JAS. 1967, 9; (3-4) : 193-97. Ref 25 KEYWORD: FISH

CONTENTS: 1. Indian puffers 2. Cranium, mandibular arch, hyoid arch, branchial arches 3. Illustrations of head skeleton are given

1033 SARKAR H. Elliptical structures in ancient India. JAS. 1965, 7; (1-2) : 83-87. Ref 24

CONTENTS: 1. Oval or elliptical ground plan 2 Sites and places are mentioned 3. Measurement and structure is described

1034 MISRA GSP. Logical and scientific method in early Buddhist text. JRAS. 1968, 1 : 54-64. Ref 56 KEYWORD: LOGIC; BUDDHIST

CONTENTS: 1. Emphasis on experience 2. Reality and logical reasoning 3. Concept of definition (dictionary type difinitions of words) 4. Classification of objects and topics under discussion 5. Mentioned various texts dealing with subject eg. Abhidhamma Pitaka, Dhammasangani, Milindapanho 6. Aggregate of matter 7. Conditioned by ignorance, habitual tendencies is consciousness (six senses)

1035 PAL RK, MITRA BN. Structural changes in some important organs of the body in Alloxan diabetes. JAS. 1968, 10; 1-4 : 41-45. Ref 14 KEYWORD: CYTOTOXIC AGENT; TOXIC AGENT-ALLOXAN

CONTENTS: 1. Alloxan (cytotoxic agent) for islets of Langerhans of pancreas, kidneys,liver,thyroid,adenohypophy & adrenal cortex 2. Diabetic & Alloxan, Experimental methods & techniques, Histopathological study

1036 LAHIRI D. Mineralogy in ancient India. IJHS. 1968, 3; (1) : 1-8. Ref 27 Howarh; Bengal Engineering College

CONTENTS: 1. Prehistoric Period (C. 4000-1800 B.C.) -Pre-Harappan Age (C.4000-1800 B.C.); -Indus valley and contemporaneous civilization (C.2500-1800 B.C.) [excavation studies] 2. Vedic period (C.3000-600 B.C.) 3. Buddhist period (600 B.C.-A.D. 800) -Arthasastra; -Ptolemy's geography (second century A.D.) 4. Later period (A.D. 800-1200) -Garuda Purana (A.D. 900): Physical properties of minerals; -Rasaratnasamuccaya (13th century A.D.): complete course of mineralogy

1037 DAS GUPTA NN, DE ML. Electron Microscope- Its past and present. IJHS. 1968, 3; (1) : 25-41. Ref 70 Calcutta; Biophysics Division, Saha Institute of Nuclear Physics, University, 37 Belgachia Road, Calcutta-37

CONTENTS: 1. Early work on electron lenses and microscopes 2. Electron microscopy in other countries 3. Commercial production of magnetic electron microscopes 4. Electrostatic microscopes

1038 SUBBARAYAPPA BV; ROY M. Matrkabhedatantram and its alchemical ideas. IJHS. 1968, 3; (1) : 42-49. Ref 35 Delhi; National Commission for the Compilation of History of Sciences of India, Bahadur Shah Zafar Marg KEYWORD: MERCURY; CHEMISTRY

CONTENTS: 1. Alchemical ideas: a) conversion processes, b)transmutation processes 2. Date of matrkabhedatantram 3. Mercurial compounds 4. Rasalinga 5. Male-female symbolism for mercury and sulphur

1039 GRAUBARD M. Rational and irrational in the History of Science. IJHS. 1968, 3; (2) : 61-79. USA; College of Liberal Arts, Natural Science Programme, University of Minnesota, Minneapolis KEYWORD: METHOD OF SCIENCE

CONTENTS: 1. Period rationalism and its poor judgement of the past 2. Growth of science to produce date, hypothese and theories : alchemy, strology, medicine 3. Struggles for knowledge and 'truth' in the past

1040 RONCHI V. Scarcely known aspect of the activity of Leonardo DaVinci in the field of optics. IJHS. 1968, 3 ; (2) : 80-90. Italy; Istituto Nazionale Di Ottica, Arcetri Firenze KEYWORD: OPTICS

CONTENTS: 1. History of optics period 1280 A.D.-1600 A.D. to present 2. 'Codices' by Leonardo da Vinci 3. Ronchi Vasco ideas/interpretation

1041 DASH B; BASU RN. Methods for sterilization and contraception in ancient and medieval INDIA. IJHS. 1968, 3; (1) : 9-24. Ref 85 Delhi Planning Commission KEYWORD: BIRTH CONTROL

CONTENTS: 1. Historical review: mentioning important texts 2. Methods for reducing child-birth (six methods) 3. Prevention of conception both by natural and artificial means (six categories) 4. Inducing abortion or miscarriage 5. Oral contraceptives for females are dealt in details thirty two contraceptives

1042 PETRI W. Colours of Lunar Eclipses according to Indian Tradition. IJHS. 1968, 3; (2) : 91-98. Ref 18 West Germany; Institut fur Geschichte der Naturwissenschaften der Universitat Munchen, 8 Munchen 22, Deutsches Museum KEYWORD: COLOUR

CONTENTS: 1. Al-Biruni observations 2. Rahu and Ketu colours 3. Nodal point of the lunar orbit 4. Moon and solar eclipses colours

1043 MAY KO. Growth and quality of the mathematical literature. ISIS. 1968, 59; 199 : 363-71. Ref 21 KEYWORD: BIBLIOMETRICS; SCIENTOMETRICS

CONTENTS: 1. From abstracting Jl how many titles appear 2. What is its quality How many titles represent new discoveries? Would it be necessary to examine. Everything in order to retrieve the information presumably contained 2. History of determinants 3. Qualitative analysis -New ideas and results, -applications, -systematization and industry, -texts and education, -duplications, -trivia 4. History of a subtopic

1044 . New Journal - Indian Journal of History of Science. Journal of the History of Biology. 1968 1(1). from Cambridge, Masschusetts. ISIS. 1968, 59; 196 : 96. KEYWORD: HISTORY OF SCIENCE - NEW JOURNALS

CONTENTS: 1. The Indian Journal of History of Science, a semi-annual publication, first appeared in May 1966

1045 SHUKLA K. Naiskramyabhumi reconstructed. VIJ. 1968, 6; 1-2 : 101-05. Ref 26 Gorakhput,; Department of Sanskrit,University of Gorkhpur KEYWORD: NUTRITION

CONTENTS: 1. Quantity in nutrition - Defects in nutrition - fine nourishment 2. Philosophy of nutrition only points mentioned

1046 DAS SR. Discovery of two plant fossil fragments from Rajbadidaiga. JAS. 1969, 11; 1-4 : 118-19. KEYWORD: FOSSIL PLANTS

CONTENTS: 1. Plant fossils belong to moraceae family 2. Plant fossils belong to Cainizoic Era

1047 PADDAYYA K. Hagargundgi a fossiliferous middle stone age site on the Bhima river (13 miles for Gulbarga in Mysore State). JAS. 1969, 11; 1-4 : 14. Ref 9 KEYWORD: FOSSIL

CONTENTS: 1. Stratigraphy 2. Animal fossils 3. Lithic industry (flake-tool assemblages)

1048 DEB S. Thermal springs of India. JAS. 1969, 11; 1-4 : 37-41. Ref 8 KEYWORD: SPRINGS-THERMAL

CONTENTS: 1. Distribution and origin of thermal springs in India 2. Utilisations of thermal waters and gases-neon & orgon in it helium,krypton etc 3. Alkaline-saline-muriatic characters of spring water 4. Sulphur-secreting facteria (Sulphurania)-in-sulphurous thermal springs of India. 5. Applications of thermal (hot) springs - heat energy to electrical,medicinal uses,tubelight industries energy"No extensive studies of Indian thermal spring gases have been done as yet"

1049 DAS AR (COMPILER). Author and subject catalogue of the botanical and allied articles published in the periodicals of the Asiatic Society of Bengal 1788-1967. JAS. 1969, 11; 1-4 : 42-62. KEYWORD: INFORMATION SCIENCE; BOTANY-SOURCE MATERIALS

CONTENTS: Periodical catalogue— 1. Asiatic Researches-vols 1-20,1788-1836 2. Journal 1st series-vols 1-75, 1832-1904 3. Journal and proceedings 2nd series-vols 1-30,1905-34 4. Journal and year Book-3rd series vols 1-24,1935-58 5. Journal (current series) vols 1-7,1959-65 6. Memoir vols 1-12,1905-1933 7. Year Book (current series) vols 1-2,1959-60,vols 4-5, 1962-63, vols 9, 1967

1050 BONGARD-LEVIN GM. Buddhist studies in the USSR. JAS. 1969, 11; 1-4 : 6-11. KEYWORD: BUDDHISM; KNOWLEDGE TRANSMISSION

CONTENTS: 1. Manuscripts & archeological discoveries

1051 BONGARD-LEVIN GM; TYOMKIN EN. New Buddhist texts from Central Asia. JAS. 1969, 11; 1-4 : 63-67. Ref 27

CONTENTS: 1. Leningrad and others

1052 BAG AK. Problem of source materials - source materials concerning astronomy and mathematics. IJHS. 1969, 4; (1-2) : 1-4. Ref 7 Calcutta; National Institute of Sciences of India, 1 Park Street, Delhi. INSA 110 002

CONTENTS: 1. Information in the list of manuscripts 2. Information required for the prupose

1053 SHASTRI TSK. Historical development of certain Hindu astronomical processes. IJHS. 1969, 4; (1-2) : 107-25. Ref 33 Madras; 10/A Krishna Puram Street, Royapettah. 14 KEYWORD: ASTRONOMY-DEVELOPMENT(TIME

CONTENTS: 1. True longitudes 2. Equation of time 3. Computation of eclipses 4. Mahapatas (sun and moon declination) and yogas

1054 DIKSHIT GS. Sivatattvaratnakara as a source for Sciences in Ancient and Medieval India. IJHS. 1969, 4; (1-2) : 11-14. Ref 7 Dharwar; Department of History, Karnatak Univeristy

CONTENTS: 1. Describes the chapters dealing with science 2. Meterology III chapter 3. Astronomy IV chapter 4. Longevity and perpetual bodily vigour V chapter 5. Town-planning, horticulture, ayurveda, mercury and mercurial, serpents and poisons VI chapter 6. Veterinary science VII chapter

1055 SHASTRI TSK. School of Aryabhata and the peculiarities thereof. IJHS. 1969, 4; (1-2) : 126-34. Ref 15 Madras; 10/A Krishna Puram Street. Royalpettah High Road. 14

CONTENTS: 1. Length for the different eons 2. Yuga calculation (i.e. 4,320,000 years approx) 3. Revolutions of the apses and nodes 4. Number of cycles of the mean or sighra motions 5. Degree of epicycles 6. True hypotenuse in computing the equation of the centre 7. Half-equation of conjunction to the mean (mercury and venus)

1056 SHASTRI TSK. System of the vatesvara siddhanta (A.D. 904, most famous astronomical work). IJHS. 1969, 4; (1-2) : 135-43. Ref 6 Madras; 10/A Krishnapurma Street, Royalpettah High Road. 14

CONTENTS: 1. Commentary on the work and their mistakes in interpretation 2.Correct interpretations by the author 3. Fundamental constants (cycles etc) 4. Yuga calculations 5. Revolution constants 6. Spastadhikara commentary (mistakes to be rectified)

1057 BHATTACHARYYA SP; SEN SN. Ahargana in Hindu astronomy. IJHS. 1969, 4; (1-2) : 144-55. Ref 10 Calcutta; History of Science in India, 1 Park Street 16 Indian Association for the Cultivation of Science. 32

CONTENTS: 1. Number of civil days elapsed from a certain epoch up to a desired date 2. Methods to calculate the mean longitudes of planets from their aharganas as given in various texts 3. Revolution numbers of sun, moon and asterisms

1058 KASHIKAR CG. Pottery in the vedic literature. IJHS. 1969, 4; (1-2) : 15-26. Ref 23 Poona; Centre of Advanced Study in Sanskrit. Poona University KEYWORD: VEDAS

CONTENTS: 1. Dates of vedas 2. Baudhayana Srautasutra (c800 BC)(Tattiriya Recension) 3. Variations according to other vedic schools of the Taittiriya Recension 4. Bharadvaja Srautasutra (c600 BC) 5. Apastamba srautasutra (c500 BC) 6. Variations according to the Maitrayani Recension 7. Variations according to the Sukla Yajurveda 8. Other pots 9. Post-vedic sources

1059 MAHADEVAN TMP. Philosophical trends in relation to History of Sciences of India Philosophical trends v. History of Sciences of India - orthodox systems. IJHS. 1969, 4; (1-2) : 27-41. Ref 11 Madras; Centre of Advanced Study in Philosophy. University of Madras, Madras KEYWORD: PHILOSOPHY AND SCIENCE

CONTENTS: 1. Metaphysics (para-vidya) 2. Four orthodox systems do make use of scientific concepts - Nyaya-vaisesika (logic, categories, atomism etc.)

1060 PANDE GC. Philosophical trends and the History of Science in India - heterodox trends. IJHS. 1969, 4; (1-2) : 42-51. Ref 32 Jaipur; Department of History and Indian Culture, University of Rajasthan KEYWORD: PHILOSOPHY AND SCIENCE

CONTENTS: 1. Relationship between philosophy and science 2. Indian philosophy of adhyatma-vidya (cosmology) and anviksiki. nyaya-vidy or (logic) 3. Heterodox systems of philosophy -Lokayata -Buddhism -Jainism

1061 CHOWDHURY KA. Archaeological plant remains for pre-and proto-historic periods as a source of History of Sciences. IJHS. 1969, 4; (1-2) : 5-10. Ref 22 Aligarh; Aligarh Muslim University KEYWORD: PALEOBOTANY

CONTENTS: 1. Stone age sites in South India 2. Shifting cultivation 3. Mohenjo-daro archaeological remains 4. Harappa archaeological remains 5. Indus-valley archaeological remains 6. Calcolithic site archaeological remains at Navdatoli-Maheshwar in Western India 7. Cereals wheat, rice, cotton

1062 GOPAL S. Social set-up, science and technology in India, social set-up science and technology in Mughal India . IJHS. 1969, 4; (1-2) : 52-58. Ref 24 Patna; Department of History, Patna University KEYWORD: MUGHAL- SCIENCE AND TECHNOLOGY

CONTENTS: 1. Fondness for 'uniqueness' suppress scientific truth 2. European traders 3. Lack of interest of business community in production technique 4. Non-emergence of vernacular prose literature for craftsmans 5. Educational curriculum was heavily weighted in favour of theology

1063 SARASWATHI TA. Mathematics development of mathematical ideas in India. IJHS. 1969, 4; (1-2) : 59-78. Ref 23 Ranchi; Women's College KEYWORD: MATHEMATICS - DEVELOPMENT

CONTENTS: 1. Indus valley excavation - Geometry 2. Sulbasutras (c.8-5th BC)- Enunciate Pythagorean theorem (significance of square, rectangle, triangles, trapezia) 3. Bakshali manuscript Pingala's Chandahsutra (c.5th B.C.-5th A.D.) 4. Aryabhata I (c.5th A.D.) 5. Brahmagupta (7th A.D.)- geometry of cyclic figures 6. Narayana (14th century A.D.) advanced cyclic figures 7. Bhaskara II - vol. and surface area of sphere 8. Madhava (14th century A.D.)- integration, infinite series for pie, arc of a circle, sine and cosine, chords of arcs

1064 BAG AK. Sine table in Ancient India. IJHS. 1969, 4; (1-2) : 79-85. Ref 19 Delhi; INSA 110 002

CONTENTS: 1. Construction of jya table (Indian sine table) 2. Sine table by different scholars since (400 A.D) is given and discussed 3. Greek Ptolemy's sine table reviewed with Indian sine table

1065 GUPTA RC. Second order interpolation in Indian mathematics up to the fifteenth century. IJHS. 1969, 4 ; (1-2) : 86-98. Ref 33 Ranchi; Birla Institute of Technology. Mesra

CONTENTS: 1. Brahmagupta's rule for equal intervals 2. Bhaskara II's form of Brahmagupta's formula and its rationale 3. Brahmagupta's rule for unequal intervals 4. Govindasvami's rules for second order interpolation 5. Mahava's Taylor series approximation 6. Forms of various rules found in the works of Paramesvara

1066 SHUKLA KS. Astronomy in Ancient and Medieval India. IJHS. 1969, 4; (1-2) : 99-106. Ref 13 Lucknow; Department of Mathematics, Lucknow University

CONTENTS: 1. Vedic astronomy, vedanga-jyotisa (c.500BC) 2. Varahamihira's panca-siddhantika 3. Astronomical constants and their calculation 4. True geocentric longitude of a planet 5. Aryabhatiya of Aryabhata I (C.A.D. 476)(Earliest work on Astronomy) 6. Brahma-sphuta-siddhanta of Brahmagupta (A.D. 628) 7. Sisya-dhi-vrddhida of Lalla (C.A.D. 749) 8. Vatesvara-siddhanta of Vatesvara (A.D. 904) 9. Siddhanta-sekhara of Sripati (C.A.D. 1039) 10. Siddhanta-siromani of Bhaskara II (A.D. 1150)

1067 KAPOOR SC. Origins of Laurent's organic classification. ISIS. 1969, 60; 204 ; 477-527. Ref 177

CONTENTS: 1. Biography an work of Laurent's 2. Influence of Dumas - Atomism, - Electrochemical Dualism, - Radicals, - Substitutions -Tradition and analogy, -Necessity of unified concepts for all branches of chemistry, -Methodological assumption, -Duma's objections to electrochemistry 3. Influence of crystallogrphy -

Baudrimont vs electrochemistry, -theory of Hauy, and the architecture of matter 4. Laurent's classificatory scheme -theory of Laurent, -methodological assumptions, - experimental basis, -application of law of substitutions, -Persistent radicals and addition reaction, -crystallography and isomorphism, -classification and rationalism, -modification of theory of fundamental and derived radicals chemical classification 5. Laurent's classification of organic compounds into multiple functional groups (1843)

1068 KENNEDY ES; UKASHAH W. Chandelier clock of Ibn Yunis. ISIS. 1969, 60; 204 : 543-45. Ref 5 KEYWORD: TIME; CLOCK - CHANDELIER

CONTENTS: The amount of oil in the lamps is to be regulated by a numerical table which is equivalent to the Babylonian system A of daylight length.

1069 HANDA D. Proto-historic Culture of Panjab. VIJ. 1969, 7; 1 : 119-29. Ref 53 KEYWORD: SAHNI DR; PROTO-HISTORY

CONTENTS: 1.In 1921 Daya Ram Sahni excavation at Harappa in 1922 by R.D. Banerji at Mohenjodaro in Larkane District sind 2. Panjab (land of five rivers) 3. Cultural sequence and things recovered mentioned

1070 PRAKASH B. Science and technology in ancient India. Social and political influences.. VIJ. 1969, 7; 1 : 143-56. Kurukshetra,; Director, Institute of Indic Studies, Kurukshetra University

CONTENTS: 1. 6th Cen. BC- 6th century AD- India witnessed growth of imperial organization (international) relations & important development of S & T in this period 2. After 6th century AD. superiority complex developed & closed their door to new idea and notion & closed the lower classes no theory & practice 3. So Brahmana has theory but did not apply, Vaisya or Suda has craft but no advantage of theory 4. S & T under went a hibernation & soon a retrogression 5. 17th C. onwards (medieval) - sway of sentiment & superstition over reason & ratiocination, decadence S & T 6. Mentioned names of literary sources of different periods

1071 SARMA KV. Raghunatha Navahasta and his contribution to sanskrit and marathi literature. VIJ. 1969, 7; 1 : 69-82. Ref 24 Hoshiarpur; Curator, Vishveshvaranand institute KEYWORD: SANSKRIT-SOURCE MATERIALS, DIETARY, MEDICINAL; MARATHI-SOURCE MATERIALS

CONTENTS: 1.Bhojanakutuhala work mentioned in Narakavarnaa - Dietary& medicinal properties of food - Hygenic properties - Household & furniture 2. Mentioned biography & his various works

1072 CHAUDHURI S (COMPILED BY). Index to the publication of the asiatic society. Vol.1.Author Index. first supplement 1954-1968. JAS. 1970 Calcutta. 1,Park Street. The Asiatic Society. 16. KEYWORD: INFORMATION SCIENCE

1073 BHATTACHARYYA S. Idea of Ghost among children. JAS. 1970, 12; 1-4 : 143.

1074 ALLCHIN FR. Piece of scale armour from shaikhan Dheri Charsada. JRAS. 1970, 2 : 113-120. Ref 10 KEYWORD: ARMOUR - PIECE

CONTENTS: 1. Peshawar area 2. Armour used during period 1st two centuries of christian era 3. Armour plates 4. Rajasthani Helmet description

1075 MALLOWAN M. Early mesopotamian link with India. JRAS. 1970, 2 : 192-94. Ref 10 KEYWORD: MESOPOTAMIA CERAMICS POTTERY; POTTERY

CONTENTS: 1. Pained ceramics and pottery works 2. Mohenjo Dara excavation reveals and its link with Mesopotamian

1076 PIGGOTT S. Copper vehicle - models in the Indus Civilization. JRAS. 1970, 2 : 200-202. Ref 9

CONTENTS: 1. Mohenjo-daro Harappa and Chanu-dara cart models study

1077 DWARAKANATH C. Some significant aspects of the origin and development of medicine in ancient India. IJHS. 1970, 5; (1) : 1-12. Ref 33 Delhi; ICMR. 110 016

CONTENTS: 1. Ayurvedic concept of the genesis of the science of life-ayuh veda - including medicine 2. Adikala life to life under urban culture 3. Rational mediçien 4. Dating of adikala -literary evidence -Archaeological grounds

1078 MISHRA BB. Human anatomy according to the agni purana (about 10-11 centries A.D.) encyclopaedic work. IJHS. 1970, 5; (1) : 101-12. Ref 41 Ranch, Government Sanskrit College

CONTENTS: 1. Embryology 2. Physiology 3. Human anatomy 4. Osteology - table showing distribution of bones in human body in different texts 5. Arthrology (Joints of bones) 6. Myology (Muscles) 7. Vascular system

1079 CHAKRAVORTY RC. Surgical principles in the sutrasthanam of the susruta samhita : management of retained foreign bodies. IJHS. 1970, 5; (1) : 113-18. Ref 34 Calcutta; Ramakrishna Mission Sevapratishthan, Chittaranjan Cancer Hospital KEYWORD: SURGERY - PRINCIPLES

CONTENTS: 1. Diagnois 2. Methods for handling 3. Surgical manoeuvres

1080 KAPIL RN. Biological science. Biology in ancient and medieval India. IJHS. 1970, 5; (1) : 119-40. Ref 105 Delhi; Department of Botany, University of Delhi

CONTENTS: 1. Prehistoric beginning 2. Post historic period 3. Biology in vedic literature 4. Post-vedic era and the classical age 5. Medieval biology in India 6. Source materials, problems and remedies

1081 DESHPANDE PJ; SHARMA KR; PRASAD GC. Contribution of susruta to the fundamentals of orthopaedic surgery. IJHS. 1970, 5; (1) : 13-35. Ref 43 Varanasi; Department of Shalya Shalakya, Postgraduate Institute of Indian Medicine, BHU. 5. KEYWORD: SURGERY - ORTHOPAEDIC

CONTENTS: 1. Classification of skeletal injuries in susrutasamhita (c.500 B.C.) comparision with modern orthopaedic surgery 2. Fracture and rehabilitation 3. Approach of susruta in fact has not changed in principles over the period of 2,500 years, in spite of the advancement of modern orthopaedic surgery 4. Types of bones

1082 CHOWDHURY KA. Wood and its use during pre-and proto-historic time. IJHS. 1970, 5; (1) : 141-43. Ref 8 Aligarh; Department of Botany, Aligarh Muslim University

CONTENTS: 1. Indus valley civilization 2. Harappan period 3. Timber utilization 4. Hastinapura - a site of copper age civilization 5. Pataliputra 6. Vedic literature

1083 VISHNU-MITTRE. Biological concepts and agriculture in ancient India. IJHS. 1970, 5; (1) : 144-61. Ref 46 Lucknow; Birbal Sahni Institute of Palaeobotany

CONTENTS: 1. Pre-aryans concepts 2. Aryans concepts -Rksamhita -Samhitas -Upanisads and the sutras 3. General Biology 4. Classification of Plants 5. Classfication of animals 6. Agriculture

1084 ROY M. Family relations of some plants in the Atharvaveda. IJHS. 1970, 5; (1) : 162-77. Ref 101 Calcutta; Indian National Science Academy, 1 Patk

Street, NCCHSI KEYWORD: BOTANY - CLASSIFICATION

CONTENTS: 1. Rational method of classification of plants in Atharvaeda 2. Classification on the basis of features, environmental conditions, anatomical structure, and medicinal uses or what the botanist call the 'Artificial Classification'

1085 RAYCHAUDHURI SP. Agriculture. History of land use in India. IJHS. 1970, 5; (1) : 178. Delhi; Land Resources Division, Planning Commission. Res: EC9 Inder Puri, N.D.-12. KEYWORD: LAND CLASSIFICATION

CONTENTS: 1. Land classification 2. Improving the productivity of the land

1086 SHUKLA MS. Arbori - Horticulture in Ancient India. IJHS. 1970, 5; (1) : 179-82. Ref 13 Varansi; College of Indology. BHU KEYWORD: PLANT DISEASE

CONTENTS: 1. Gardens maintenance 2. Methods of propagation of plants in Arthasastra and Brhatsmhita 3. Plant diseases - Nature and treatment

1087 MUKHERJEE SK; NATH N. Horticultural science in ancient India. IJHS. 1970, 5; (1) : 183-86. Ref 10 Delhi; Division of Horticulture, IARI.12. Calcutta; Botanical Survey of India. 14 Madan Street, Calcutta.13. KEYWORD: HORTICULTURE

CONTENTS: 1. Propagation by cuttings and grafting 2. Promote inflorescence and fruiting 3. Planting distance 4. Construction of a garden (kitchen garden) 5. Moghul gardens (in Kashmir, Punjab, Delhi and Agra)

1088 LAHIRI D. Dynamic Geology in pre-historic India and vedic literature. IJHS. 1970, 5; (1) : 188-96. Ref 56 Howrah; Bengal Engineering College

CONTENTS: 1. Desert i.e. the sea of sand (Described in Vayupurana as Balukarnava) 2. Protecting the surface of the earth 3. Philosophical side of dynamic geology - water (Varuna, agni (sun) Earthquakes (incarnation of visnu) 4. Soak pits construction 5. Water percolation in roofs, walls 6. Mohenjo-daro and Harappa remains

1089 SIKDAR JC. Nine Physical concepts. Jaina Atomic theory. IJHS. 1970, 5; (2) : 199-218. Ref 142 Ahmedabad; Research Officer, L.D. Institute of Indology. 9.

CONTENTS: 1. Conception of atom 2. Analysis of concept of ultimate atom (Paramanu) from the point of view of the definition of Pudgola (matter) and its properties 3. Vibration and motion of Paramanu 4. Speed of motion of Paramanu 5. Motion of atom by touching other pieces of matter (Sprstagati of Paramanu) 6. Resistance of ultimate atom (Pratighata of Paramanu Pudgala) 7. Combination of ultimate atoms (Paramanus) 8. General rules of combination of particles of matter

1090 CHAKRAVARTHY GN. Concept of the structure of space-time : a comparative estimate between the western and the vedic view. IJHS. 1970, 5; (2) : 219-28. Ref 28 Mysore; St. Philomena's College KEYWORD: TIME-SPACE

CONTENTS: 1. Cosmogenic theories - Mechanical and structural contents 2. Four stages in the development of the theory of cosmology in the West 4. Five-dimensional universe 4. Vedic concepts concerning the solar symbol, cosmic wheel, cosmic tree and lotus in their spatial as well as temporal aspects

1091 BHARDWAJ HC. Alchemy Chemistry. Problem of advent of copper in India. IJHS. 1970, 5; (2) : 229-37. Ref 37 Varanasi; College of Indology. BHU. 5. KEYWORD: HARAPPA

CONTENTS: 1. Genesis of copper metallurgy in Ancient India 2. Location of probable copper mines exploited by Harappans 3. Impurity pattern of Harappan copper objects and copper ores 4. Origin of copper metallurgy in India

1092 AGARWAL DP. Technology. Metal technology of the Harappa culture and its socio-economic implications. IJHS. 1970, 5; (2) : 238-52. Ref 40 Bombay; Tata Institute of Fundamental Research. 5. KEYWORD: METALLURGY

CONTENTS: 1. Chemical analyses 2. Alloying 3. Metal forging techniques 3. Ecology and technology

1093 RAO BKG. Development of technology during the iron age in South India (C.1000B.C. to the beginning of the historical times). IJHS. 1970, 5; (2) : 253-71. Ref 18 Mysore; Department of Ancient History and Archaeology, Univ. of Mysore

CONTENTS: 1. Ceramics 2. Metallurgy 3. Iron 4. Copper and bronze 5. Gold 6. Beads 7. Glass and paste 8. Stone working 9. Architecture 10. Agriculture 11. Dress

1094 CHOUDHURY M. Techniques of colouring glass and ceramic materials in ancient and medieval India. IJHS. 1970, 5 ; (2) : 272-80. Ref 49 Calcutta; NCCHS. 1 Park Stree. KEYWORD: COLOUR - GLASS AND CERAMIC; CERAMICS

CONTENTS: 1. Table showing specimens of coloured and multicoloured glass 2. Glazing of ceramic materials 3. Preparation of glazing

1095 GOVIND V. Some aspects of glass manufacturing in ancient India. IJHS. 1970, 5; (2) : 281-308. Ref 169 Delhi; INSA

CONTENTS: 1. Literary sources 2. Foreign travel accounts (c. first century A.D. To A.D. 902) 3. Archaeological sources 3. Technology of glass manufacturing 4. Chemical analysis of glass specimens

1096 MAHALINGAM TV. Engineering. South Indian temple - medium of construction . IJHS. 1970, 5; (2) : 309-14. Ref 9 Madras; Department of Ancient History and Archaeology, University of Madras KEYWORD: ENGINEERING - CIVIL

CONTENTS: 1. Rock architecture (rock-cut techniques) 2. All-stone temples 3. All-brick tradition

1097 SRINIVASAN TM. Brief account of the ancient irrigation engineering systems prevalent in South India. IJHS. 1970, 5; (2) : 315-25. Ref 51 Delhi; NCCHSI. INSA. National Institute of Oceanography, Panjim, Goa

CONTENTS: 1. Nomenclature 2. Treatises 3. Engineers 3. Testimony of skill 4. Method of construction. ponds, tank 5. Constituents of a good tank 6. Faults or dosas 7. Construction of reservoirs (Big and small tanks) 8. Construction of embankments 9. Trees on the bank of tanks 10. Construction of slucies

1098 FILLIOZAT J. Transmission of science ideas and techniques influence of mediterranean culture areas on Indian science. IJHS. 1970, 5; (2) : 326-31. Pondicherry; Institut Francais d'Indologie

CONTENTS: 1. Astronomy and mathematics (yuga, zodiacal) 2. Medicine 3. Chemistry 4. Arts and techniques

1099 SEN SN. Influence of Indian science on other culture areas. IJHS. 1970, 5; (2) : 332-46. Ref 21 Calcutta; Indian Association for the Cultivation of Science. 32 KEYWORD: SCIENCE - INFLUENCE

CONTENTS: 1. Medicine - Poisons 2. Astronomy - Ayurveda

1100 VERMA RL. Growth of Greco-Arabian medicine in medieval India. IJHS. 1970, 5; (2) : 347-63. Ref 25 Delhi. NCCHSI. INSA Persian &

Arabic Scholar, Department of Anatomy and the History of Medicine. AIIMS. Delhi. 16. KEYWORD: MEDICINE - GRECO-ARABIAN

CONTENTS: 1. Foundation of a hospital 2. Unani Tibb in Bahmani Kingdom 3. Golden period in Medical history of Gujarat 4. Lodi period 5. Medical aid under Adil Shahi period 6. Medical relief during Qutab Shahi period 7. Treatment of haematuria by Simple drugs 8. Mughal period 9. Relations of Hakims and Vaidyas 10. Contribution of Tibb-e-Unani to Indian medicine

1101 SINGH LM; THAKRAL KK; DESHPANDE PJ. Susruta's contributions to the fundamentals of surgery. IJHS. 1970, 5; (1) : 36-50. Ref 24 Varanasi; Department of Shalya Shalakya, Postgraduate Institute of Indian Medicine, Banaras Hindu Univ. 5. KEYWORD: SURGERY - PLASTIC; SURGERY - WOUND

CONTENTS: 1. Training of surgery 2. Wound management 3. Pre-operative preparation 4. Operation 5. Post-operative care 6. Prevention of infection 7. Plastic surgery 8. Concept of thermal injuries

1102 SHARMA PV. Date of Dhanwantari Nighantu. IJHS. 1970, 5 ; (2) : 364-70. Ref 36 Varanasi; Dept. of Dravyaguna, Postgraduate Institute of Indian Medicine. BHU. 5.

CONTENTS: 1. External evidences 2. Internal evidences

1103 SHUKLA HC. Ideas of scientific measurement in basic principles of Ayurveda with special reference to somatometry. IJHS. 1970, 5; (2) : 371-78. Ref 23 Varanasi; Dept. of Basic Principles. Postgraduate Institute of Indian Medicine. BHU. 5. KEYWORD: MEASUREMENT; SOMATOMERTIC

CONTENTS: 1. Measurement in human biology 2. Angulipramana - a unit of measurement by Ayurveda 3. Indices - for morphological classification 3. Morphological basis of the factors like strength, happiness, longevity and prosperity 4. Table I showing measurement of body parts by Caraka Samhita, Susruta Samhita and Astanga Samgraha

1104 SRINIVASAN TM. Water-lifting devices in ancient India: their origin and mechanisms. (from the earliest times to C.A.D. 1000). IJHS. 1970, 5; (2) : 379-89. Ref 41 Goa; NCCHSI. National Institute of Oceanography, Panjim

CONTENTS: 1. Intermittent or discontinous water-supply from stream and canals 2. Semi-mechanical devices, i.e. the balanced bucket 3. Continuous water-supply by water-lifting machinery

1105 GAUR DS, GUPTA LP. Theory of pancamahabhuta with special reference to ayurveda. IJHS. 1970, 5; (1) : 51-67. Ref 27 Varanasi; Department of Basic Principles. Postgraduate Institute of Indian Medicine, BHU. 5. KEYWORD: MATTER

CONTENTS: 1. Physical reals (Bhutas), defintion, evolution 2. Pancabhuta doctrine of matter and the ayurveda 3. Genesis of the Atoms

1106 MITRA J. Methodology for experimental research in ancient India. IJHS. 1970, 5; 1 : 68-75. Ref 23 Varanasi; Post-graduate Institute of Indian Medicine, Faculty of Medical Sciences. BHU.5

CONTENTS: 1. Contribution by lower animals 2. Method of producing diseases on experimental basis 3. Experimentation of poisoned food and drinks on lower animals 4. Antidotes for diseases produced on experimental basis 5. Recipes for producing discolourations of the skin 6. Test of poisoned food and other materials 7. Trial of drugs on human beings

1107 GURUMUTHRY S. Medical science and dispensaries in ancient South India as gleaned from Epigraphy. IJHS. 1970, 5; (1) : 76-79. Ref 11 Madras; Department of Ancient History and

Archaeology, Univ. of Madras. 5. KEYWORD: EPIGRAPHY - MEDICINE

CONTENTS: 1. Sangam literature in medicine 2. Colas refer to the dispensary 3. Institutions, remuneration paid, study and teaching of medicine

1108 MURTHY RSS. Brahmanas on medicine and biological sciences. IJHS. 1970, 5; (1) : 80-85. Ref 40 Mysore; Department of Sanskrit. University of Mysore KEYWORD: MEDICINE - BRAHMANAS

CONTENTS: 1. Medicine, hygiene 2. Biological sciences - Physiology and anatomy of human being -Embryology -Botany -Zoology

1109 RAY P. Medicine - as it evolved in ancient and medieval India. IJHS. 1970, 5; (1) : 86-100. Ref 48 Calcutta; 50/1 Hindustan Park. 29

CONTENTS: 1. Chronological stages 2. Pre-vedic age (c.1000 B.C.-1500 B.C.)) Belief in planets, magic etc. 3. Vedic age (c.1500 B.C.-500 B.C.). Humoral theory 4. Post-vedic or ayurvedic age (c.500 B.C.-A.D. 600) 5. Early medieval age - age of alchemy and tantras(c. 600 AD-1300 AD) 6. Late medieval age (c.A.D. 1300-A.D. 1600) - Commentaries, systematization, stagnation 7. Decline of medicine in India reasons

1110 . Museum and library of the Institute of History of Medicine and Medical Research of New Delhi. ISIS. 1970 , 61; 206 : 113.

CONTENTS: 1. Inaugurated on 14 Feb. 1970 by the Prime Minister, Mrs. Indira Gandhi. The Institute comprises departments of theory and philosophy of medicine, rural and domestic medicine, medical education, psychological medicine, health and sanitation, food and nutrition, and pharmacy. The Institute solicits contributions to the library in the form of duplicate books and periodicals and catalogues, for which shipping charges will be borne by the Institute. The address is Tughalqabad, New Delhi-44, India

1111 SARMA KV (ED & TRANSL). Golasara of Gargya-Kerala: Nilakantha Somayaji (V.I.Series-47). VIJ. 1970, 8; 1-2 : 1-28. Hoshiarpur, Curator, Vishvesharanand institute of Sanskrit & Indological Studies KEYWORD: KERALA

CONTENTS: 1.Astronomy in Kerala - Golasara 2. Biography & works of Nilakantha Somayaji 3. Planets - oribits - revolutions - Apses - Latitudes - Diameters of sun, moon & Earth - motion in terms of signs 4. Situation of celestial sphere 5. Pythogoras - R Sines - computation of Tabuler R Sines - Arc & Circumference

1112 SHARMA BR. Cakra The symbol of eternity through the ages. VIJ. 1970, 8; 1-2 : 135-52. Ref 19 Tirupati; Director, Kendriya Sanskrit Vidyapeeth KEYWORD: TIME

CONTENTS: 1.Cakra as infinity ultimate Truth, & continous motion 2. Concept of time 3. Vedic, post-vedic, Buddhist Literature description of cakra

1113 MEHRA KL. History of Masa- pulse in India and its cultural significance. VIJ. 1970, 8; 1-2 : 217-26. Ref 62 U.P. Jhansi, Head of Division of plant improvement, Indian Grassland & fodder Research Institute KEYWORD: PULSES

CONTENTS: 1.Urud Dal (Masa-Pulse) nomenclature 2. Archaeological records 3. Historical records vedic period (before 800 BC) - Sutra period (800-300 BC) - Early Buddhist and Jain Period(500-300 BC) - Maurya and sunga period (300 BC-75 AD)Epics and Manusmrti (4th cent. BC. - 5th cent. AD) - Kusana and saka Satavahana period (75-300 AD) - Gupta period (C 300 - 750 AD) - Post Gupta period (after 750 AD) 4. Food preparations & cultivation and cultivation value of masa-pulse

1114 BHIDE VV. Vedic concept of yuga consisting of five years. VIJ. 1970, 8; 1-2 : 74-81. Ref 37 Poona; Vaidika Samsodhana Mandala Pin 30

CONTENTS: 1.Concept of Yuga - in Rgveda etc. 2. Gaturmasya sacrifices of four and five years 3. Cycle of five years

1115 O'FLAHERTY WD. Submarine mare in the mythology of Siva. JRAS. 1971, 1 : 9-27. Ref 136 KEYWORD: FIRE; WATER

CONTENTS: 1. Background symbolism of fire and water in India 2. Fire and water in mythology of Siva 3. Fiery mare: Samjna and control of tejas 4. Symbolism of horse 5. Origin of submrine mare-fire (Aurva)

1116 CHAUDHURI S. English Trade in Bengal Raw silk, (1650-1720). JAS. 1971, 13; 1-4 : 153-65. Ref 45 KEYWORD: SILK-RAW

CONTENTS: 1. Tables indicating price, Quality, total export value of Raw silk from Bengal given 2. Historical account is given of Raw silk trade in Bengal

1117 TIKADER BK. Description of some little known spiders from India of the genus Miagrammopes cambridge (Uloboridae). JAS. 1971, 13; 1-4 : 172-77.

CONTENTS: 1. Five new species of Uloborid spiders, belonging to genus Miagrammopes cambridge 2. Characters, of genus, General Description, Cephalothorax, Abdomen,Holotype,Type-locality 1. Miagrammopes poonaensis sp.nov 2. Miagrammopes indico sp.nov 3. Miagrammopes sutherlandi sp.nov 4. Miagrammopes gravelyi sp.nov 5. Miagrammopes Kirkeensis sp.nov, Illustrations are given

1118 LOEWE M. Spices and silk : aspects of world trade in the first seven centuries of the christian era. JRAS. 1971, 2 : 166-79. Ref 12 KEYWORD: TRADE - SPICES; TRADE - SILK CONTENTS: 1. China, description mainly, Rome and Han 2. Indian spices - Indian Route etc.

1119 DESHPANDE MN. Archaeological sources for the reconstruction of History of Sciences of India. IJHS. 1971, 6; (1) : 1-22. Ref 51 Delhi; Archaeological Survey of India

CONTENTS: 1. Prehistoric science and technology - early Palaeolithic - Mesolithic- middle Palaeolithic -Neolithic 2. Protohistoric science and technology 3. Science and technology of the historical period

1120 CHINTAMANI BM; SUBBARAYAPPA BV. History of Science in India - Pali Sources. IJHS. 1971, 6; (1) : 102-12. Ref 51 Delhi; NCCHSI. INSA

CONTENTS: 1. Atomism 2. Medicine 3. Astronomical ideas 4. Measurements 5. Dyeing methods 6. Architectural

1121 MUKHERJI V. Short history of the Meson Theory form 1943 to 1948. Part II. IJHS. 1971, 6; (2) : 117-34. Ref 44 Kharagpur; Department of Humanities and Social Sciences, IIT.

CONTENTS: 1. Divergence difficulty and nature of coupling 2. Meson-pair theory followed up 3. Two-mesons, mixed meson, point model etc.

1122 RAI RN. Extant Siddhanta Sekhara - an error in one of its sine values (11th century A.D.). IJHS. 1971, 6; (2) : 135-38. Ref 14 Delhi; INSA KEYWORD: SINE

CONTENTS: 1. Sine value for 78 45' is observed to be erroneous by 2' 2. Reasons for erroneous sine value

1123 RAO EGK. Exploration of undergound water springs according to the ancient Hindus. IJHS. 1971, 6; (2) : 139-46. Ref 15 Kotekar; Mangalore KEYWORD: SPRING

CONTENTS: 1. Importance of termites as a clue to underground springs 2. Association of termites and vegetation

1124 RAI RN. Ardharatrika system of Aryabhata I. IJHS. 1971, 6; (2) : 147-52. Ref 24 Delhi; INSA

CONTENTS: 1. Midnight system of reckoning 2. Calculation of ksepa quantities 3. Asronomical tables references

1125 SHARMA GP; SHARMA PV. Sivadasa Sen- A scholar commentator on Indian medicine of later medieval period (15th cent. A.D.). IJHS. 1971, 6; (2) : 153-67. Ref 14 Varanasi, Department of Dravyaguna, Faculty of Medical Sciences, BHU

CONTENTS: 1. Life and works 2. Date of Sivadasa Sen 3. Contributions in the field of Materia medica - basic concepts, description of drugs

1126 CHINTAMANI BM. Notices of thirteen MSS in Prakrt with special references to their scientific and technological contents. IJHS. 1971, 6; (2) : 168-72. Delhi; NCCHSI. INSA

CONTENTS: 1. Prakrt language 2. Thirteen manuscripts on S&T- their chapter, name and of what it deal, length of the chapter etc. have been described

1127 SENGUPTA R. Influence of certain Harappan architectural features on some texts of early-historic period. IJHS. 1971, 6; (1) : 23-26. Ref 20 Delhi; Archaeological Survey of India KEYWORD: MOHENJO-DARO

CONTENTS: 1. Cullavagga and Mahavagga of the Vinaya texts refer hot-bath tank and bath of Mojenjo-daro 2. Bituminous mortar, use of hollowed tree-trunk for conveying water, use of a windlass

1128 RAI RN. Karanasara of Vatesvara. IJHS. 1971, 6; (1) : 27-35. Ref 15 Delhi; Indian National Science Academy

CONTENTS: 1. Vatesvara Siddhant 2. Ksepa year of Karanasara 3. Revolutions of the Great Bear 4. Angular size of the earth's shadow 5. Year of composition of Karanasara 6. Birth-place of vatesvara

1129 PANIKKAR NK; SRINIVASAN TM. Concept of Tides in ancient India. IJHS. 1971, 6; (1) : 36-50. Ref 42 Goa; National Institute of Oceanography, Miramar, Panjim Delhi; NCCHSI. Marine Science Unit. INSA

CONTENTS: 1. Observations without the concept of Tides- Proto-historic period 2. Observation and concept of Tides - concept in Vedic period, Epic period, period of buddhist works, Puranic period, modern parallel - works of Kalidasa, period of Sangam literary work etc. 3. Tidal range 4. Effects of Tides 5. Utilization of Tides

1130 GUPTA RC. Fractional parts of Aryabhata's sines and certain rules found in Govindasvami's Bhasya on the Mahabhaskariya. IJHS. 1971, 6; (1) : 51-59. Ref 9 Ranchi; Department of Mathematics, Birla Institute of Technology. Mesra KEYWORD: SINE - DIFFERENCES

CONTENTS: 1. Fractional parts of Aryabhata's sine-differences 2. An approximate rule concerning the last Tabular sine-differences 3. A crude rule for computing Tabular sine-differences in the third sign (60 to 90 ) 4. Govindasvami's modified rule for computing tabular sine-differences in the third sign

1131 CAROZZI AV. Lucien Cayeux- A challenger of the principle of uniformity in Geology. IJHS. 1971, 6; (1) : 60-66. Ref 5 USA; Department of Geology, University of Illinois, Urbana, Illinois

CONTENTS: 1. Sediments - 'ancient' and 'present' causes- statements and conclusions of Cayeux- critics of Cayeux- Laffitte and Rutten

1132 SHARMA PV. Trimalla Bhatta- His date and works with special reference to his materia medica in one hundred verses. IJHS. 1971, 6; (1) : 67-74. Ref 19 Varanasi; Dept. of Dravyaguna Postgraduate Institute of Indian Medicine, BHU

CONTENTS: 1. Dravyaguna-satasloki and its different chapters name description 2. Other similar work

1133 MUKHERJI V. Short history of the Meson theory from 1935 to 1943. Part I. IJHS. 1971, 6; (1) : 75-101. Ref 70 Kharagpur; Department of Humanities and Social Sciences, Indian Institute of Technology

CONTENTS: 1. Fermi field and nuclear forces 2. Electron- positron pair theory of nuclear forces 3. Groping for a new concept (Mid-1937) 4. Yukawa hypothesis 5. Advocacy of Yukawa hypothesis 6. Invocation of the neutral meason 7. American Physical Society and Yukawa hypothesis 8. Eddington's 'Fundamental Equation' and Yukawa hypothesis 9. Problem of Meason mass 10. Bhabha's Neutral Meason Theory 11. Behte's Netural Meason Theory, coupling etc.

1134 WALKER AR. Aw Ha Khuve : the Lahu Nyi rite for the recall of a wandering sout. JRAS. 1972, 1 : 16-29. KEYWORD: THAILAND; MYSTICISM; SOUL

CONTENTS: 1. Lahu texts and translation in English to call soul 2. Theory of multiple souls 3. Given recall rites account

1135 MARR JR. Examination of some plant-names and identitites in India. JRAS. 1972, 1 : 40-56. Ref 126 KEYWORD: BOTANY; FIG

CONTENTS: 1. Burkill Isaac his specialization, literary works are mentioned 2. Tree 'Acer', fig (plantanus) 3. Ta. Palam in English, Tamil and other views 4. Black Pepper, pepper-vine, Areca palm, Acacia, Rice, Ginger, Akill, Plant names in Timil Poetry, flower, scarlet Ixora 5. Bibliography giving 56 sources

1136 SARMA KV. History of the Kerala school of Hindu astronomy in perspective (to be continued) (V.I. series - 55). VIJ. 1972, 10; 1-2 : 1-96. Hoshiarpur, ; Reader in Sanskrit, Vishveshvaranand Institute of Sanskrit and Indological studies, Panjab University KEYWORD: SCHOOL-KERALA

CONTENTS: 1. Salient features of Kerala astronomy 2. Anticipation of modern mathematical discoveries by Kerala astronomers 3. Trends in Kerala Literature of Jyotisa 4. Kerala astronomers & their works 5. Bibliography of Kerala & Kerala-based Jyotisa literature

1137 MEHRA KL. History of Masura-pulse in India and its cultural significance. VIJ. 1972, 10; 1-2 : 131-39. Ref 62 Jhansi (UP) Head of Division of Plant improvement, Indian Grassland and Fodder Research institute.. KEYWORD: PULSES

CONTENTS: 1. Masur Dal (Masura-pulse) Various names 2. Archaeological records 3. Historical records vedic period (before 800 BC) - Early Buddhist and Jain period (500-300 BC) - Maurya and sunga period (300 BC - 75 AD) - Kusana and saka satavahana periods (75-300 AD) - Gupta period (C-300 -750 AD) - Past-Gupta period (after 750 AD) 4. Cultivation & religious values & as food

1138 BAJAJ PN. Numerals - a gift of Hindu scholars. VIJ. 1972, 10; 1-2 : 140-41. Ref 13 U.S.A., Wichita State University

CONTENTS: 1.Qutations of various scholars for Hindu-numerals 2.Hindu numerals picked up by Arabs & transmitted it to Europe

1139 IYER SV. Contribution of Kerala to sanskrit sastra literature. VIJ. 1972, 10; 1-2 : 62-79. Ref 129 Trivandrum; Professor of Sanskrit, University of Kerala KEYWORD: SANSKRIT-SOURCE MATERIALS

CONTENTS: 1.Mentioned Ayurveda & Jyotisa sources from Kerala

1140 MUKERJEE SK. Fern-flora of the plains of North Bengal. JAS. 1972, 14; 2-4 : 111-31. Ref 15

CONTENTS: 1. Twenty families of fern-flora mentioned giving Habits & Habitat & examples of plants & location

1141 RAMAPRABHU T. Observations on the autecology of Ceratophyllum demersum(L) with notes on its control. JAS. 1972, 14; 2-4 : 149-62. Ref 39

CONTENTS: 1. Physico-chemical feature of environmental of certophyllum 2. Biological factors includes competition with other plants 3. Description of Habitat,morphology and reproduction ecology in plants Ceratophyllum

1142 MITRA E. Contributions to our knowledge of Inidan freshwater plants. 6. on some aspects of the habit,external morphology reproduction and autecology of spirodela polyrrhiza linni and Lemma trisulca Linn. JAS. 1972, 14; 2-4 : 29-51. Ref 30 Barrackpore, Central Inland Fisheries Research Institute

CONTENTS: 1. Chronological study of persons in the field are mentioned 2. External morphology 3. Studies under natural conditions, Experimental studies 4. Reproduction, vegetative,sexual 5. Physico-chemical conditions of existence, field studies, Laboratory experiments (three) 6. Relation of the plants to fish 7. Six tables are given.indicating chemical Analysis of water, water quality, growth in different substratum etc.

1143 JANA BB, SARKAR HL. Hydrobiology of selected thermal springs of Rajgir, Bihar. JAS. 1972, 14; 1-2 : 60-69. Ref 34 KEYWORD: SPRINGS-THERMAL

CONTENTS: 1. Study area include eight springs thermal 2. Physico-chemical characteristics of thermal springs 3. Physico-chemical,temperature, H+ ions concentration,free CO2 (carbon dioxide),Carbonate alkalinity,Bicarbonate alkalinity,chloride, phosphate,silica,dissolved oxygen etc 4. Bacteriological, polate count, presumptive coliform organisms,confirmative coliform organism,viable yeasts and molds

1144 MITRA E; BANERJEE AC; BANERJEE MK. Control of weeds by copper sulphate treatment in a sewage fed fish pond and post-treatment observations. JAS. 1972, 14; 2-4 : 91-102. Ref 7 KEYWORD: FISH

CONTENTS: Noted submerged plants- 1.Vallisneria spiralis L,Hydrilla verticillata(Linn.F) and rooted floating plant-Eichhornia crassipes solms are studied 2. Post-treatment observations, Higher aquatic plants, manuring of treated pond,chemical aspects, plankton,Bottombiota, data are given in tables

1145 TROLL CW. Note on an early topographical work of Sayyid Ahmad Khan : Asar Al-Sanadid. JRAS. 1972, 2 : 135-46. Ref 33 KEYWORD: BUILDING - BRICKS

CONTENTS: 1. Bricks in building in different ages in India 2. Changes in editions of Asar AL-Sanadid : gives buildings and architectural items in Delhi

1146 RAI RN. Sine values of the Vatesvara Siddhanta. IJHS. 1972, 7; (1) : 1-15. Ref 62 Delhi; INSA

CONTENTS: 1. Method of calculating the sine (Jya) values - relevant stanza and its translation 2. Actual values alongwith the emended text have been given 3. R sin values of ninety-six angles at intervals of 56'15" taking R3437'44"

1147 SHARMA SD. Maxima and minima of Tithis. IJHS. 1972, 7; (2) : 115-18. Ref 6 Patiala; Department of Physics. Punjab University

CONTENTS: 1. Perturbation theory 2. Surya Siddhanta - maxima & minima of tithis 3. Theory - discrepancies in Surya- Siddhanta tithis calculation 4. Extreme durations of all tithis according to Drk - School

1148 RAMAN VV. Relativity in the early twenties - many - sided reactions to a great theory. IJHS.

1972, 7; (2) : 119-45. Ref 167 USA; Physics Department. Rochester Institute of Technology. Rochester

CONTENTS: 1. High praises and compliments 2. Objections and criticisms 3. Philosophy of relativity theory 4. Relativits

1149 MUKERJI V. Historical note - the meson mass value in the history of the Yukawa theory. IJHS. 1972, 7; (2) : 146-52. Ref 28 Kharagpur; Department of Humanities and Social Sciences, IIT

CONTENTS: 1. Meson mass value 2. Cosmic-ray meson and Yukawa particle reconciliation and the meson mass value

1150 PANIKKAR NK; SRINIVASAN TM. Kappal Sattiram- A Tamil treatise on ship-building during the seventeenth century A.D.. IJHS. 1972, 7; (1) : 16-26. Ref 15 Delhi; NCCHSI. Marine Science Unit. INSA

CONTENTS: 1. Danish in Tarangambadi (Tranquebar) 2. Manuscript - Kappal Sattiram 3. Measurements of a cubit 4. How to estimate the characteristics of a ship 5. Propitious day fixed for launching the ship 6. Installation of ship's mast 7. Tutelary deity of the ship 8. Astrological instructions 9. Measurements of masts and sails 10. Sizes of anchors 11. Measurements of British ships

1151 RAI RN. Calculation of Ahargana in the Vatesvara Siddhanta. IJHS. 1972, 7; (1) : 27-37. Ref 18 Delhi; INSA

CONTENTS: 1. Methods to calculate the Ahargana (the number of civil days from a certain epoch to another date) 2. Method of Brahmadeva Ganaka to calculate Ahargana 3. Pulisa Siddhanta's Method to calculate Ahargana 4. Other methods of Vatesvara 5. Calculation of solar days from civil days

1152 SHARMA AL; SEERWANI AB; SHASTRY VR. Botany in the vedas (part I) - a critical study of references to plants, etc., and their uses. IJHS. 1972, 7; (1) : 38-43. Ref 26 Jagadalpur; Government Degree College, Beena (M.P.) Ujjain; School of Studies in Chemistry, Vikram University KEYWORD: VEDAS - BOTANY

CONTENTS: 1. Main plants and the comparison of their properties as has been revealed due to recent findings with vedic literature 2. Asvattha, Khadira, Kustha, Soma, Palasa, Nyagrodha, Pippali, Bilva, Udumbar, Apamarga, their medicinal properties

1153 SEN SN. Scientific works in sanskrit, translated into foreign languages and vice-versa in the 18th and 19th century A.D.. IJHS. 1972, 7; (1) : 44-70. Ref 20 Calcutta; Indian Association for the Cultivation of Science, Jadavpur KEYWORD: KNOWLEDGE TRANSMISSION; SANSKRIT - S & T WORKS

CONTENTS: 1. Astronomy, Mathematics and Astrology 2. Medicine 3. Translation of scientific works into sanskrit (10 approx.) 4. Table showing 52 sanskrit works, names of translators, languages of translations, references etc.

1154 BAG AK. Notes on the sine table in ancient India. IJHS. 1972, 7; (1) : 71-74. Ref 3 New Delhi; Indian National Science Academy

CONTENTS: 1. Radius (trijya) in Pancasiddhantika should be 120' and not 120' 1" which follows by its edited work 2. sanskrit verses and their english translation- sine aries, sine in Taurus, sine in Gemini 3. Varahamihira's and Ptolemy's values of R sin 0

1155 GUPTA RC. Early Indian on second order sine differences. IJHS. 1972, 7; (2) : 81-86. Ref 10 Ranchi; Assistant Professor of mathematics, Birla Institute of Technology, P.O. Mera KEYWORD: SINE

CONTENTS: 1. Proportionality factor, forms of the rule F(D1-D2)/D1 2. Nilakantha Somayaji's Aryabhatiya - Bhasya, proof of the rule

1156 SHARMA PV; SHARMA GP. Jejjata (9th century AD) and his information about Indians drugs. IJHS. 1972, 7; (2) : 87-98. Ref 23 Varanasi; Department of Dravyaguna, BHU. 5

CONTENTS: 1. Date -external evidences (from other authors work), internal evidences (from his description) 2. Place 3. Works 4. Contributions as a commentator 5. Contributions in the field of drugs

1157 BISWAS TK; DEBNATH PK. Asoka (Saraca Indica Linn.) - a cultural and scientific evaluation. IJHS. 1972, 7; (2) : 99-114. Ref 77 Varanasi; Bharat Kala Bhawan. BHU Varanasi; Department of Pharmacology, Institute of Medicnal Sciences. BHU KEYWORD: ASOKA - MEDICINE; DRUG

CONTENTS: 1. Place of Asoka tree in cultural history of India 2. Scientific evaluation - botanic description, macroscopical structure, microscopical, vernacular name of Asoka 3. Asoka in ayurvedic system of medicine, and in unani system of medicine 4. Chemical component of Asoka 5. Therapeutic uses of Asoka 6. Pharmacological investigation on Asoka 7. Adulterant of Asoka 8. Indian pharmacoepial preparation

1158 SARMA KV (ED & TRANSL). Computation of True Moon by Madhava of sangamagrama (V.I. Series - 62). VIJ. 1973, 11 : 1-56. Hoshiarpur, ; Reader in Sanskrit Vishveshvaranand institute of Sanskrit and indological studies KEYWORD: MOON

CONTENTS: 1.Moon,Sentence-Numbers 2. Zero-correction to longitudes 3. Correction for the Equation of time 2. Lump-days 3. Sphutacandapti of Madhava 4. Text & Translation given

1159 SARMA KV. History of the Kerala school of Hindu Astronomy: in perspective (Continued from 1972,10 sup 1-96)(V.I. series 55). VIJ. 1973, 11 : 97-208. KEYWORD: SCHOOL-KERALA

CONTENTS: 1. Some chapters only the bibliograbhy is present in this volume 2. Subject index given

1160 SASTRY TSK. Some mis-interpretations and omissions by Thibaut and Sudhakara Dvivedi in the Pancasiddhantika of varahamihira. VIJ. 1973, 11 : 107-18. Ref 4 Madras; Retd. Prof. of Sanskrit, 10-A, Sri Krishnapuram St. KEYWORD: DVIVEDI S; ASTRONOMY; DAY-SOLAR, CIVIL

CONTENTS: 1. Twenty six errors & their correction are made by TSKS 2. Solar day, civil days 3. Mean of sunset 4. Computation of two yogas, moon's nodes, latitudes 5. Tithis & naksatras, diameter of the circle 6. Sun's, moon's Shadow computation 7. Eclipses 8. Orbits of sun & moon 9. Diameter of sun & moon 10. venus, jupiter & saturn.

1161 SARMA KV. Decennial Index to vishveshvarnanand Indological Journal (vols 1-10, 1963-72)(V.1.Series-60). VIJ. 1973, 11 : 1-32. KEYWORD: INFORMATION SCIENCE

CONTENTS: 1.Author Index, subject Index, index of books reviewed, index of illustrations 2. Text author wise alphabetical - paper - ob. notice - Lit. Rev. - Supplements

1162 CHOWDHURY AKR (ROY CHOWDHURY AMIYA K). Trepanation is ancient India. JAS. 1973, 15; 1-4 : 203-04.

CONTENTS: 1. Trepanation of skull 2. Medicne & diseases in ancient India (before the vedic period 1500 B.C.) 3. Mohenjo-dara, Harappa & Lothal sites cemeteries studies for human skeleton

1163 RADHKRISHNAN S; BHUYAN BR. Observations on the incidence of weeds and seasonal fluctuations in abundance of Lemna minor Linn. Azolla pinnata R.Br. and Salvinia cucullata Roxb in ponds in sibsagar, Assam. JAS. 1973, 15; 1-4 : 35-47. Ref 24

CONTENTS: 1. Submerged, emersed and free floating macrophytes 2. Material & methods, Area, periods, sample 3. Observations in seven tables indicating,Average no. & wt. of plants per half, sq metre quadrant,physico-chemical conditions,Zoo phyto planlkton per litre of water etc.

1164 JAIN LC. Set theory in Jain school of mathematics. IJHS. 1973, 8; (1-2) : 1-27. Ref 114 Sehore; Department of mathematics, Govt. Postgraduate College M.P. KEYWORD: JAIN - MATHEMATICS

CONTENTS: 1. Classification of sets (14 classes of sets) 2. System of set operations 3. Comparability 4. Sequences locating all the sets 5. Symbolism for set theory

1165 SWAMINATHAN KD. Jyotisa in Kerala. IJHS. 1973, 8; (1-2) : 28-36. Ref 41 Aligarh; Centre of Advanced Study, Department of History, AMU

CONTENTS: 1. Scholars and their works 2. Bhaskara, Paramesvaran Namputiri, Madhava, Comatiri of Kelallur, Comatiri of Putumana, Sripati, Parakkad Namputiri and Acyuta Pisaroti

1166 SHARMA JN; ARORA RB. Arthritis in ancient Indian literature. IJHS. 1973, 8; (1-2) : 37-42. Ref 22 Delhi; Department of Pharmacology, All India Institute of Medical Sciences. 110016 KEYWORD: CARAK SAMHITA; ATHARVAVEDA

CONTENTS: 1. Carak Samhitas and Atharvaveda 2. Sign and symptoms of arthritis 3. Arthritis is born due to ( Tridiscordance theory (tridosa)) 4. Treatment of arthritis

1167 SHUKLA KS. Use of hypotenuse in the computation of the equation of the centre under the epicyclic theory in the school of Aryabhata I. IJHS. 1973, 8; (1-2) : 43-57. Ref 20 Lucknow; Department of mathematics, University of Lucknow

CONTENTS: 1. Tabulated manda epicycles, true or actual manda epicycles, and computation of the equation of the centre 2. Views of astronomers of the school of Aryabhata I - Bhaskara I (629), Govinda Svami (c.800-850), Paramesvara (1430), Nilakantha (c.1500), Suryadeva Yajva (b.1191), Putumana Somayaji (1732) 3. Views of astronomers of other schools - Brahmagupta, Sripati (c.1039), Aditya Pratapa, Suryasiddhanta school 4. Use of hypotenuse under the eccentric theory indispensable 5. Exceptions, Use of true manda epicycle

1168 TIWARI KK; MITRA TR. Thirteenth century Indian reference to plant Nematodes. IJHS. 1973, 8; (1-2) : 58-60. Ref 3 Calcutta; Zoological Survey of India KEYWORD: NEMATODES - PLANT

CONTENTS: 1. Sarngadhara - Paddhati's chapter Upavanavinoda refer to plant nematodes with description 2. Related stanzas and its english translation

1169 SWAMY BGL. Sources for a history of plant sciences in India I. epigraphy. IJHS. 1973, 8; (1-2) : 61-98. Ref 11 Madras; Department of botany, Presidency College KEYWORD: BOTANY

CONTENTS: 1. Land and its vegetation - cultivated plants (nine plants) 2. Public gardens and temple gardens 3. Plants products 4. Totem plants 5. Fuel 6. Poetic embellishment 7. Plants and profession

1170 ALUR KR. Faunal studies in archaeology. IJHS. 1973, 8; (1-2) : 99-113. Ref 16 Mysore; Kusum Kunj, Gandhi Chowk. Dharwar KEYWORD: FAUNA

CONTENTS: 1. Excavation of faunal remain at different centres 2. Evolution of the Manus and its significance 3. De ction from the evolution of the Manus 4. Prevalence of wild hunting and pastoral cultures

1171 ADAMSON PB. Anatomical and pathological terms in Akkadian : Part I. JRAS. 1974 : 102-06. KEYWORD: ANATOMY - TERMS; PATHOLOGY - TERMS

1172 MEHTA RN. Sandstone mines of Nathakua. VIJ. 1974, 12; 1-2 : 202-04. KEYWORD: MINE-SANDSTONE

CONTENTS: 1. Uses of sandstones 2. Knowledge of mining techinique by remains

1173 PATYAL HC. Significance of the aegle marmelos (Bilva) in the veda. VIJ. 1974, 12; 1-2 : 261-72. Ref 26

CONTENTS: 1.Wood-apple tree (bilva), religious & medicinal uses 2. Word bilva-derivation 3. Habit of the Tree 4. Implements & vessels made of the Aegel Marmelos 5. Medicinal significance. Rituals 6. Aegle marmelos as the healing one

1174 RAHURKAR VG. Use and control of Liquor in ancient India. VIJ. 1974, 12; 1-2 : 286-300. Ref 3 KEYWORD: SOMA

CONTENTS: 1.Process of preparation of Soma (Rgveda) 2. Result of drinking soma 3. Medicinal value of madya (wine)

1175 SHASTRI AM. Varahamihira and Bhadrabahu. VIJ. 1974, 12 ; 1-2 : 361-83. Ref 84

CONTENTS: 1.Bhadrabahu-samhita & Brhat-samhita relations discussed 2. Others works are listed

1176 BISWAS TK; DEBNATH PK. Asvattha (Ficus Religiosa Linn): a cultural and medicinal observation . VIJ. 1974, 12; 1-2 : 39-47. Ref 40 KEYWORD: DRUG-FICUS

CONTENTS: 1. Cultivation, worship places in India 2. Roots & barks, seeds, leaves as medicine

1177 STERNBACH L. Camphor in India. VIJ. 1974, 12; 1-2 : 425-67. Ref 161

CONTENTS: 1. Camphor from camphor trees - order Lauraceae 2. Legends about camphor 3. Camphor in sanskrit 4. Use of camphor (63 uses) - medicines - food

1178 KHARADZE EK. Copernicus and modern science. IJHS. 1974 , 8; (1) : 1-7. USSR; Academy of Sciences, Georgian KEYWORD: RELATIVITY

CONTENTS: 1. Application of the principle of relativity 2. Kinematic method 3. Conception of elliptical orbits

1179 SUNDARAMOORTHY G. Contribution of the cult of sacrifice to the development of Indian astronomy. IJHS. 1974, 9; (1) : 100-06. Ref 49 madurai; Madurai University.2

CONTENTS: 1. Period required for sacrifices 2. Ritual calendar 3. Time certainity by natural objects (sun., moon, planets etc.) 4. Astronomical knowledge in vedas 5. Sacred time

1180 SEN SN. Epicyclic eccentric Planetary theories in ancient and medieval Indian Astronomy. IJHS. 1974, 9; (1) : 107-21. Ref 24 Calcutta; Indian Association for the Cultivation of Science, Jadavpur.32 KEYWORD: GEOSTATIC MODEL; HELIOSTATIC MODEL

CONTENTS: 1. Equivalence of geostatic and heliostatic models 2. Geometric models 3. Concept of equant 4. Planetary positions - methods of calculations, use of sine function, table 5. Development among Greeks

1181 TRAUTMAN A. Copernicus and modern physics and cosmology. IJHS. 1974, 9; (1) : 11-13. Ref 3 Poland; Warsaw.Polish Acadmey of Sciences

CONTENTS: 1. Copernicus ideas to tackle the problems 2. Present research and its dependibility on the past some time proves wrong (there is no generation of new ideas except we well study the foundation of those classic theories 3. Relativistic quantum theories, analogus to quantum electrodynamics 4. Newton's law of universal attraction based on Copernicus ideas 5. Copernican principle and the cosmology

1182 ANSARI SMR. Arab antecedents of Copernican method. IJHS. 1974, 9; (1) : 122. Aligarh; Physics Department, AMU KEYWORD: PTOLEMY

CONTENTS: 1. Ptolemy (2 AD) Planetary theory 2. Copernican planetary model governed ptolemy's ideas 3. Ibn ash-shatir(1304-1375/6):secondary epicycle in copernican method 4. Al-Tusi (120-1274) and Copernican model 5. Qutub-al-din al- shirazi (1236-1311) 6. Neugebauer O

1183 PANDE MC. Pre-Copernican Astronomy in Europe. IJHS. 1974, 9; (1) : 123-37. Ref 7 Nainital; Uttar Pradesh State Observatory KEYWORD: EUROPE - ASTRONOMY

CONTENTS: 1. Real reason for beginnings of astronomy 2. Greek astronomy - Pythagorean picture of universe 3. Puzzles that remained unsolved - variations in brightness of planets, inequality of the terrestrial seasons, riddles of lunar motion, planetary motions 4. Shortcomings of ancient astronomy 5. Post-ptolemaic astronomy

1184 KAPREKAR DR. Copernicus Magic Square. IJHS. 1974, 9; (1) : 138-39. Nasik; 311 Devlali Camp, Devlali

CONTENTS: 1. Sum from all sides 594

1185 JOSHI SC. Nicolaus Copernicus(1473-1543) and the commentariolus. IJHS. 1974, 9; (1) : 14-24. Ref 7 Nainital; Uttar Pradesh State Observatory

CONTENTS: 1. Commentariolus (Heliocentric Universe):assumptions 2. Arrangement of spheres 3. Apparent motion of the sun 4. Motions of the moon 5. Three superior planets Saturn, Jupiter and Mars 6. Venus - motion in latitude 7. Mercury - motion in latitude

1186 GUPTA RC. Sines and Cosines of multiple arcs as given by Kamalakara. IJHS. 1974, 9; (2) : 143-50. Ref 6 Ranchi; Department of mathematics, Birla Institute of Technology P.O. Mesra KEYWORD: ARC - MULTIPLE

CONTENTS: 1. Formulas for double arc, triple arc, quadruple arc, quintuple arc and derivations

1187 NARLIKAR JV. Copernicus and modern astronomy. IJHS. 1974, 9; (2) : 151-57. Bombay; Jawaharlal Nehru Fellow, Tata Institute of Fundamental Research, Homi Bhabha Road.400005 KEYWORD: HELIOCENTRIC THEORY

CONTENTS: 1. Motion of heavenly bodies - venus 2. Geocentric picture of planetary motion 3. Heliocentric picture of planetary motion 4. Typical planetary orbit - Copernicus, Ptolemy

1188 KULKARNI RP. Note on the examination of soil for foundation of buildings and of township in ancient/medieval India. IJHS. 1974, 9; (2) : 158-63. Ref 57 Nasik; Maharashtra Engineering Research Institute KEYWORD: BUILDINGS - SOIL

CONTENTS: 1. Different types of sites 2. Examination of soil (colour, smell, taste, sound, temperature etc.) 3. Detailed description of field tests 4. Procedure of soil tests

1189 GUPTA RC. Addition and subtraction theorems for the sine and the cosine in medieval India. IJHS. 1974, 9; (2) : 164-77. Ref 18 Ranchi; Department of Mathematics, Birla Institute of Technology.P.O. Mesra

CONTENTS: 1. Statement of theorems 2. Method based on theory of indeterminate analysis 3. Geometrical derivation as given in NAB 4. Geometrical proof quoted in Marici commentary on Jyotpatti by Munisvara (1638) 5. Proofs found in Siddhanta Tattva-Viveka commentary

1190 MISRA A. Consciousness in plants. IJHS. 1974, 9; (2) : 178-84. Ref 21 Darbhanga; Botany Department, Mithila University, C.Msc. College.846004 KEYWORD: SENSITIVITY - PLANT

CONTENTS: 1. Sensitivity in plants - Physiology, psychology, molecular approach 2. Philosophy of consciousness in plants 3. Veda to Bose (1917)

1191 MARZA VD; TARNAVSCHI IT. Darwin's theory of unity of reacting mechanisms in plants and animals: its present day importance. IJHS. 1974, 9; (2) : 185-220. Ref 73 Iasi; Laboratory of Histology, Institute of Medicine and Pharmacy Rumania; Laboratory of Morphology of Plants, University of Bucharest KEYWORD: DARWIN - UNITY OF REACTING MECHANISMS

CONTENTS: 1. Darwin's forerunners in study of plant movements and analysis of his two works on this problem 2. Climbing plant and power of movement in plant - history and content 3. Analysis of two works - power of movements in plants, movements and habits of climbing plants

1192 CHANNABASAPPA MN. Note on Colebrooke's translation of a stanza from Bhaskara's Lilavati. IJHS. 1974, 9; (2) : 221-23. Ref 6 Karnataka; Department of Mathematics, Karanataka Regional Engineering College

CONTENTS: 1. Auhtor explained the wrong translation made by Colebrooke about Stri

1193 GHOSH SK. Nicolaus Copernicus and his heliocentric theory. IJHS. 1974, 9; (1) : 25-30. Ref 5 Santiniketan; Department of physics, Visva Bharti University

CONTENTS: 1. Early life of Copernicus 2. Copernicus astronomy - theory 3. Copernicus' arguments in preface of "De Revolutionibus" 4. Orginality of Copernicus (debts to Ptolemy) 5. Copernicus' theory and afterwards

1194 KUPPANNASASTRY TS. Main Characteristics of hindu astronomy in the period corresponding to pre-copernican European astronomy. IJHS. 1974, 9; (1) : 31-44. Madras; 10 A Srikrishnapuram, Royapettah High Road. 14 KEYWORD: PRE-COPERNICAN ASTRONOMY CONTENTS: 1. System of measuring longitudes 2. Cucles in Yugas and Kalpas 3. Concept of precession of the equinoxes 4. Eccentric and epicyclic theories of planetary motion 5. Concept of the sidereal day 6. Certain specialities in the use of trigonometry in astronomical work

1195 RAI RN. Prime Meridian in Indian astronomy. IJHS. 1974 , 9; (1) : 45-50. Ref 20 Delhi; 10 A/4 Shakti Nagar

CONTENTS: 1. Location of cities on prime meridian 2. Longitudes and latitudes of cities on prime meridian

1196 CHATTERJEE B. Glimpse of Aryabhata's theory of rotation of earth. IJHS. 1974, 9; (1) : 51-55. Ref 17 Delhi; Indian National Science Academy KEYWORD: EARTH ROTATION

CONTENTS: 1. Aryabhatiyam - Rotation of earth (1st chapter) 2. Critics - Varahamihira, Brahmagupta, Lalla, Sripati, Bhaskaracarya II 3. Prthudaka

1197 PENDSE CG. Copernican Helicentric Theory(b.1473 Feb 19, d.1543 May 24) and Indian astronomy. IJHS. 1974, 9 ; (1) : 56-61. Poona; Maharashtra Association for the Cultivation of Science. 4

CONTENTS: 1. Copernicus, Tycho Brahe and Kepler 2. Indian astronomy and Aryabhata

1198 SHUKLA KS. Pancasiddhantika of Varahamihira(1). IJHS. 1974, 9; (1) : 62-76. Ref 11 Lucknow; Department of mathematics. University of Lucknow KEYWORD: ASTRONOMY

CONTENTS: 1. Visnucandra and Romaka criticized by Paulisa 2. Declination table of Varahamihira 3. Fifth correction for mercury and venus in the old Suryasiddhanta 4. Traditional correction of the Paulisa school for the longitude of the moon's ascending node

1199 DHAVALE DG. Kapitthaka of Varahamihira. IJHS. 1974, 9; (1) : 77-78. Ref 4 Poona; 850/4, Shivaji Nagar

CONTENTS: 1. Location of Kapitthaka (place name)

1200 GANGADHARAN N. Pre-Copernican Astronomy as known to the Puranas. IJHS. 1974, 9; (1) : 79-85. Ref 31 Madras; Department of sanskrit, University of Madras KEYWORD: POLE STAR

CONTENTS: 1. Movement of sun 2. Position and role of pole-star 3. Distances of planets from earth

1201 SMAK J. Copernicus and his theory. IJHS. 1974, 9; (1) : 8-10. Poland.; Warsaw. Institute of Astronomy, Polish Acadmey

CONTENTS: 1. General intellectual and cultural background on which copernicus appeared 2. Heliocentric theory of Copernicus 3. Signification of the work of Copernicus (applications) - Copernican revolution

1202 GUPTA RC. Solution of the astronomical triangle as found in the Tantrasamgraha (AD 1500). IJHS. 1974, 9; (1) : 86-99. Ref 7 Ranchi; Department of mathematics, Birla Institute of Technology, Mesra

CONTENTS: 1. Spherical Triangle 2. Ten cases for solving astronomical triangle

1203 SHARMA PV. Pseudo-Harita Samhita. IJHS. 1975, 10; (1) : 1-8. Ref 50 Varanasi; Department of Dravyaguna, BHU KEYWORD: HARITA SAMHITA - PSEUDO

CONTENTS: 1. Title of Harita Samhita 2. Contents - peculiarities of subject matter, language and style, date

1204 SHASTRI AM. Sanskrit literature known to Al-Biruni. IJHS. 1975, 10; (2) : 111-38. Ref 97 Nagpur; Department of Ancient Indian History and Culture, Nagpur University

CONTENTS: 1. Sanskrit works refered in Al-Biruni works 2. Vedas 3. Puranas -18, Epics, Dharmasastra (smrtis), philosophy, grammar, metrics, astronomy and astrology (46 authors and their works), medicine, fables (pancatantra, Al-Biruni translated it into Arabic), alchemy

1205 GHAYASUDDIN. Varahamihira, the best sanskrit sources of Al-Biruni on Indian jyotisa. IJHS. 1975, 10; (2) : 139-52. Ref 48 Hyderabad; Henry Martyn Institute

CONTENTS: 1. Varahamihira's life and works 2. Three branches of jyotisa 3. Pancasiddhantika, Pulisasiddhanta, Brhat-samhita 4. Degree of star - Al-Kawakib (Grahas), Al-Buruj (Rasis), Al-Buyut (Grhas)

1206 NAYAR BK. Al-Biruni and the authorities on sanskrit prosody. IJHS. 1975, 10; (2) : 153-58. Ref 31 Delhi; INSA

CONTENTS: 1. Chapter XIII of Al-Biruni's India, where in Al-Biruni mentioned sanskrit authorities

1207 RAMAN VV. Permeation of thermodynamics into nineteenth century chemistry. IJHS. 1975, 10; 1 : 16-37. Ref 139 USA; Professor of Physics and Humanities, Rochester Institute of Technology, 1 Lomb Memorial Drive, Rochester, New York

CONTENTS: 1. Role of heat in chemistry 2. Affinity and thermochemistry 3. Energy conservation principle and chemistry 4. Second law of thermodynamics 5. Difficulties with Thmsen-Berthelot principle 6. Duhem's thermodynamic potential 7. Nernst and third law 8. Thermodynamics and physical chemistry

1208 CHATTERJEE B. Al-Biruni and Brahmagupta. IJHS. 1975, 10; (2) : 161-65. Ref 3 Delhi; Adya Jha Hostel, C3, SDA Hauz Khas

CONTENTS: 1. Quotations from Brahmasphuta siddhanta and Khandakhadyaka of Brahmagupta (7th century AD) used by Al-Biruni in his "India"

1209 RAI RN. Al-Biruni and Indian eras. IJHS. 1975, 10; (2) : 166-73. Ref 20 Delhi; INSA

CONTENTS: 1. Historical eras -existence of Brahman, Kalpa, manvantava, caturyuga, Kaliyuga, Pandava-kala 2. Eras in actual use -Sri Harsa, Vikramaditya, Saka, Valabha and Gupta 3. Era of astronomers

1210 BAG AK. Al-Biruni on Indian arithmetic. IJHS. 1975, 10 ; (2) : 174-84. Ref 29 Delhi; INSA

CONTENTS: 1. Numerals 2. Decimal place-value system 3. Pati and use of dust numerals 4. Indian arithmetic in Arabia 5. Rule of three 6. Fraction

1211 SEN SN. Al-Biruni on the determination of latitudes and longitudes in India. IJHS. 1975, 10; (2) : 185-97. Ref 46 Calcutta; Indian Association for the Cultivation of Science

CONTENTS: 1. Latitudes - "Tarikh al-Hind" (calculation) 2. Longitudes - "India" (calculation)

1212 ANSARI SMR. Physical research of Al-Biruni. IJHS. 1975 , 10; (2) : 198-217. Ref 50 Aligarh; Physics Department,AMU

CONTENTS: 1. Determination of specific gravity 2. Physical world - view and methodology, manuscripts data calculations and discussion of results, nature of heat and light, philosophy and methodology

1213 ROY S. Al-Biruni and Hindu speculations on Gravitation. IJHS. 1975, 10; (2) : 218-23. Ref 16 Delhi; Department of Economics, Delhi University

CONTENTS: 1. Terrestrial speculations -Aryabhata, Varahamihira, Bhaskara and their commetators 2. Al-Biruni's own reactions to these speculations are examined

1214 MANIKAR VB. Metrology in Al-Biruni's "India". IJHS. 1975, 10; 2 : 224-29. Ref 9 Delhi; Director, Weights and Measures, Ministry of Industry and Civil Supplies KEYWORD: TIME

CONTENTS: 1. Yojana - Base-unit for astronomy 2. Units of time 3. Weights and dry measures

1215 ALAVI SMZ. Al-Biruni's contribution to physical geography. IJHS. 1975, 10; 2 : 230-34. Ref 13 Aligarh; AMU KEYWORD: GEOGRAPHY - PHYSICAL

CONTENTS: 1. Sculpturing of earth's surface 2. Origin of landforms 3. Climatic changes 4. Origin of lakes and inland seas 5. Earth's equilibrium - origin of ocean basin, tidal phenomena, concept of antipodes 6. Physical geography of India - northern mountain chain, origin of great India plain, physical setting: frontiers, rivers, mountains and climate

1216 PANIKKAR NK; SRINIVASAN TM. Al-Biruni and the theory of tides. IJHS. 1975, 10; 2 : 235-41. Ref 27 Goa; History of Marine Sciences Unit of INSA, National Institute of Oceanography, Panjim KEYWORD: TIDES - THEORY

CONTENTS: 1. Constancy of volume of ocean waters 2. Theory of tides 3. Beliefs and legends 4. Tidal ranges 5. Somanath and the "Ebb and Flow"

1217 AHMAD SM. Al-Biruni as a synthesizer and transmitter of scientific knowledge. IJHS. 1975, 10; (2) : 244-48. Ref 14 Aligarh; Centre for West Asian Studies. AMU KEYWORD: KNOWLEDGE TRANSMISSION

CONTENTS: 1. Growth of science in -India, Islamic East, Europes up to (10th Century AD) 2. Position of Al-Biruni as a scientist and his role

1218 NAYAR BK. Al-Biruni and science communication in Sanskrit. IJHS. 1975, 10; (2) : 249-52. Ref 10 Delhi; INSA

CONTENTS: 1. Poetic presentation of scientific knowledge so that being learnt by heart 2. Metrical form 3. Advantages and disadvantages of poetic rendering of authoritative works - Al-Biruni 4. Use of synonyms

1219 ASIMOV MS. Al-Biruni's astronomical treatise in the Dari language. IJHS. 1975, 10; (2) : 254-56. USSR; Academy of Science of Tadjikistan, Dushanbeh

CONTENTS: 1. Dari language (Parsiyi-i Dari) emerged, alongside Arabic as international language of arts and science 2. Al-Biruni's works in Dari language. Description of chapters -the book of learning the elements of astrology, 'Astrology'

1220 HABIBI AH. Where was Al-Biruni situtated?. IJHS. 1975, 10; (2) : 257-58. Afghanistan; Tarith Tolana, Afghan Historical Society, Kabul

CONTENTS: 1. Birun place 2. Meaning of Birun 3. Geographic situation of Birun

1221 NAYAR BK. Al-Biruni and the arithmetical sequence of the sanskrit ganas. IJHS. 1975, 10; 2 : 259-70. Ref 18 Delhi; INSA

CONTENTS: 1. Al-Biruni's "India" - chapter XIII deals with grammatrical and metrical literature of Hindus 2. Long and short letters or sounds 3. Doublets and triplets 4. Calculation of representation in sequence 5. Modes of representations 6. Prastara of metre having one, two, three and four letters 7. Binary system 8. Application of rule quoted by Al-Biruni 9. Al-Biruni and metrology 10. Translation of Al-Biruni's India

1222 BHAR MR. Al-Biruni's treatment of the Laghujataka and comets:a critique. IJHS. 1975, 10; 2 : 271-76. Ref 6 Delhi; Department of sanskrit, Hindu College, University of Delhi

CONTENTS: 1. Tables of comets 2. Comparative scientific thinking in astronomy and astrology

1223 GUPTA RC. Circumference of the Jambudvipa in Jain Cosmography. IJHS. 1975, 10; 1 : 38-46. Ref 20 Ranchi; Assistant Professor of Mathematics, Birla Institute of Technology, P.O. Mesra

CONTENTS: 1. Jain value of circumference 2. How circumference was obtained

1224 ANSARI SMR. Quinquecentenary of Nicolaus Copernicus: a report. IJHS. 1975, 10; 1 : 47-54. Ref 14 Aligarh; Physics Department, AMU

CONTENTS: 1. Author attended Copernicus birth celebration at Poland and giving here a report of the same

1225 KULKARNI RP. Soil stabilization by early Indian methods. IJHS. 1975, 10; (1) : 9-15. Ref 26 Nasik; Maharashtra Engineering Research Institute

CONTENTS: 1. Soil stabilization of prehistoric and vedic time 2. Soil stabilization as given in treatise of Silpa Sastra - four methods - Compaction, rearrangement and addition of soil particles, chemical reactions, thermal processes

1226 AHMAD M; BEHARI R; SUBBARAYAPPA BV. Al-Biruni: an introduction to his life and writing on the Indian science. IJHS. 1975, 10; 2 : 98-110. Ref 33 Aligarh; Director, Centre of West/Asian Studies, AMU Delhi; 80 Darya Ganj Delhi; Supervisor, NCCHSI

CONTENTS: 1. Life of Al-Biruni 2. Mathematical and astronomical concepts 3. Physical Sciences 4. Geography of India 5. Al-Biruni's works on India

1227 LAL SK. Krtya. VIJ. 1975, 13; 1-2 : 163-72. Ref 18 KEYWORD: MYSTICISM; MAGIC - BLACK

CONTENTS: 1. Black & witchcraft, sorcery 2. Female friend of witchcraft 3. Post vedic literature, puranas 4. Fire, Religious stories

1228 MEHRA KL. History of Mudga-pulse in India and its cultural significance. VIJ. 1975, 13; 1-2 : 190-97. Ref 66 KEYWORD: PULSES

CONTENTS: 1.Mung Dal (Mudga-pulse) different names 2. Archaeological records 3. Historical Records - vedic period (before 800 BC) - Early Buddhist & Jain Period (500-300 BC) - Maurya & Sunga period (300-75 BC) - Epics & Manusmrti (4th Cent. BC. - 5th Cent. AD) - Kusana and Saka satavahana period (75-300 AD) - Gupta period (300-750 AD) - Post gupta period (after 750 AD) 4. Cultivation & as food & religious beliefs.

1229 SHARMA S. Denotata of Time in Himachali. VIJ. 1975, 13 ; 1-2 : 318-29. Ref 1 KEYWORD: HIMACHAL

CONTENTS: 1.Consciousness of time in ritual practice/religious belief 2. Linguistic peculiarities to the terms of time 3. Classification of time - Ages - Historical times - Temporal changes - Sacred times - stages of life - Calendar - operation of day and night -spatial analysis - Cries of birds and beasts - indicators of day - indicators of night - Grammatical categories - Regimentation - proverbs

1230 KUMAR V. Luxurious Living of Indian buddhists. VIJ. 1975, 13; 1-2 : 415-17. Ref 10 KEYWORD: TEA; BUDDHIST

CONTENTS: 1. Salted Tea-Preparation 2. Buddhism lost its hold over the Indian masses

1231 SHARMA SD. Systematics of nuclear quadrupole moments using ancient India mathematics. VIJ. 1975, 13; 1-2 : 418-21. Ref 4 KEYWORD: TRANSITION-MULTIPOLE

CONTENTS: 1. Positive deformation and assymetry of nuclei 2. Theory 3. Q – Qoc + Qop and Brahmagupta equation 4.Multipole transitions

1232 CHOPRA DV. Contributions to the development of mathematics in ancient India . VIJ. 1975, 13; 1-2 : 72-81. Ref 21 KEYWORD: MATHEMATICS - DEVELOPMENT

CONTENTS: 1. Zero contribution 2. Nine digits & place value 3. Permutations 4. Geometric progression, trigonometry, indeterminate analysis, arithmetic

1233 CHOUDHURY J. Menstrual cycles of some urban and rural girls of West Bengal. JAS. 1975, 17; 1-4 : 1-16.

CONTENTS: 1. menstrual cycle in humans 2. Analysis of data-stature, of women-height,weight,constitution Index,menstrual cycle length in rural & urban women 3. Age at menarche 't' test between women groups

1234 PATEL S. Skin tanning variation and Helix-folding among the tibetans at Chandragiri,Ganjam District,Orissa. JAS. 1975, 17; 1-4 : 17-19. Ref 12 KEYWORD: ANTHROPOLOGY

CONTENTS: 1. Skin reflectances in male & female 2. Tanning coefficient,frequency of Helix-folding

1235 MITRA E. Contributions to our knowledge of India freshwater plants. 7. On some aspects of the habit,life history and autecology of Vallisneria spiralis L. JAS. 1975, 17; 1-4 : 30-49. Ref 14

CONTENTS: 1. Habit, External morphology Root, Stem,leaves 2. Reproduction,vegetative , sexual 3. Germination of seeds 4. Physico-chemical conditions 5. Tolerance of draght condition, low water,temperate,growth of plants in varying of substratum

1236 CHONDAR SL. Length weight and volume relationship of Gudusia Chapra (Hom). JAS. 1975, 17; 1-4 : 55-65. Ref 12 KEYWORD: ANTHROPOLOGY

CONTENTS: 1. L-W & V relationship of Gudusia Chapra by least square method from keetham,Agra 2. 't' test for significant departure of b values of regression of relationship from isometric growth value etc 3. Allometric law & cube law for L,W& V relationships

1237 SEN AK. Indigo uprising of Bengal & Harish Chandra Mukherjee. JAS. 1975, 17; 1-4 : 88-95. Ref 40 KEYWORD: MUKHERJEE HC

CONTENTS: 1. Socio-economic condition of labours plantation & factory etc

1238 LAL BB. Indus script : some observations based on archaeology. JRAS. 1975, 2 : 173-77. Ref 27

CONTENTS: 1. Origin of script 2. Direction of writing 3. Use of Indus seal 4. Observations on occurrence of seals 5. V-like sign 6. Survival of Indus script 7. Ochre colour ware

1239 GILLISPIE CC. Note on the Society's prizes. ISIS. 1975 , 66; 234 KEYWORD: HISTORY OF SCIENCE - PRIZE

CONTENTS: 1. The Henry Schuman Prize 2. Pfizer Award 3. Sarton Medal

1240 MAZUMDAR PMH. Johannes Muller on the blood, the Lymph, and the chyle. ISIS. 1975, 66; 232 : 242-53. Ref 35 Maryland, Institute of the History of Medicine, The Johns Hokkins University, Baltimore, USA

CONTENTS: 1. Description of clotting of blood and views of others.

1241 GUPTA RC. Sine of eighteen degrees in India upto the eighteenth century. IJHS. 1976, 11; (1) : 1-10. Ref 15 Ranch; Depatment of Mathematics, Birla Institute of Technology, P.O. Mesra

CONTENTS: 1. Greek trigonometry and Aryabhatiya 2. Exact value of the sine of 18 degrees 3. Derivations (Profs of the formula) in -Marici commentary by Mnisvara (AD 1638), Siddhanta-tattva-viveka by Kamalakara (AD1658), Siddhanta-sara-kaustubha by Jagannatha

1242 SWAMY BGL. Sources for a history of plant sciences in India II The RG Vedic soma plant. IJHS. 1976, 11; (1) : 11-32. Ref 90 Madras; ; Department of Botany, Presidency College KEYWORD: RGVEDA

CONTENTS: 1. Effects of soma drinking 2. Backward-tracing endeavours on the identification of soma 3. Post-RG-vedic substitutes for soma 4. Commentaries, lexicons and medical texts 5. On the soma-Haoma analogy 6. Intoxicating versus non-intoxicating quality 7. Status of soma in the Ritual 8. Unemphasized undercurrent in soma research 9. Forward-tracing endeavours 10. Soma, the 'wild-goose' 11. Wasso's fly-agaric theory 12. Cannabis (Bhang) theory

1243 CHANNABASAPPA MN. Square root formula in the Bakhshali manuscript. IJHS. 1976, 11; (2) : 112-124. Ref 40 Karnataka; Department of Mathematics, Karnataka Regional Engineering College, Surathal

CONTENTS: 1. Correct deciphering of the Bakhshali sutra 2. Kaye's translation of the sutra 3. Datta's translation of the Bakhshali sutra 4. CN Srinivasiengar's translation 5. Correct translation of the Bakhshali sutra 6. Plausible method of the derivation of the Bakhshali formula

1244 MITRA R; KAPOOR LD. Kamala - The national flower of India - its ancient history and uses in Indian medicine. IJHS. 1976, 11; (2) : 125-132. Ref 31 Lucknow; Central Council for Research in Indian Medicine and Homoeopathy Unit, National Botanic Gardens KEYWORD: LOTUS FLOWER - MEDICINE

CONTENTS: 1. Indian lotus - sacred plant 2. Application in art and literature 3. Nativity 4. Kamala - drugs properties, uses, description, distribution

1245 AGARWAL DP; MARGRBANDHU C; SHEKAR NC. Ancient copper workings: some new C dates. IJHS. 1976, 11; (2) : 133-136. Ref 10 Ahmedabad; Physical Research Laboratory, Baroda Archaeological Survey of India, Nagpur, Mineral Exploration Corporation Ltd. Seminary Hills

CONTENTS: 1. Evidence of ancient workings 2. Historical background 3. Emergence of town along copper belt 4. 14 C. Dates of ancient copper workings - Table

1246 CHAUDHURI M. Ship-building in the Yuktikalpataru and Samarangana sutradhara. IJHS. 1976, 11; (2) : 137-147. Ref 47 Calcutta; Indian National Science Academy I Park Street

CONTENTS: 1. King Bhoja a stanzas and their translation are given for the following aspects of ship building 2. Season for the ship-building 2. Woods for the construction of ship 3. Tying of iron-nail to a sea-going vessel 4. Classification of ships 5. Names and measurements of the Samanya (Ordinary) type of vessels 6. Two types of special (visesa) ship Dirgha and Unnata names and measurement of Dirgha and Unnata ships 7. Bhojas view about Dirgha type of ships and unnata 8. Painting of the ship 9. Decoration of ship 10. Three types of ships with cabins 11. Characteristics of the royal ship according to Bhoja 12. Despicable water-vessels (Jaghanya Jalayanani)

1247 SRINIVASAN TM. Measurement of rainfall in ancient India. IJHS. 1976, 11; (2) : 148-157. Ref 41 Delhi; National Commission for the Compilation of History of Sciences in India, Indian National Science Academy

CONTENTS: 1. Rain gauge (Varsamana) 2. Material of the gauge 3. Selection of site 4. Seasons of rainfall 5. Quantity and distribution of rainfall 6. Forecast of rainfall

1248 SWAMY BGL. Sources for a history of plant science in India III. The Maritime trade of ancient Tamils in plant products- A critique. IJHS. 1976, 11; (1) : 33-48. Ref 30 Madras; Department of Botany, Presidency College

CONTENTS: 1. Ports of the Tamil country 2. Plant products (20 plants) 3. Voyages of the ancient Tamils

1249 RAI RN. Some observations on Vrddha-Vasistha Siddhant. IJHS. 1976, 11; (1) : 49-53. Ref 5 Delhi; 10 A/4, Shakti Nagar.7.

CONTENTS: 1. Revolution numbers of the node and the apogee of the moon 2. Ahargana calculations 3. Jya value of 60 4. Revolutions in a yuga-table 5. Jya and utkramajya value-table

1250 BAG AK. Madhava's sien and cosine series. IJHS. 1976, 11; (1) : 54-57. Ref 7 Calcutta; Indian National Science Academy I Park Street KEYWORD: SINE SERIES; COSINE SERIES

CONTENTS: 1. Madhava (1350-1410) twenty-four sine values- table 2. Tantr-samgraha (1510AD) stanza and its English translation 3. Calculations of values 4 Date with European men, Newton, Demoivre and Euler

1251 JAIN LC. Kinematic motion of Astral Peal and counter bodies in Trilokasara. IJHS. 1976, 11; (1) : 58-74. Ref 81 M.P.; Sehore, Professor of Mathematics, Govt. Post Graduate College

CONTENTS: 1. Celestial diagram 2. Motion of Naksatras 3. Motion of the sun 4. Motion of the moon

1252 SINGH RH; SINHA BN. Ayurvedic concept of the psychosomatic basic of health and disease. IJHS. 1976, 11; (1) : 75-80. Ref 18 Varanasi; Department of Kayachikitsa Institute of Medical

Sciences, BHU KEYWORD: AYURVEDA; DISEASES

CONTENTS: 1. Correlation between the features of the mind and body 2. Stress, cause of disease 3. Organ neurosis (psychosomatic disease) 4. Ayurveda views -fundamental casuse of disease, prajnapuradha, asatmyendriyarthasamyoga

1253 JAIN LC. Certain mathematical topics of the Dhavala texts: Jaina school of mathematics. IJHS. 1976, 11; (2) : 85-11. Ref 54 Khandwa; Professor of Mathematics, S.N. Govt. Poist Graduate College

CONTENTS: 1. Virasenacarya's Dhavala commentary (9th century AD) of Satkhandagama by Puspanta and Bhutabali (C. second century AD) 2. Logarithmic treatment -finite and transfinite sets 3. Measure and analytical methods -measure relative to fluent, time, quater and becoming, methods of cut, division, reduction, measure, reason, explanation, and abstraction

1254 SARMA KV (ED & TRANSL). Computations concerning moon's shadow of Nilakantha Somayaji(Panjab University indological series-6). VIJ. 1976, 14; 1 : 1-31. Hosiarpur; Acting Director, Vishveshvaranand Vishva Bandhu Institute of Sanskrit and Indological Studies Punjab University V.V.B.I.S. & I.S. KEYWORD: SHADOW-MOON; TIME

CONTENTS: 1. Chayaganita (shadow computations) in Hindu astronomy (Kerala) 2. Biography & works of Nilakantha 3. Candracchayaganitam with Translation & Text 4. Shadow from Time 5. Time from Shadow 6. Four main R sines 7. Moon's latitude, shadow etc.

1255 SARMA KV (ED & TRANSL). Mirror of the laws of Astronomy Siddhanata-darpana of Nilakantha Somayaji: with auto-commentary (Panjab Univ. Indological Series 7). VIJ. 1976, 14; 2 : 1-54. Hoshiarpur, Acting Director, V.V.B.I.S. & I.S., Panjab Univ.

CONTENTS: 1.Biography & works of Nilakantha 2.Siddhanta -Darapana with commentary 3. Revolutions of planets - Diameters of earth moon & sun - Epicyles of apses & of conjunction - Measures of aeons - position of ecliptic - orbit of stars - Equinoxes - Definition of sine etc - declination, lunor crescent - Geocentric positions of a planet - Ecentric and orbital circles 4. Text & Translation given

1256 SARMA KV. Astronomy in India - Vedic period. VIJ. 1976 , 14; 1 : 133-52. Ref 12 Hoshiarpur; Acting Director, Vishveshvaranand Institute, Punjab University

CONTENTS: 1. Pre-scientific and sceintific astronomy its sources. 2. Vedic period : Nature of the source material 3. Evolution of the science during the vedic period 4. Genesis of the universe 5. Configuration of the Universe 6. Vedic astronomers 7. Sun, Moon 8. Earth, spherical & self supporting 9. Cause of day & night 10. Units of Time: Kalpa & Yuga 11. Year & month 12. Day-Division 13. Ayanas [Sun moves southwords for six months and northwords for six (months) 14. Equinoxes & solstices 15. Stars, Eclipses, planets 16. Meteors & comets

1257 SARMA KV. Scientific basis of Indian astrology. VIJ. 1976, 14; 2 : 309-12. Ref 1

CONTENTS: 1.Litrature, sources on astrology 2. Description of astrology given

1258 MISHRA M. Etymology of parjanya'rain-cloud'. VIJ. 1976 , 14; 1 : 5-6. Ref 7

CONTENTS: 1.Meaning & History of the word 'Parjanya' rain-cloud

1259 MULHOLLAND J (MISS). Soma an attempt to classify the plant and the drug. JAS. 1976, 18; 1-4 : 47-55. Ref 63

CONTENTS: 1. Rgveda description of soma-meaning,properties of drink soma, preparation of soma 2. Filtration of soma

1260 GHOSH AK, DUTTA RM. History of the teaching of Botanyin Bengal,(Including Bihar and Orissa) (1800-1918). JAS. 1976, 18; 1-4 : 7-46. Ref 36 KEYWORD: EDUCATION

CONTENTS: 1. Fort william college 2. Serampore college 1835-1857,calcutta medical college 1857-1887, 4. Krishnath college, Berhampore (Late Berhampore college) 5. Hughli Mohsin college (Late Hughli college) 6. Patna college 7. Ravenshan college, cutack 8. Rajshahi college..25. Colleges 1919- study of botany,name of teachers etc. syllabus of college are given for B.A., M.A. in Science

1261 KAPIL RN; BHATNAGAR AK. Portuguese contributions to Indian botany. ISIS. 1976, 67; 238 : 449-52. Ref 40 Delhi, Department of Botany, University of Delhi

CONTENTS: 1. Trade of east-west 2. Garcia da Orta (Portuguese) work "Coloquios dos simples e drogas he coucas medicinais da India compostos pello Doutor Garci da Orta" 3. Describes plant mentioned in the work. 4. Christoral Acosta (1574) work's "Trata de las drogas. Y medicinas de las Indias orientalis con sus plantas for Christoval Acosta"

1262 BAG AK. Method of integral solution of indeterminate equations of the type : BYAXC in ancient and medieval India. IJHS. 1977, 12; (1) : 1-16. Ref 32 Delhi; Indian National Science Academy. 110 002

CONTENTS: 1. Aryabhata I (b476AD)-knowledge of continued fraction 1.1 Solution of Yax+c/b in +ve integers. Total number of quotients of mutual division (n) Case I when n is even, Case II when n is odd 1.2. Solution of xby+c/a in +ve integers 2. Bhaskar I (600 AD) - solution of Yax-c/b 3. Scholars from seventh to sixteenth century (Bhaskara II (b.1150) 4. Treatment in the Karanapaddhati (16th century) and Yuktibhasa (1500-1600). Stanzas and their translations 5. Transmission to other nations

1263 SHARMA ML. Indian astronomy at the time of aryabhata. IJHS. 1977, 12; (2) : 100-05. Ref 8 Varanasi; Department of Jyotisa, Sampurnanand Sanskrit University

CONTENTS: 1. Five siddhantas of Panca siddhantika : description of each siddhantas 2. Existed knowledge at about the time of Aryabhata -knowledge of mathematics, knowledge of gola or celestial sphere, knowledge of astronomical instrument, actual observation 3. Astronomical knowledge, later used in all siddhantas 4. Problems remained yet unsolved - use of a yuga, beginning of mahayuga, vara, mean civil-day

1264 LISHK SS; SHARMA SD. Role of pre-aryabhata Jaina school of astronomy in the development of siddhantic astronomy. IJHS. 1977, 12; (2) : 106-13. Ref 34 Patiala; Department of Physics, Punjabi University KEYWORD: TIME

CONTENTS: 1. Planetary motion 2. Measurement of celestial angular distances 3. Use of Zigzag and linear step functions 4. Shifting of first point of Zodiacal circumference form winter solstice to vernal equinox 5. Latitudinal motion, notion of declination, notion of obliquity of ecliptic, graduation of zodiacal circumference 6. Heliacal motion 7. Lunar occultations 8. Subdivisions of time 9. Absence of certain elements of siddhantic astronomy

1265 PETRI W. Moving reference systems in the Aryabhatiya. IJHS. 1977, 12; (2) : 114-19. Germany (Fed. Rep); Institut fuer Geschichte der Naturwissenschaften der Universitaet, Deutsches Museum, D-8000 Muenchen KEYWORD: ASTERISM

CONTENTS: 1. Alleged 60 years revolution period of the asterism 2. No refernce system beyond the stars 3. Four elements only 4. "Buddhist" interpretation of Nimitta 5. Earth-wind 6. Self-consistency of Aryabhata's astronomy

1266 MERCIER R. Newly discovered mathematical relations between Greek and Indian astronomy. IJHS. 1977, 12; (2) : 120-26. U.K.; Department of Mathematics, Southampton

University KEYWORD: MATHEMATICS - GREEK AND INDIAN

CONTENTS: 1. Calendars preparation 2. Tropical coordinates 3. Sidereal coordinates 4. Babylonian - Indian sidereas astronomy 5. Continuation of Indian Astronomy in Arabic and Latin 6. Star map

1267 SIKDAR JC. Eclipses of the sun and moon according to Jaina astronomy. IJHS. 1977, 12; (2) : 127-36. Ref 62 Ahmedabad; L.D. Institute of Indology. 9.

CONTENTS: 1. Causes of eclipses 2. Phases of the moon 3. Limit and frequency of the lunar and solar eclipses 4. Reoccurrence of eclipses : the saros 5. Eclipses of the moon 6. Eclipses of the sun 7. Effects of total lunar and solar eclipses

1268 JAIN LC. Aryabhata I and Yativrsabha- a study in kalpa and Meru. IJHS. 1977, 12; (2) : 137-46. Ref 33 Khandwa; S.N., Govt. P.G. College, Professor of Mathematics

CONTENTS: Stanzas and their translations 1. Parallel Prakrit and Sanskrit verses concerning periods 2. Parallel verses concerning meru 3. Space and time concepts

1269 BEHARI R. Aryabhata as a mathematician. IJHS. 1977, 12 ; (2) : 147-49. Delhi; 1 Market Road. 110 001

CONTENTS: 1. Decimal place value system 2. Zero 3. Value of pie 4. Indeterminate equation

1270 FORBES, EG. Mesopotamian and Greek influences on ancient Indian astronomy and on the work of Aryabhata. IJHS. 1977, 12; (2) : 150-60. Ref 55 Scotland; University of Edinburgh, Edinburg KEYWORD: KNOWLEDGE TRANSMISSION

CONTENTS: 1. Calendrical system 2. Planetary motion 3. Method of calculating planetary distances 4. Manda and Sighra epicyle construction 5. Eccentric and epicycle- different models of planetary motion 6. Spherical (Gola) trigonometry

1271 DHANI SL. Manvantara theory of evolution of solar system and Aryabhata. IJHS. 1977, 12; (2) : 161-66. Chandigarh

CONTENTS: 1. Cosmic eras: Brahma Visnu and Rudra 2. Phases of evolution (Manvantaras) 3. Manvantara theory of evolution of solar system -matter evolved into solar system, era of darkness (beginning of day and night), formation for oceans and mountains, life-man, sun to dissolve after 2347 million years

1272 VOLODARSKY A. Mathematical achievements of Aryabhata. IJHS. 1977, 12; (2) : 167-72. Ref 8 USSR. Moscow; Institute of the History of Science and Technology

CONTENTS: 1. Life and works of Aryabhata 2. Commentaries to the work 3. Decimal place-value system 4. Square-root and cube-root 5. Alphabetic numeration 6. Rule of three a/bc/x (how to find x when three are given) 7. Linear equation 8. Problem of messengers (Calculate the time of meeting of 2 planets moving in opposite or same direction 9. Quadratic equation 10. Arithmetical progression 11. Indeterminate equations (Theory of numbers) 12. Area of plane figure 13. Value of pie 14. Volume of a pyramid 15. Pythagoreas theorem 16. Trigonometric problems

1273 JAIN BS. Ganita-sara-samgraha of Mahavira (C.850AD). IJHS. 1977, 12; (1) : 17-32. Ref 5 Delhi; Department of Mathematics, Correspondence Courses, University of Delhi

CONTENTS: 1. Rules and translations on 2. Minor methods of cubing 3. Treatment of geometric series 4. Unit fraction 5. Rational right triangles -right triangles having a given hypotenuse, problems involving areas and sides, problems involving sides but not areas, pairs of rectangles

1274 SCHNEIDER I. Contributions of the sceptic philosophers arcesilas and carneades to the development of an inductive logic compared with the Jaina-logic. IJHS. 1977, 12; (2) : 173-80. Ref 11

Germany (Fed. Rep.); Institute for Geschichte der Naturwissenschaften der Universitat Munchen, D8 Munchn 26, Deutsches Museum KEYWORD: ARISTOTLE

CONTENTS: 1. Probability concept of Aristotle 2. Sensual experience for the recognition 3. Absolute knowledge 4. Sceptical philosophy 5. Presentation with respect to the probable 6. Jain logic 7. Syadvada 8. Three kinds of probability

1275 SHUKLA KS. Glimpses from the aryabhata-siddhanta. IJHS. 1977, 12; (2) : 181-86. Ref 11 Lucknow; Department of Mathematics, University of Lucknow KEYWORD: TIME

CONTENTS: 1. Time from shadow 2. Possibility of an eclipse 3. New moon or full moon 4. Observation of the planets 5. Distances of mercury and venus at rising or setting 6. Rising and setting of canopus (Agastya) 7. Shadow instruments 8. Water-clocks

1276 LEVIN GMB. Aryabhata and Lokayatas. IJHS. 1977, 12; (2) : 187-93. Ref 26 USSR; Moscow, Institute of Oriental Studies, Armyansky Peru

CONTENTS: 1. Gupta period and Aryabhata 2. Al-Biruni cited Aryabhata 3. Lokayata (Materialist school of ancient Indian philosophy) -concept of four elements -sense- preception is the only source of authentic knowledge about the world 4. Eclipses concept Aryabhata's concept of the earth's rotation

1277 SARMA KV. Tradition of aryabhatiya in Kerala: revision of planetary parameters. IJHS. 1977, 12; (2) : 194-99. Ref 19 Hoshiarpur; Vishveshvaranand Inst. P.U. Sadhu Ashram

CONTENTS: 1. Aryabhata's astronomy in Kerala 2. Astronomic constants 3. Number of revolutions 4. Planetary revolutions 5. Drk system of Paramesvara (AD1431-32) 6. Revision by Nilakantha (AD1443-1560) 7. Revision in Drkkarana 8. Comparative study of the revisions

1278 GUPTA RC. Mathematical rules form the aryabhatiya. IJHS. 1977, 12; (2) : 200-06. Ref 22 Ranchi; Department of Mathematics, Birla Institute of Technology, P.O. Mesra

CONTENTS: 1. Rules for area of a triangle 2. Rules for volumes of a pyramid 3. Rules for volume of a sphere 4. Altitude of sun at any time

1279 BILLARD R. Aryabhata and Indian astronomy ( an outline of an unexpected insight ). IJHS. 1977, 12; (2) : 207-24. France; Ecole Francaise d' Extreme Orient, 22, avenue du President Wilson, 75116 Paris

CONTENTS: 1. Yugabhagana and yuga canon 2. Radiography of - 'Almagest's canon by Claudius Ptolemy -Yuga canon defined by Aryabhata -'S' of Almagest's canon - 'S' of Aryabhatiya's canon -Yuga speculations (Common multiples of the diverse revolutions)

1280 MUKHERJEE RN. Back ground to the discovery of the symbol for Zero. IJHS. 1977, 12; (2) : 225-31. Ref 61 Bihar; Department of Mathematics, Purnea College, Purnea

CONTENTS: 1. Representation of zero 2. Spiritual symbols for zero 3. Physical symbols for zero 4. Vedic literature 5. Double meaning of zero in vedic literature

1281 ELFERING K. Area of a triangle and volume of a pyramid as well as the area of a circle and the surface of the hemisphere in the mathematics of Aryabhata I. IJHS. 1977, 12; (2) : 232-36. Germany (West); Institute of History of Science, University of Munich. WG

CONTENTS: 1. Text stanzas and their translations

1282 KHAN MS. Aryabhata I and Al-Biruni. IJHS. 1977, 12; (2) : 237-44. Ref 54 Calcutta KEYWORD: ASTRONOMY

CONTENTS: 1. Greek and Indian work on astronomy 2. Kitab al-Hind or Indica of Al-Biruni

3. Sources of Al-Biruni's information about Aryabhata's works 4. Difficulties Al-Biruni faced in studying the subject (Aryabhatiya) 5. Error commited by others in their writings in studying Aryabhata

1283 LISHK SS; SHARMA SD. Season determination through the science of sciatherics in Jaina school of Astronomy. IJHS. 1977, 12; (1) : 33-44. Ref 30 Patiala; Department of physics, Punjabi University. Punjab

CONTENTS: 1. Theory in Aitareya Brahmana satapatha Brahmana 2. Jambudvipa prajnapti-variation of shadow-length text and its explanation 3. Other explicit references -Suryaprajnapti, Candraprajnapti 4. Table of shadow-lengths for season-determination 5. Gnomonic text of a Babylonian tablet (noon-shadow-length)

1284 BEHARI R; JAIN BS. Mathematical contributions of ancient indian mathematicians as given in the works of Bhaskaracarya II (12th Century AD). IJHS. 1977, 12; (1) : 45-56. Ref 8 Delhi; University of Delhi

CONTENTS: 1. Bhaskara II life and works 2. Cyclic method of solving the indeterminate quadratic equation -Brahmagupta's Lemma, Bhaskara, two examples on Brahmagupta's samasa operation, four examples on Cakravala method 3. Bhaskara's idea of 'zero' as an infinitesimal and the idea of 'infinity' 4. Bhaskara's work Lilavati

1285 JAIN LC. Divergent sequences locating transfinite sets in trilokasara. IJHS. 1977, 12; (1) : 57-75. Ref 50 Sehore; Govt. P.G. College, Professor of Mathematics KEYWORD: SET-TRANSFINITE

CONTENTS: 1. Nuber-measure: Sequences 2. Measure: classification and notation 3. Forms of the sequences (S1 to S14) 4. Analysis of the sequence S1 5. Complementary sequences (S2 and S3) 6. Structural Analysis of the sequence S12, S13 and S14

1286 YANO M. Three types of Hindu sine tables. IJHS. 1977, 12; (2) : 83-89. Ref 9 Japan; Kyoto Sangyo University, Kyoto

CONTENTS: Sine Tables 1. Type A:R120 2. Type B:R3438 3. Type C;R3270 4. Derivation of sines

1287 SHARMA ML. Aryabhata's contribution to Indian astronomy. IJHS. 1977, 12; (2) : 90-99. Ref 34 Varanasi; Department of Jyotisa, Sampurnanand Sanskrit University

CONTENTS: 1. Aryabhatiyam - Description of the verses and three chapters 2. Cycles of revolutions (Bhaganas of planets within a Mahayuga) 3. Position of apogee of planets 4. Circumference of Epicycles 5. Correction for geometric positions and equation of centre after Aryabhata's work

1288 SARMA KV (ED & TRANSL). True Longitude computation on the sphere of Zodiac (Panjab Univ. I.S.-8. VIJ. 1977, 15; 1 : 1-41. Hoshiarpur, Acting Director. VVBIS & I.S.,Panjab Univ. KEYWORD: ASTRONOMY

CONTENTS: 1.Rasigolasphutaniti - edition & translation 2.Biography & works of Acyuta Pisarati 3. Sphere of Zodiac - Essence of an eclipse - conjunction - Refutation of Determination - Correction for Reduction to the Ecliptic

1289 SARMA KV (ED & TRANSL) . Investigations on astronomical theories by Nilakantha Somayaji (Panjab Univer I.S. - 11). VIJ. 1977, 15; 2 : 1-90. Ref 170 Hoshiarpur; Acting Director, VVBIS & IS, Panjab University

CONTENTS: 1.Jyotirmimamisa - text only 2. Biography & work of Nilakantha 3.Contents of Jyotirmimamisa correction through eclipses - vedic authority for inference in astronomy - True motion, position etc, of planets - Relation of the sides and hypotenuse - Reduction of angular distances 4. Nilakantha's esteam for Aryabhata 5. Significant of Karana texts

**1290** SARMA KV. Aryabhata and the revision of planetary parameters in the astronomical tradition of Kerala. VIJ. 1977, 15; 1 : 125-39. Ref 27

CONTENTS: 1. Aryabhata and scientific Indian astronomy 2. Aryabhathya 3. Earth's rotation 4. Opposition to Aryabhata 5. Decline & resurgence of the system 6. Astronomical tradition of Kerala 7. Aryabhata's astronomy in Kerala 8. Parahita system of Haridatta (AD. 684) 9. Drk System of Paramesvara (AD 1431) 10 Revision by Nilakantha (AD 1444-1545) 11. Revision in Drkharana & their comparative study of revision 12. Table of parameters of astronomers.

**1291** PATYAL HC. Atharvanic practices with roots of plants (Mulakarmans or Mulakriyas). VIJ. 1977, 15; 1 : 13-19. Ref 18 Poona; Deccan College,6 KEYWORD: ROOTS-PLANET; MYSTICISM

CONTENTS: 1.Counter-practices againts magic 2. Dharma - Texts views on Magic or Sorcery 3. Meaning of the terms Mulakarmans and mulakriyas.

**1292** GUPTA MN. Vedic metre in Indus Script. VIJ. 1977, 15; 2 : 188-96. Ref 11 KEYWORD: METRE - VEDIC

CONTENTS: 1. Metrical technicalities 2. Numeral notation for metrical Technicalities in Indus script 3. Indus Seals, Kalibangan seals in relation to vedic metre 4. Language Study of Indus Seal

**1293** ARORA RK. Concept of Human body in the Adigranth. VIJ. 1977, 15; 2 : 216-28. Ref 26 Chandigarh; Dept. of History, Directorate of correspondance course, Panjab University

CONTENTS: 1. Body as the abode of God 2. Microcosm & Macrocosm 3. Body & two opposite forces 3. Tenth door 4. Truth & whole earth, slag in body 5. Religious & philosophical description of body.

**1294** KRISHAN Y. Astronomical Revolution in India about AD. 400 and its implications.. VIJ. 1977, 15; 2 : 265-84. Ref 57 Delhi; C.II/55, Dr. Zakir Hussain Marg

CONTENTS: 1.Vedanga - Jyotisa pertain to Rgveda & Yajureda 2. Siddhanta - Jyotisa - descriptions of there 3. Naksatradarsa - Jyotisa description 4. Religious rites & astronomy - law of karma 5. Pauranic Literature - description - Markandeya - Purana (3rd C AD) - Narada - Purana (AD 875-1000) - Garuda - Purana (AD 800-1050) - Padma-purana (AD 900-1400) - Agni-Purana(9th C AD) - Matsya-Purana (650-1000 AD) Related to astronical aspects (worship of planets)

**1295** GHOSH SK. New ways of developing speech in the mute (Mukam karoti Bachalam). JAS. 1977, 19; 3-4 : 1-12.

CONTENTS: 1. Causes of mutism 2. Visible-gesticulational language 3. Audiological education 4. Communication dissorders 5. Prelingual training

**1296** GHOSH SC, SHAH SC. Estheriella Taschi sp.Nov.: a new triassic conchostracan from the panchet formation of East-Bokaro coalfield,Bihar. JAS. 1977, 19; 1-2 : 14-18. Ref 15

CONTENTS: 1.Fossiliferous bed systematic description (classification of species) 2. Description of species

**1297** PINGLE U (URMIL); REDDY NVR. Late Carboniferous glacial conglomerate (Talchir), a rich factor site for abbevillio-Acheulean hand-Acheulean hand-axe culture in central India. JAS. 1977, 19; 3-4 : 65-72. Ref 11

CONTENTS: 1. Paleolithic sites along Talchir glacial conglomerate, implements from site Talchir 2. Typological classification of implements in site (Talchir) 3. Types of material used for implements in sities Talchir

**1298** YAMADA I. Vijnaptimatrata of Vasubandhu. JRAS. 1977, 2 : 158-76. Ref 49 KEYWORD:

RESEARCH METHOD; SCIENTIFIC METHOD

CONTENTS: 1. Observations, conceptualisation, verbal expression 2. Modes of transformation, Consciousness, eye, ear, nose, tongue, body, mind 3. Ultimte reality 4. Soteriology, realization of ultimate and cognition without conceptualisation

1299 HARRIES HC. Sixteenth-Century dissemination of the coconut palm. ISIS. 1977, 68; 244 : 605-06. Ref 10 Jamaica; Coconut Industry Board, P.O. Box 204, Kingston

CONTENTS: 1. Report coconut palm in July 1501 earlier than 1563 as reported by Kapil RN & Bhatnagar AK in ISIS 1976. 67, 449-52 2. Give the detail account of Portugese travel trade

1300 WALKER AR. Lahu Nui (Red Lahu) funerary chants : two Lahu texts with a brief ethnographic introduction. JRAS. 1978 : 163-70. Ref 19 KEYWORD: THAILAND; ANTHROPOLOGY; MYSTICISM; SOUL

CONTENTS: 1. View of death and after life 2. Funerary practice text and translation in English

1301 . Silpasastra works in Sri Lanka. JRAS. 1978, 1 : 73. Ref 21 KEYWORD: TOWN-PLANNING

CONTENTS: 1. Mentioned Indian sanskrit on Silpasastra 2. Sarajo timalai (Tamil names) gives, reference to construction of houses, villages cities, town etc., site judgement 3. Sinhales sanskrit works in Sri Lanka

1302 MAJUMDAR PK. Ganita Kaumudi and the continued fraction. IJHS. 1978, 13; (1) : 1-5. Ref 7 Calcutta, Central Library, Research Block, Calcutta Univeristy

CONTENTS: 1. Indeterminate equation (Nx2 + K2Y2)- Continued fraction integral solution 2. Original sanskrit verses with translation

1303 MAJUMDAR PK. Rationale of Bhaskara I's method for solving ax+cby. IJHS. 1978, 13; (1) : 11-17. Ref 6 Calcutta; Central Library, Calcutta University

CONTENTS: 1. Indeterminate equation (by ax-c)-method continued fraction. Integral solution 2. Original sanskrit verses and its english translation with modern interpretation 3. Few lines about the continued fraction 4. Bhaskara I's rule 5. Rationale of the rule (ax-cby)

1304 PRASAD KN; MURTHY SRN. Dating the quaternary and human civilisation. IJHS. 1978, 13; (2) : 113-16. Ref 6 Hyderabad; Geological Survey of India, Southern Region KEYWORD: QUATERNARY PERIOD

CONTENTS: 1. Human evolution 2. Vedic period and civilisation 3. Configuration of the Himalayas during quaternary period and evolution

1305 KULKARNI RP. Geometry as known to the people of indus civilization. IJHS. 1978, 13; (2) : 117-24. Ref 2 Nasik; B-6/2 MERIQuarters, Maharashtra Engineering Institute KEYWORD: INDUS CIVILIZATION - GEOMETRY

CONTENTS: 1. Mensuration knowledge 2. Geometrical knowledge - properties of rectilinear figures - properties of a circle 3. Area and volume of rectilinear figures 4. Pythagoras theorem 5. Knowledge of drawing similar figures

1306 GUPTA RC. Indian values of the sinus totus. IJHS. 1978 , 13; (2) : 125-41. Ref 64 Ranchi; Department of Mathematics, Birla Institute of Technology, P.O. Mesra KEYWORD: SINE VALUES

CONTENTS: 1. Values of sine totus (R) [ RC/2 pie ] total or complete sine 2. Table of Indian sinus totus toti is given 3. Smaller values of R 4. Transmission of Indian values of the sinus totus 5. List of sanskrit terms for sinus totus given 35 items

1307 GERSHENOWITZ H. Treatment of lamarckism as found in forty-one college textbooks. IJHS. 1978, 13; (2) : 144-50. Ref 41 USA;

Department of History and Philosophy of Science, Glassboro State College, Glassboro, N.J.

CONTENTS: 1. 41 College textbook analysis for lamarckism (1950-74) (a) Role of lamarck in development of evolutionary theory (b) Presence or absence of statement of his postulates (c) Positive or negative illustration of the postulates (d) evaluation of his postulates

1308 SHASTRI, TSK. Epoch of the Romaka siddhanta in the panca siddantika and the epoch longitudes of the sun and moon in the vasistha-paulisa. IJHS. 1978, 13; (2) : 151-58. Moscow; 10/A Krishna Puram Street, Royapettah -14 KEYWORD: SUN LONGITUDES; MOON LONGITUDES

1309 MURTHY SRN. Geological evidences in support of the antiquity of some ancient Indian events. IJHS. 1978, 13; (1) : 18-22. Ref 5 Hyderabad; Geological Survey of India KEYWORD: TECTONICS PLATE, GEOLOGY

CONTENTS: 1. Antiquity of the time of Surya Mayasura (Suryasiddhanta) and Ramayana by theory of Plate Tectonics

1310 GERSHENOWITZ H. Gavin de beer and the neo-lamarckians. IJHS. 1978, 13; (1) : 23-27. Ref 11 USA; Department of History and Philosophy of Science, Glassboro State College, Glassboro, J.J. KEYWORD: ORIGIN OF SPECIES

CONTENTS: 1. Reflections of a Darwinian 2. Embryology and evolution 3. Embryos and ancestors 4. Galatea 5. Theories of evolution described by Fisher's their examination disscussed

1311 SINGH RH; SRIVASTAVA PK. Nephrology in ancient Indian system of medicine. IJHS. 1978, 13; (1) : 28-31. Ref 6 Varanasi, Division of Postgraduate Studies in Indian Medicine and Nephrology, Institute of Medical Science. BHU KEYWORD: MEDICINE NEPHROLOGY

CONTENTS: 1. Anatomical considerations 2. Physiological considerations 3. Clinicopathological considerations 4. Pathophysiological consideration 5. Therapeutic considerations

1312 KULKARNI RP. Value of pie known to Sulbasutrakaras. IJHS. 1978, 13; (1) : 32-41. Ref 15 Nasik; B-6/2, MERI Quarters, Maharashtra Engineering Research Institute-422 004 KEYWORD: PIE

CONTENTS: 1. Pie value, historical background -Indus civilization -vedic period -Brahmana period 2. Value of pie as deduced from sulbasutras

1313 JAIN LC. Spiro-elliptic motion of the sun implicit in the Tiloyapannatti. IJHS. 1978, 13; (1) : 42-49. Ref 15 Sehore; Govt. P.G. College, Professor of Mathematics KEYWORD: SUN SPIRO-ELLIPTIC MOTION

CONTENTS: 1. Data regarding motion of the sun 2. Geometry of the implied path of the sun 3. Dynamical laws of the tacit and closed-spiral motion, envisaged in Jaina School

1314 MAHDIHASSAN S. Triphala and its arabic and Chinese synonyms. IJHS. 1978, 13; (1) : 50-55. Ref 18 Pakistan; SD 34, Block A, North Nazimabad. 33. KEYWORD: MEDICINAL PLANTS, DRUGS

CONTENTS: 1. Drug (Triphala) its importance 2. Botanical name, different species and uses in diseases

1315 GERSHENOWITZ H. George Gaylord Simpson and Lamarck. IJHS. 1978, 13; (1) : 56-61. Ref 19 USA; Department of History and Philsophy of Science, Glassboro State College, Glassboro, N.J. KEYWORD: SIMPSON GG

CONTENTS: 1. Process of inhertance of acquired characteristics 2. Lamarckians demonstrated in laboratory 3. Simpson GG ideas toward Lamarck

1316 MAJUMDAR PK. Extant siddhanta sarvabhauma an error in the sine of one-third part of an angle. IJHS. 1978, 13 ; (1) : 6-10. Ref 6 Calcutta; Central Library, Calcutta University KEYWORD: TRIGONOMETRY

CONTENTS: 1. Trigonometrical formulae, of sin A/3

1317 ANSARI SMR. Establishment of observatories and the socio-economic conditions of scientific work in nineteenth century India. IJHS. 1978, 13; (1) : 62-71. Ref 28 Aligarh; Physics Department, AMU

CONTENTS: 1. History of observatories and astronomical work done there 2. Madras observatory (1792) 3. Lucknow observatory (1832) 4. Socio-economic conditions for scientific work in 19th century India

1318 MURTHY SRN. Critical evaluation of mineralogical aspects of some sanskrit texts. IJHS. 1978, 13; (2) : 77-82. Ref 4 Hyderabad; Geological Survey of India, Southern Region. 500 001 KEYWORD: MINERALOGY -GEMS -ORES -MEDICINE; GEM STONE

CONTENTS: 1. Gem-stones (list of sanskrit names of a well known gems 2. Medicinal uses of minerals 3. Occurrence of metallic ores

1319 ROY M. Dyes in ancient and medieval India. IJHS. 1978, 13; (2) : 83-112. Ref 214 Calcutta; 80B, Vivekananda Road. 700 006

CONTENTS: 1. Red dye-producing substances 2. Yellow dye-producing substances 3. Blue dye-producing substances 4. Black dye-producing substances 2. Tables are given indicating dye-producing substances, name and period, english and botanical names, colour-producing part of the substance, extraction of colouring matter, nature of the dye, uses (long-lasting, mordant dye, fast, fugitive etc.)

1320 CHOPRA DV. Number PI (Pie) in ancient India. VIJ. 1978 , 16; 1 : 104-111. Ref 18 USA; Wichita State university, Wichita (Kansas 67208) KEYWORD: PIE

CONTENTS: 1.Introdution to mathematical achievements of an ancient Indians 2. History of Pie in world 3. Aryabhata I (b476 AD) - Gave approx Value of Pie (two mathematicians & astronomers) II (10th C. AD) 4. Umasvati (150 BC) mentioned C denotes circumference of a circle 5. Varahamihitra, (505 AD) Bhaskara. 6. Nilakantha 7. S Ramanujan(1887-1920) in 1914 gave several noteworthy approximation evaluating Pie 8. Foreigners also mentioned.

1321 SARMA KV. Mathematical rationale in Kerala texts on Astronomy and mathematics. VIJ. 1978, 16; 1 : 112-18. Ref 6

CONTENTS: 1.Sankara Variyar (AD 1500-60, 2. Kriyakramakari of sankara 3. Exposition of rationale - Multiplication - summation of series etc. 4. Yuktidipika of sankara 5. Depiction of the rationales

1322 MAHDIHASSAN S. Earliest Divisions of time and the associated observations of Heavenly bodies. VIJ. 1978, 16; 1 : 127-31. Ref 3

CONTENTS: 1.Divisions of time as day, week, month & year on the basis of Heavenly body 2.Counting board of the ancients (fingers & thumb) time into 7 days & 7 nights 3. Babylonians symbols all the units of time, 4. Time-god or chronos, being worshipped in ancient Mesopotamia.

1323 ARORA RK. Concept of Nadis in the Adigranth. VIJ 1978 , 16; 1 : 55-66. Ref 40 Chandigarh; Reader, correpondence courses, Panjab University KEYWORD: PULSE; NERVE PULSE

CONTENTS: 1. Nerve (Nadis), artery 2. Three nadis 3. Susumna nadi 4. Sadhana of nadis 5. Literature sources & their views

1324 MUKHOPADHYAY B. On the significance of Soma. VIJ. 1978 , 16; 1 : 6-9. Ref 15 Burdwan,; Lecturer, Department of Sanskrit, Burdwan University

CONTENTS: 1. Soma: Inebriating juice of plants 2. Soma: Plant bearing soma 4. Soma: Elixir of life & delight 5. Soma: The god 6. vedic period description

1325 ADAMSON PB. Anatomical and pathological terms in Akkadian : Part II. JRAS. 1979, 1 : 2-8. KEYWORD: ANATOMY - TERMS; PATHOLOGY - TERMS

CONTENTS:

1326 LISHK SS; SHARMA SD. Zodiacal circumference as graduated in Jaina astronomy. IJHS. 1979, 14; (1) : 1-5. Ref 26 Punjab. Patiala; Department of Physics, Punjabi University KEYWORD: ASTRONOMY JAINA; JAINA ASTRONOMY

CONTENTS: 1. Asterism and zodiacal stretch list in a table. 28 items 2. Asterism and zodiacal stretch in celestial parts in a table 3. Asterisms and their days and nights with months of the year

1327 GERSHENOWITZ HARRY. Influence of Lamarckism on the development of freud's psychoanalytic theory. IJHS. 1979, 14; (2) : 105-113. Ref 25 USA; Department of Life Sciences, Glassboro State College, Glassboro 08028

CONTENTS: 1. Lamarckism and social science theory 2. Criticism of Freud's use of Lamarkian theory 3. Phylogenetic link between oedipus complex & fear of castration 4. Cannibalistic act as phylogenic basis for war throughout history 5. Phylogenetic pattern of phantasy world 6. Psychoanalytic theory

1328 GERSSHENOWITZ 'Z H. John Burroughs : Right wing Neo-Lamarckian. IJHS. 1979, 14; (2) : 114-21. Ref 20 USA; Department of Life Sciences, Glassboro State Colege, Glassboro 08023 KEYWORD: BURROUGHS J

CONTENTS: 1. Biography of Burroughs J. 2. Mystical interpretation of man's soul in great cosmos 3. Organic evolution of Charles Darwin

1329 SIKDAR JC. Indian concepts of matter-part I. IJHS. 1979, 14; (2) : 122-33. Ref 77 Ahmedabad; L D Institute of Indology. 9.

CONTENTS: 1. Concept of matter by i) Sankhya ii) Nyaya-vaisesika 2. Bauddha 4. Jains

1330 MURTHY SRN. Vagbhata on medicinal uses of gems. IJHS. 1979, 14; (2) : 134-38. Ref 3 Hyderabad; Geological Survey of India, Southern Region. 500 001 KEYWORD: GEMS - MEDICINE; MEDICINE - GEMS

CONTENTS: 1. A list is given for gems. their sanskrit name and modern name, classification etc. 2. Rasaratnasamuccaya (Sanskrit text) on the medicinal uses of gems

1331 SINGH RS. Botanical identity and a critical appreciation of Maluva Lata as evinced in the Buddhistic Pali literature . IJHS. 1979, 14; (2) : 139-43. Ref 6 Varanasi; Raga-Shastra Dept. Institute of Medical Sciences, BHU. 5.

CONTENTS: 1. Habit and habitat of maluva lata (plant) 2. Botanical name, parts of plants and its uses in medicine

1332 MURTHY SRN. Geological aspects of the Surya siddhanta. IJHS. 1979, 14; (2) : 144-49. Ref 11 Hyderabad; Geological Survey of India, Southern Regional Office KEYWORD: GEOLOGY

CONTENTS: 1. Brief contents of text 2. Geography and geological implications of Suryasiddhanta 3. Studies on the text and plate theory of earth

1333 SASTRY TSK. Vasistha-Paulisa venus in the Pancasiddhantika of Varahamihira. IJHS. 1979, 14;

(2) : 150-54. Madras; 10 A Krishnapuram Street. 600 014

CONTENTS: 1. Thibaut and Drivedi and Neugebauer and Pingree studies on Pancasiddhantika studies of star-planets, Jupiter and Saturn equation of conjunction 2. Motion on the synodic circle computed is given

1335 GERSHENOWITZ HARRY. Chauncey Wright's view on Lamarck. IJHS. 1979, 14; (1) : 16-24. Ref 24 USA, NJ; Department of History and Philosophy of Science, Glassboro State College, Glassboro

CONTENTS: 1. Wright's structure of science 2. Darwinian and evolution mind of man and animal brain 3. Modes of natural section 4. Wright's ideas about Darwin's and Lamarck's theories of evolution

1336 GERSHENOWITZ HARRY. George Henslow : True Darwinist. IJHS. 1979, 14; (1) : 25-30. Ref 20 USA, NJ; Department of Life Science, Glassboro State College, Glassboro 08028

CONTENTS: 1. Henslow did experiments and verify the neo-Lamarckian-Darwinian theory

1337 JAIN LC. System theory in Jain school of mathematics. IJHS. 1979, 14; (1) : 31-65. Ref 81 Khandwa; Department of Mathematics, S.N. Govt. P.G. College KEYWORD: MATHEMATICS - JAINA; JAINA

CONTENTS: 1. Chronology of source material indicating period, contributor work 2. Cognate terminology : English Synonym and prakrit term 3. Mathematicophysical concepts and system theoretic principles 4. Role of yoga and kasaya 5. Structural details of karma (action) system i) Timeset ii) Input values, functions and their relations iii) Measures in input and relations iv) Output 5. Elements of operational Algorithm operational phases (Ten) 6. Cybernatics

1338 GUPTA RC. Munisvara's modification of Brahmagupta's Rule for second order interpolation. IJHS. 1979, 14; (1) : 66-72. Ref 4 Ranchi; Department of Mathematics, Birla Institute of Technology, P.O. Mersa

CONTENTS: 1. Symbols 2. Brahmagupta's rule 3. Munisvara's modification of the rule

1339 MURTHY SRN. Occurrence of Cinnabar in Rasarnavakalpa. IJHS. 1979, 14; (2) : 83-86. Ref 12 Hyderabad; Geological Survey of India, Southern Regional Office. 500 001 KEYWORD: MERCURY ORE

CONTENTS: 1. Region of occurrence of Cinnabar 2. Examination of ores for Cinnabar and associated mercury ores

1340 MUKHERJI V. Some historical aspects of Jagadis Chandra Bose's microwave research during 1895-1900. IJHS. 1979 , 14; (2) : 87-104. Ref 16 Kharagpur; Indian Institute of Technology. 721 302 KEYWORD: BOSE JC

CONTENTS: 1. Galena detector 2. Notable lines of research (Collected Physical papers) of 1927 Bose published by Longmans, Green and Co. 3. Bose's work and the International background

1341 KAMAVISDAR SS. Evidences of the inorganic substances in amalgamation process during Indian alchemy.. IJHS. 1979, 14(2) : 155-56. Ref 7 MP, Department of Chemistry, Govt. PG College, Ambikapur-49700 KEYWORD: MERCURY

CONTENTS: 1. Purifiction & the potentiation of Mercury 2. Metal transforming or transmuting 3. Solidification of mercury

1342 SARMA KV (ED) WITH INTRODUCTION. Rationals of Hindu Astronomy : Part I (Punjab University Indological Series-24). VIJ. 1979, 17; 1-2 : 1-124.

CONTENTS: 1. Location of astronomical tracts, individual works, authorship of the tracts 2. Ganitayuktayah - Moon's shadow, Midday shadow - Elevation of moon's hours - partallax in longitude, latitudes of planets - declination - shadows - Planets, computations - lump-day of zero correction-rationale - Motion of planetery latitude-rationalistic exegeses 3. Original text only

1343 MARGABANDHU C. Early historic Rangmahal culture: its origin and diffusion in north-west Rajasthan (an archaeological synthesis). VIJ. 1979, 17; 1-2 : 206-15. Ref 32 Baroda, Superintending Archaeologist Archaeological survey of India, Western circle KEYWORD: DRY LAND-ANCIENT CULTURE

CONTENTS: 1. Explorations of 1951-52 by Ghosh A & 1952-54 by Hanna Rydh 2. Dried up river valleys in northern part of Bikaner 3. Study of river channel to reconstruct the environmental & past climatic conditions of this area. 4. Origin & growth of the culture & its description in sources.

1344 GUPTA RC. Time-altitude and altitude-azimuth equations in Hindu astronomy. VIJ. 1979, 17; 1-2 : 256-63. Ref 13 Ranchi,; Associate Professor of Mathematics, Birla Institute of Technology, Mesra.

CONTENTS: 1.Rule from the Aryabhatiya I(b.AD. 476). (due to extreme brerity, elliptic language and poetical form, it is often difficult to identify the actual meaning of the rules) 2. Finding time from given altitude of the sun 3. Finding azimuth from altitude 4. The kona-sanku (corner-gnomon)

1345 MAHDIHASSAN S. Distillation assembly of Pottery in ancient India with a single item of special construction.. VIJ. 1979, 17; 1-2 : 264-66. Ref 8 Pakistan; Karachi, S.D. 34-Block A. North Nazimabad. Pin 33 KEYWORD: DISTILLATION-POTTERY

CONTENTS: 1.Distillation assembly of pottery at Taxilla (about 300 AD.) 2. Uses of Distillation pottery 3. Description of distillation assembly of pottery 4. Muller RFG idea on distillation in India before Islam.

1346 MAJUMDAR PK. Extant siddhanta Tattvaviveka-an error in the addition theorem for the cosine. JAS. 1979, 21; 1-2 : 23-27. Ref 5 KEYWORD: TRIGONOMETRY

CONTENTS: 1. Trigonometrical formula cos (L+B) some error 2. Explanation of formula cos (L+B) some error 3. Kamalakara's (C.1658 AD) Siddhanta tattvaviveka

1347 . New ISIS prize. ISIS. 1979, 70; 251 : 150. KEYWORD: HISTORY OF SCIENCE - PRIZE

CONTENTS: 1. The History of Science Society announces the establishment of a new prize to encourage the publication in ISIS of origianl research of the highest standard. Consisting of $200 and a certificate, this prize will be given annually, on the recommendation of the committee on ISIS, to the author of the beat article in ISIS in the three years prior to the year of the award. Articles published in ISIS between march 1978 and December 1980 inclusive, will be eligible for the first annual award, to be eligible for the first annual award to be presented in 1981. The Zeithlin - Ver Brugge Prize for the best article in ISIS in the three years preceding of the year of award."ISIS 1979.70(254).583". The Pfizer Award, for the best book in the History of Science published in the previous year. The Sarton medal for distinguished contributions to the cause of furthering the History of Science. The Schuman Prize for best essay by an undergraduate or graduate student in History of Science Society, Inc. The Society was established to foster interest in the History of Science and its social and cultural relations, to provide a forum for disussion and to promote scholarly research in the History of Science. The Society pursues these objectives by the publication of its journal ISIS, by the support and subvention of other forms of scholarly publication; by the organization of annual meeting and other programs; by the award of medals and prizes.

1348 JAYAWARDENE SA. Magda Whitrow (Editor). ISIS cumulative Bibliography. A Bibliography of the History of Science formed from ISIS Critical Bibliographies 1-90, 1913-65. I-IV Volumes. ISIS. 1979, 70; 251 : 160-63. Ref 8 England. Science Museum, South Kensington, SW7 5NH, London KEYWORD: INFORMATION SCIENCE

1349 FULLER JC. Calendarical system in Tamilnadu (South India). JRAS. 1980, 1 : 52-63. Ref 18 KEYWORD: CALENDAR; ALMANACS

CONTENTS: 1. Calendarical system currently used in Tamilnadu 2. Principles of Tamil Almanacs 3. Planets and days of week 4. Racis and Tamil months 5. Naksattirams/tithis: determines timing of events/festivals

1350 VERMA IN. Calico printing in India. IJHS. 1980, 15; (1) : 1-5. Ref 5 Delhi; Y-64A Hauz Khas. 110 016 KEYWORD: PRINTING-CALICO

CONTENTS: 1. Centres in India for Calico printing 2. Farrukhabad Calico printing -method printing, pattern, colouring

1351 KUMAR DEEPAK. Patterns of colonial science in India. IJHS. 1980, 15; (1) : 105-113. Ref 53 Kurukshetra; Department of History Kurukshetra University KEYWORD: SCIENCE-COLONIAL

CONTENTS: 1. Science and imperialism 2. Reslut-oriented research in applied science (Production Science) 3. Botanical, zoological, coal and mineral resources survey aims to get maximum economic which illustrate use of science for colonial purposes 4. History of colonical science in 18th, 19th and 20th century in India

1352 GERSHENOWITZ H. Monsieur Poliakov's recent attack upon Lamarck. IJHS. 1980, 15; (1) : 114-19. Ref 14 USA, New Jersey; Department of Life Sciences, Glassboro State College Glassboro. 28.

CONTENTS: 1. Social neo-Lamarckians, quotations from his work "some observation with regard to man 'Philosophie Zoologique' 2. Lamarck supporter of Napoleon 3. Social Darwinism besides material- physical phenomenon

1353 SHUKLA KS. Hindu geometry by Bibhtibhusan Datta and Abadesh Narayan Singh 'Revised by. IJHS. 1980, 15; (2) : 121-188. Ref 278 Lucknow; Department of Mathematics and Astronomy, Lucknow University KEYWORD: DATTA B; SINGH AN

CONTENTS: 1. Origin of Hindu Geometry 2. Hindu names for Geometry 3. Technical terms in Geometry 4. Propositions of early geometry 5. Measurement of triangles 6. Measurement of Quadrilaterals 7. Measurement of Segment of circle 8. Measurement of volumes 9. Measurement heights and distance 10. Squaring the circle 11. Figures like, drum, corn, crescent moom, thunder-bolt etc.

1354 SIDDIQI T. Unani medicine in India during the Delhi sultanate. IJHS. 1980, 15; (1) : 18-24. Ref 15 Delhi; Institute of History of Medicine and Medical Research, Tughlaqgabad. 110 062 KEYWORD: MEDICINE - UNANI

CONTENTS: 1. Lahore, first centre of unani medicine 2. Slave dynasty, khali dynasty, Tughluq dynasty, Majmu-i Diya'i source of majmu'a-i Diyai, Lodi dynasty 3. Unani medicine in Gujarat

1355 MURTHY SRN. Vedic river Saraswati A myth or fact- A geological approach. IJHS. 1980, 15; (2) : 189-192. Ref 6 Hyderabad; Geological Sruvey of India, Southern Regional Office.1. KEYWORD: GEOLOGY

CONTENTS: 1. Saraswati river was a live river in geological part 2. Palaeochannels and offshore silts of Saraswati river 3. Diamond content base at Saraswati river

1356 LISHK SS; SHARMA SD. Standarization of time-unit muhurta through the science of Sciatherics in atharva vedanga jyotisa. IJHS. 1980, 15; (2) : 193-203. Ref 29 Patiala; Department of Physics, Punjabi University. 2.

CONTENTS: 1. Theory of time- lava (time required for 12 times blinking an eye) length of muhurta 30 lavas 1 kala .. etc. 2. Shadow-lengths and latitude at different time at the time of sunrise 3. Determination of seasons

1357 GERSHENOWITZ H. Napoleon and LamarcK. IJHS. 1980, 15; (2) : 204-209. Ref 12 New Jersey; Department of Life Sciences, Glassboro State Collge Glassboro. 28.

CONTENTS: 1. Napoleon disgrace Lamarck's philosophic zoologique 2. Georges Cuvier's catastrophic theory put him czar of Science and education ministries by Napoleon 3. Luvier and Lamarck interaction

1358 KAMAVISDARSS Analytical stuides in the evidence regarding chemico-culture in the history of Indian medicine in ancient period - Allium series. IJHS. 1980 , 15; (2) : 210-222. Ref 27 Rajnandgaon; Government P.G. College KEYWORD: MEDICINE - CHEMICO - CULTURE

CONTENTS: 1. Physico-chemical observations on action of allicin inplestruction of works 2. Experimentally produced diseases are given indicating, symptoms, causing factor and preparations etc. by consuming for a month or forthnight 3. Use of Allium Sativum (Plant)

1359 MAHDIHASSAN S. Comparative study of the early system of Indian cosmology and the Tridosa humoral doctrine. IJHS. 1980, 15; (2) : 223-229. Ref 11 Karachi; SD 34 Block, 'A', North Nazimabad KEYWORD: HUMORS

CONTENTS: 1. Chandogya Upanisad recognizes a cosmology as cosmic element Heat, water and earth 2. Human physiological functions, respiration, body temperature, intake and elimination of water 3. Oxidation -reduction, operations in body , hydroxylation, dehydration

1360 BAGCHI S; GHOSE AK. History of mining in India- Circa 1400-1800 and technology status. IJHS. 1980, 15; (1) : 25-29. Ref 8 Dhanbad; Central Mining Research Station. Dhanbad. Indian School of Mines KEYWORD: GEM STONE

CONTENTS: 1. Precious stones 2. Mining for gold, copper, lead, zinc, silver, iron 3. Fuel mineral 4. Technology overview

1361 SANGAR SP. Indian silk fabrics in the seventeenth century. IJHS. 1980, 15; (1) : 30-43. Ref 154 Chandigarh; Panjab University

CONTENTS: 1. Indian silk manufacturing industries in Bengal and other centres 2. Techniques of silk manufacturing 3. Classification of silk good 4. Silk fabrics quality contents and manufacturing techniques 5. Export and import of silk goods from India

1362 GUPTA RC. The Marici commentary on the Jyotpatti. IJHS. 1980, 15; (1) : 44-49. Ref 16 Ranchi (Bihar); Department of Mathematics, Birla Institute of Technology

CONTENTS: 1. Trigonometry- ancient Indian 2. Jagannatha's sanskrit translation (18th C) of Arabic version of Ptolemy's Almagest (C.ISO AD) 3. Marici commentary does not mention Aryabhata school in trigonometry 4. Derivation of Sin 36 & 18 5. Addition and subtraction theorems for the Indian sine

1363 GHORI SA. Impact of modern European astronomy on Raja Jai Singh. IJHS. 1980, 15; (1) : 50-57. Ref 34 Aligarh; Ghani Building, Near Jama Masjid

CONTENTS: 1. Jai Singh's biography 2. Before the building of the observatory 3. During the building and working of the observatory 4. After the observatory was completed 5. Traces of modern astronomical discoveries in Jai Singh's table (i) Theoretical: Kepler's "first law" (planetary motion by a combination of uniform circular motions) (ii) Practical: use of telescope and modern discoveries 6. Fixed stars

1364 NAQVI HK. Colour making and dyeing of cotton textiles in medieval Hindustan. IJHS. 1980, 15; (1) : 58-70. Ref 35 Delhi; Department of Indian History and Culture, Jamia College, Jamia Milia Islamia, Jamia Nagar. 25. KEYWORD: TEXTILE COTTON; DYE

CONTENTS: 1. Table given, indicating processes of dyeing as detailed in Nuskha Khulasatul Mujarrebat (20th chapt on dyeing and printing) (before C.1766) give 32 items indicating colour obtained, ingredients used, process followed and reference etc. 2. Modes of printing and painting of fabrics

1365 SIKDAR JC. Jaina Alchemy. IJHS. 1980, 15; (1) : 6-17. Ref 118 Ahmedabad; L.D. Institute of Indology. 9. KEYWORD: DRUG-METALS

CONTENTS: 1. Philosophy of mercury, Iatro-chemical period of India (1300-1550 AD) 2. Chemico-metallic processes (gold, silver, iron, copper, mercury, pearl)diseases and medical receipes process of killing metals (mercury), processes of occultism

1366 VERMA RL. Medical trends in Kashmir during Zain-Ul-Abidin's reign (1420-70 A.D.). IJHS. 1980, 15; (1) : 71-78. Ref 9 Delhi; Department of History of Medicine, All India Institute of Medical Sciences. 16.

CONTENTS: 1. Shri Karpura Bhat (accomplished Indian medical doctors) 2. Mansur bin Muhd wrote (Kifaya-e-Mansari deals with general principles of medicine, drugs and methods of treatment), Tashrih-e-Mansuri (collection of anatomical knowledge of Unani physicians) 2. Zain-Ul-abidin's biography, his apptitude in medicine

1367 BAG AK. Indian literature on mathematics during 1400-1800 A.D.. IJHS. 1980, 15; (1) : 79-93. Ref 8 Delhi; Indian National Science Academy.2. KEYWORD: MATHEMATICS - SOURCES

CONTENTS: 1. Works in sanskrit literature 2. Tradition and trend of mathematical literature in Sanskrit 3. Indo-persian/Arabic literature 4. Trend of Indo-Persian literature

1368 BEHARI K; GOVIND V. Survey of historical astrolabes of Delhi. IJHS. 1980, 15; (1) : 94-104. Ref 20 Delhi; C-4/F/91 Janak Puri, Delhi, University Grants Commission 35 Ferozshah Road

CONTENTS: 1. Description of various astrolabes found at National Museum, Delhi, Redfort Museum, Delhi, are given 2. Armillary astrolabes and Planisphere astrolabes 3. Astrolabe makers and astrolabe making centres in India

1369 SHUKLA KS. Series with fractional number of terms. VIJ. 1980, 18; 1-2 : 475-81. Ref 3 Lucknow (UP); Professor of mathematics, Lucknow university KEYWORD: ARITHMETIC

CONTENTS: 1. Bakhshali Manuscript (C. 300 AD), Brahma-sphuta-siddhanta, Acarya Mahavira (9th C. AD) & Acarya Sridhora occurrence of series with fractional terms 2. Geometrical interpretation 3. Paradoxical situation-series with fractional terms 4. Symbolical interpretation 5. Non-equivalance of interpretation

1370 BAJAJ PN. Numerals an invention of the Hindus. VIJ. 1980, 18; 1-2 : 482-84. Ref 22 Wichita, USA; Deptt. of mathematics, Wichita state University KEYWORD: ZERO

CONTENTS: 1. Numerals by Hindu, carried by Arabs to Europe 2. Notion of Zero 3. Tower of Bramah at Benares, - ancient Indian use of exponents

1371 GUPTA RC. Citrabhanu (1475-1500 AD) : a little known mathematician of medieval India . VIJ. 1980, 18; 1-2 : 485-90. Ref 10 Ranchi (Bihar);

Associate Professors of mathematics, Birla Institute of Technology,Mesra KEYWORD: CITRABHANU-MATHEMATICIAN

CONTENTS: 1. Citrabhanu-ignored his inclusion by many source 2. His works & their description 3. Mentioned twenty one pairs of simultaneous equations in two unknowns(psay x & y)

1372 CHAKRAVARTY AK. Works of Samuel Davis on Hindu Astronomy. VIJ. 1980, 18; 1-2 : 491-95. Midnapur,(W. Bengal), ; M.R. College, Mahishadal - 721628 KEYWORD: ECLIPSES

CONTENTS: 1. 18th Century Indian astronomy. fall in study 2. Study Hindu method computation of eclipse and computation of Jovian years of the Sixty year cycle papers published in philosophical transactions in 1780 & 1791 respectively

1373 CHHABRA BC. Navy in ancient India. VIJ. 1980, 198; 1-2 : 381-84. Bangalore; 494, Indiranagar Ist stage, Karnataka-560038

CONTENTS: 1. Mention of ships & vessels in ancient sources & - Ramayana - Rajavalliya - Mahavamsa - Fresco paintings at Ajanta (1500 AD) - Prince vijayasimha (543 BC) - Stone inscription (related to Bengal) - coins of Satavahana ruler of Ambaha, - Pallava, Chalukyas etc.

1374 CHITTICK N. Indian relations with east Africa before the arrival of the portuguese. JRAS. 1980, 2 : 117-27. Ref 43 KEYWORD: BOTANY

CONTENTS: 1. Mining - technique for crushing, concentration, tools etc. 2. Food plants to East Africa from India 3. Before Islam 4. Islamic period 5. Archaeological evidences

1375 DERRETT JDM. Scrupulousness and a Hindu-Jain contact. JRAS. 1980, 2 : 144-67. Ref 114 KEYWORD: JAINA

CONTENTS: 1. Trade amongst Hindu and Jain

1376 CHONDAR SL. Osteology of Gudusia Chapra (Pisces:Clupeidae). II. the vertebral column and the appendicular skeleton. JAS. 1980, 22; 3-4 : 32-48. Ref 15 Barrackpore, Inland Fisheries Trading Centre,Central Institute of Fisheries,Education KEYWORD: FISH

CONTENTS: 1. Materials & methods 2. Description-vertebral column,pectoral girdle & fin,pelvic girdle & fins,Median fins,Dorsal fin,Caudal fin,Ribs etc.

1377 MAITRA JN. Coronary Occlusion. JAS. 1980, 22; 1-2 : 43-44.

CONTENTS: 1.Heart attack - atheromatosis in arteries 2.Athermona production - nature and illustrations

1378 CHONDAR SL. Curved-normal vertebral mutants in Cirrhinus mrigala. JAS. 1980, 22; 1-2 : 47-50. Ref 7 KEYWORD: DEFORMITY - SKELETON

CONTENTS: 1.Deformed morphological characters of the fry C. mrigala 2.Deformity in skeleton - Cirrhinus mrigala

1379 LEMAY H. Stars and human sexuality: some medieval scientific views. ISIS. 1980, 71; 256 : 127-37. Ref 54 KEYWORD: SEXUALITY - STARS AND HUMAN

CONTENTS: 1. Virginity 2. Sexual productivities 3. Marital fidelity and illegitimate progeny 4. Conception and fetal development

1380 BOSE AK. Aphrodisiacs- a psychosocial perspective. IJHS. 1981, 16; (1) : 100-103. Ref 12 Calcutta; NRS Medical College and Hospital - 14 KEYWORD: SEX STIMULANTS

CONTENTS: 1. Sex stimulants (Aphrodisiacs) 2. Works on sex life, Kama Sutra by Vatsayana (3rd C. AD) Ananga Ranga by Kalyanamala (about 1450-1526 A.D.) etc, Koka Sastra etc

'Lizzat-al-Nisa in Arabic 2. Psychosocial reaction on sex by way of poetry, paintings and sculpture, Rasamanjari (by Bhanudatta), Khajuraho and Konarak sculptures, kangra pictures which show sexuality with dignity and eseteric charm 3. Ratirahasya (by Kokkoka) 4. Two kinds of aphrodisiacs (Vajikaranas) (i) Mechanical or external methods (ii) Medicinal or oral preparation to excite sexual passions 5. Use of smell and beverages aphrodisiac

1381 DUBE CB; KANSAL CM. Diseases due to deficiencies of vital principles in the body. IJHS. 1981, 16; (1) : 104-07. Ref 7 Lucknow; P.G. Department of Kayachikitsa, Lucknow University KEYWORD: DIET; DISEASES - DEFICIENCIES IN DIET

CONTENTS: 1. 18 types of deficiencey or ksaya and their reactions are given 2. A balance diet list is given 3. Diseases due to deficiency and diseases due to excessive nutrition are given

1382 THAKRAL KK. Techniques for extraction of foreign bodies from war wounds in medieval India. IJHS. 1981, 16; (1) : 11-16. Ref 10 Lucknow; Department of Shalya Shalakaya, State Ayurvedic College-226 004 KEYWORD: WOUNDS -WAR, FOREIGN BODIES

CONTENTS: 1. Causative agents 2. Methods of diagnosis of foreign bodies 3. Techniques for extraction of foreign bodies

1383 MAJUMDAR PK. Rationale of Brahmagupta's method of solving ax+cby. IJHS. 1981, 16; (2) : 111-17. Ref 3 60 Talpukur Road, P.O. Naihati, 24 Parganas KEYWORD: INDETERMINATE EQUATION

CONTENTS: 1. Indeterminate equation byax+c 2. Rule 3. Rationale of the method

1384 SHARMA SD. Pluto and a transplutonian planet as predicted by Venkatesha Ketakara. IJHS. 1981, 16; (2) : 118-29. Ref 42 Department of Physics, Punjabi University, Patiala - 147 002 KEYWORD: SAPTAGRAHI

CONTENTS: 1. Saptagrahi

1385 GERSHENOWITZ H. John Wesley powell: Staunch neo-lamarckian. IJHS. 1981, 16; (2) : 130-38. Ref 23 Department of Life Sciences, Glassboro State College, Glassboro, New Jersey-08028 KEYWORD: DARWIN

CONTENTS: 1. Lamarck's law 2. Principle of Laissez-faire 3. Darwinian concepts

1386 KHAN AJ. Survey of the concepts and measures developed by the greco-arab physicians related with the prevention and treatment of the infections and epidemic diseases. IJHS. 1981, 16; (2) : 139-44. Ref 12 A & U Tibbia College, Karol Bagh, New Delhi KEYWORD: DISEASES -INFECTIONS, EPIDEMIC; UNANI MEDICINE

CONTENTS: 1. Unani system of medicine 2. Use of drug of 1st-4th degree 3. Use of cardiotonics and protection of vital heat 4. Anti-dotal drugs

1387 GARG NK; NIGAM SK. Interaction in chemistry and medicine between Indian and Europe in 18th-19th century. IJHS. 1981, 16; (2) : 145-55. Ref 12 Lucknow; Central Drug Research Institute, National Botanical Research Institute KEYWORD: KNOWLEDGE TRANSMISSION

CONTENTS: 1. Metals 2. Fermentation 3. Indian system of medicine and plant chemistry 4. Dyes and colouring matters 5. Socio political conditions 6. Implantation of western science and education 7. 19 plants used by Ayurveda and unani system of medicine 8. Species and their constituents 9. Synthetic dyes discovered during 19th C. 10. Educational institutes in 19th C

1388 SANCHEZ, JL. Carlos J. Finlay and the conception of contagion. IJHS. 1981, 16; (2) : 156-68. Ref 21 New Delhi; B 56 Greater Kailash,

New Delhi - 110 048 KEYWORD: YELLOW FEVER

CONTENTS: 1. History of medicine 2. Yellow fever 3. Infecto - contagious diseases

1389 MITRA R. Bakula- A reputed drug of ayurveda, its history, uses in Indian medicine. IJHS. 1981, 16; (2) : 169-80. Ref 68 Lucknow; Central Council for Research in Ayurveda and Siddha, National Botanical Research Institute, Lucknow -226 001 KEYWORD: BOTANY

CONTENTS: 1. Correct botanical nomenclature 2. Botanical description 3. Distribution 4. Cultivation and collection 5. Properties and uses in Indian medicine 6. Dosage, economic uses 7. Modern researches 8. Pharmacognosy of the drug

1390 CHAUHAN DKS; SINGH RN. Contribution of medieval India to ayurvedic materia medica. IJHS. 1981, 16; (1) : 17-21. Ref 20 Lucknow; Department of Pharmacognosy, Department of Dravyaguna, State Ayurvedic College KEYWORD: MATERIA MEDICA-AYURVEDIC

CONTENTS: 1. Nighantus and commentaries 2. Common and selected drugs added to Ayurvedic Materia Medica - listed 11 drugs

1391 THATTE DG; TIWARI SP; TIWARI GP. Techniques of venupuncture (Siravedha) in India in 18th century. IJHS. 1981, 16; (2) : 181-88. Ref 6 Lucknow; Department of Sharir, State Ayurvedic College

CONTENTS: 1. Anatomical sites of venupuncture (Siravedha) 2. Choice of season and time 3. Posture 4. Procedure of venupuncture 5. After care

1392 HEGDE KTM. Scientific basis and technology of ancient Indian copper and iron; metallurgy. IJHS. 1981, 16; (2) : 189-201. Ref 29 Vadodara; Department of Archaeology and Ancient History, Maharaja Sayajirao University of Boroda. 390 002

CONTENTS: 1. Chalcolithic culture 2. Pre-Harappans, post-Harappan 3. Chemical analysis 4. Smelting of copper and iron

1393 SANGAR SP. Intoxicants in mughal India. IJHS. 1981, 16 ; (2) : 202-14. Ref 79 Chandigarh; 3093/-28 D Chandigarh-160 002

CONTENTS: 1. Opium 2. Post 3. Bhang 4. Dhatura 5. Tobacco

1394 ABRAHAM G. Gnomon in early Indian astronomy. IJHS. 1981, 16; (2) : 215-18. Ref 6 Madras; S66 Anna Nagar, Madras-600 040 KEYWORD: TRIGONOMETRY

CONTENTS: 1. Yavanajataka and pancasiddhantika 2. Trigonometric formula 3. Arthasastra

1395 SINGH RS; SINGH AN. On the identity and economico-medicinal uses of hastikarnapalasa (Leea marcrophylla roxb, family : Ampelidanceae) as evinced in the ancient (Sanskrit) texts and traditions. IJHS. 1981, 16; (2) : 219-22. Ref 15 Varanasi; Department of Rasa Shastra, Institute of Medical Sciences, Benaras Hindu University, Varanasi 221 005 KEYWORD: DRUG

1396 SIDDIQI T. Unani medicine in India. 1524 to 1605 AD. IJHS. 1981, 16; (1) : 22-25. Ref 9 Delhi; Institute of History of Medicine and Medical Research. Tughlaqabad, P.O. Madangir-110 062 KEYWORD: MEDICINE - UNANI

CONTENTS: 1. Mughal soldiers-change of climate (in India) 2. Babur's physician Atka Bakhsi 3. Well-known physicians and surgeons and their works of Babur's court 4. Education of physicians

1397 MAHDIHASSAN S. Parisrut the earliest distilled liquor of vedic times or of about 1500BC. IJHS. 1981, 16; (2) : 223-29. Ref 14 Karachi; SD 34-Block A, North Nazimabad, Karachi 33 KEYWORD: BEVERAGES

CONTENTS: 1. Beverages of vedic times 2. Soma (juice of ephedra) 3. Sura (millet beer) Su-la 4. Parisrut (distilled liquor)

1398 SIDDIQI T. Two eminent physicians (of Unani medicine) during Shah Jahan's reign. IJHS. 1981, 16; (1) : 26-30. Ref 5 Delhi; Institute of History of Medicine and Medical Research Tughlaqabad, P.O. Madangir-110 062 KEYWORD: MEDICINE - UNANI

CONTENTS: 1. Hakeem Amanallah Khan his works and the description there in 2. Hakeem Nur Al-Din Muhammad biography and his work

1399 SHARMA PV. Contributions of Sarngadhara in the field of materia medica and pharmacy. IJHS. 1981, 16; (1) : 3-10. Ref 72 Varanasi; Department of Dravyaguna, Faculty of Indian Medicine, BHU

CONTENTS: 1. Contributions in Materia Medica (i) New indigenous drugs (ii) New therapeutic uses of indigenous drugs (iii) Propagation of rural medicine (iv) Use of animal products (v) New therapeutic techniques (vi) Formulation (vii) Poisons and psychotropic drugs (viii) Use of metallic and mercurial preparation (ix) Use of unani drugs (x) Sexological medicine 2. Contribution in pharmacy -Pharmaceutical forms grouping

1400 CHATURVEDI GN; TIWARI SK; RAI NP. Medicinal use of opium and cannabis in medieval India. IJHS. 1981, 16; (1) : 31-35. Ref 11 Varanasi; Department of Kayachikitsa, Institute of Medical Sciences. BHU.5.

CONTENTS: 1. Opium and its powder preparation (Akarakarabhadi) drug for sexual hyperactive delay ejaculation 2. Other preparations of opium, karpurarasa and ahiphenasava (drugs) for acute gastro-enteritis 3. Cannabis indica (Sativa) powder (Jatiphatadi churna) formula given 4. Opium and cannabis their historical origin/perspective

1401 MISHRA SK; BHATIA BR; BHAT PSN; DWIVEDI RB. Ideas of integration as a process of evolution of Indian system of medicine in the medieval period. IJHS. 1981, 16; (1) : 36-40. Ref 10 Jamnagar; Department of Kayachikitsa, Institute of Postgraduate Teaching and Research, Gujarat Ayurvedia Univ.-361 001 KEYWORD: MEDICINE -EVOLUTION OF SYSTEMS

CONTENTS: 1. Controversies and evolution through the medieval period 2. Interaction between Ayurveda and Arab medicine 3. Stagnation in the process of evolution

1402 SINGH RS. Contribution of Unani materia medica to the identification of vedic plants with special reference to Usana (Name of plant). IJHS. 1981, 16; (1) : 41-46. Ref 25 Varanasi; Department of Rasa-Shastra, Institute of Medical Sciences. BHU.5. KEYWORD: MATERIA MEDICA - UNANI

CONTENTS: 1. Preparation of soma, from usana 2. Characteristics of Usana (plant) botanical description 3. Alkalin water from usana 4. Usana location in other parts of the world

1403 KHAN MS. Arabic source for the history of ancient Indian medicine. IJHS. 1981, 16; (1) : 47-56. Ref 47 Calcutta; Flat no.304, Wellsley Mansion, 44A Wellesley Street.16. KEYWORD: MEDICINE - ARABIC SOURCES

CONTENTS: 1. Translation of Indian scientific works 2. Method of translation 3. Indian physicians and their works

1404 KHAN AB; TARIQ M; AFAQ SH; ASIF M. Poisons and antidotes in unani system of medicine. IJHS. 1981, 16; (1) : 57-63. Ref 8 Aligarh; Post Graduate Pharmacological Laboratory, AK Tibbiya Collge, AMU KEYWORD: MEDICINE - UNANI; UNANI MEDICINE

CONTENTS: 1. List of poisons and antidotes are given 56 and 48 items respectively 2. Concept of unani and modern system of medicine regarding the poisons and antidotes are discussed 3. Mechanism of action given

1405 MAHDIHASSAN S. Alchemy and its fundamental terms in Greek, Arabic, Sanskrit and Chinese. IJHS. 1981, 16; (1) : 64-76. Ref 27 Pakistan, Karachi; SD 34 Block A, North Nazimabad-33 KEYWORD: LONGEVITY

CONTENTS: 1. Elixir and ferment-gold 2. Drug of longevity 3. Energizer plant ephedra 4. Longevity-cum-resurrection (Gold pulverized with herbal juice) 5. Sanskrit- Rasa/Soma-Rasa 6. Indian Rasayana began with herbalism 7. Alexandria and Chinese pharmacy 8. Arabic terms for Elixir 9. List terms constituting a system of alchemy in Chinese, Arabic, Greek and Sanskrit

1406 RAI NP; TIWARI SK; UPADHYA SD; CHATURVEDI GN. Origin and development of pluse examination in medieval India. IJHS. 1981, 16; (1) : 77-88. Ref 11 Varanasi; Department of Kayachikitsa, Institute of Medical Sciences, BHU.5.

CONTENTS: 1. Sarngadhara samhita, Bhava Prakasa, Yoga Ratnakara describing pluse examination (anatomical position, pathological state) 2. Tables indicating types of pulse, characteristics of pulse, simile to movement of animals, relation to fingers 3. Pulse in physiological conditions and mental states 4. Pulse indicating bad prognosis describes, periods indicating death, characteristics of the pulse 5. Avicenna's contribution to pulse examination (description of main features of the pulse)

1407 SINGH BN. Contribution of Madanapala Nighantu to the knowledge of Indian materia medica with particular reference to fig(Anjira) . IJHS. 1981, 16; (1) : 89-94. Ref 20 Lucknow; Department of Dravyaguna, State Ayurvedic Collge, Lucknow University KEYWORD: BOTANY; FICUS CARICA LINN

CONTENTS: 1. Botanical description of Fig (Ficus Carica Linn) 2. Fruit of fig tree historical perspective 3. Three categories of Anjira Group of plants 4. Wild fig, cultivated fig, country fig or gular mentioned 5. Synonyms of the fig in Iranian or Persian names in Quran

1408 ISRAILI AH. Humoral theory of Unani Tibb. IJHS. 1981, 16; (1) : 95-99. Ref 12 Delhi, Hamdard (Wakf) Laboratories, Hamdard Marg KEYWORD: HIPPOCRATIC THEORY

CONTENTS: 1. Concept of disease in Hippocratic theory of humours 2. Four humours in number : blood, phlegm, yellow bile and black bile and their functions 2. Avicenna work Qanun fil Tib (Canon of medicine) dealt with the concept of humours in detail 3. Humoral imbalance

1409 ADAMSON PB. Anatomical and pathological terms in Akkadian : Part III. JRAS. 1981, 2 : 125-32. Ref 54 KEYWORD: ANATOMY - TERMS; PATHOLOGY - TERMS

CONTENTS: 1. Guraru (Kuraru) diseass affecting head 2. Congestion/stricture 3. Cardiac pain, vertebra, breast bone, phalanges, tumour in skin

1410 MAJUMDAR PK. Bijaganitam of Bhaskara II and the continued fraction. JAS. 1981, 23; 1-2 : 124-36. Ref 12

CONTENTS: 1.Indeterminate equation of second degree 2.Varga Prakrit and Cakravala (cyclic method) 3.Brahmagupta integral solution of Nx2 + 1Y2 4.Bhaskara II continued fraction 5. Transliterated and its interpretation given

1411 MAJUMDAR PK. Use of successive convergents of the continued fraction in the works of Brahmagupta and Bhaskara II. JAS. 1981, 23; 3-4 : 99-109. Ref 19

CONTENTS: 1. Brahmasphuta Siddhanta,mean mars,budhasighra phala,moon's node,apogee of mars,mean mercury,sun & venus 2. Khandakhradyaka,Sun's equations etc 3. Siddhanta Siromonih,Continued fraction , true angular diameter at moment of eclipse,Sun & Moon motion.

1412 GULLARD P. Charles Babbage Institute for the History of Information Processing (CBI). ISIS. 1981, 72; 262 : 262-64. KEYWORD: HISTORY OF SCIENCE - INTERNATIONAL; INFORMATION SCIENCE

CONTENTS: 1. Program 2. Future Plans 3. Problems and opportunities.

1413 CLAGETT M (ED) . Critical problems in the History of Science: Proceedigs of the institute for the History of Science at the University of Wisconsin. September 1-11, 1957. Mcadison: Univ. Wisconsin Press 1959-1962 : 1969. ISIS. 1981, 72; 262 : 267-83. KEYWORD: HISTORY OF SCIENCE - CRITICAL PROBLEMS

CONTENTS: 1. Rubwick M, Coleman W, Sylla E, Daston L

1414 SEN SN. Tieffenthaler on Latitudes and longitudes in India - An Eighteenth Century Study of Geographical coordinates. IJHS. 1982, 17; (1) : 1-17. Ref 45 Narendrapur; 24 Parganas. Ramakrishna Mission Palli, P.O. Narendrapur 743 508

CONTENTS: 1. Extension of India - breadth and length at different places 2. Data on latitudes and longitudes from the works of oriental geographers and foreign travellers 3. First meridian for reckoning longitudes : Vary 4. Tieffenthaler's stressed more on determining the latitudes 5. Computed the longitude differece between two places from the route mileage

1415 JAIN BL. Development of life sciences in India in eighteenth-nineteenth century. IJHS. 1982, 17; (1) : 114-31. Ref 67 Jaipur; Department of Botany. University of Rajasthan. 302 004 KEYWORD: FLORA; FAUNA

CONTENTS: 1. Survey of Indian flora by East India company and made collections and publications 2. Exploration of Indian Fauna done by few 3. Development of medical sciences since Ayurveda text 4. Progress of life sciences in 19th century in Botany 5. Proliferation of botanic gardens 6. Explorations of cryptogamic flora 7. Surveys of forest flora and start of Indian forest service 8. Tea garden and industries 9. Development of animal science, medical sciences, veterinary science, Homoeopathic system 10. Museums and all India surveys 11. Education in life sciences

1416 SIKDAR JC. Phirangiroga(Syphilis) and its management as described in Vaidyaka samgraha an old Gujarati manuscript of an unknown author (18th century A.D.). IJHS. 1982, 17; (1) : 132-53. Ref 121 Gujarat; L. D. Institute of Indology

CONTENTS: 1. Describes primary stage -Etiology 2. Symptomology. First stage, II and III stages - cure- mercurial remedies 3. Vaidyaka samgraha chapter 1 is given describing remedy for Phirangiroga (Syphilis)

1417 GANGADHARAN N. State of Ayurveda in the eighteenth and nineteenth centuries. IJHS. 1982, 17; (1) : 154-63. Ref 38 Madras; Lecturer in Sanskrit, University of Madras

CONTENTS: 1. Works belonging to 18th century and 19th century 2. Some great exponents 3. Western studies on Ayurveda - Plastic surgery, inoculation , against smallpox 4. Teaching of Ayurveda at - Sanskrit College at Calcutta by British rulers

1418 MURTY KS. Geological sciences in India in the 18th-19th century. IJHS. 1982, 17; (1) : 164-78. Ref 65 Nagpur; Univeristy Department of Geology, Law College Compound. 440 001 KEYWORD: TRIGONOMETRIC SURVEY; GEOLOGICAL SURVEY

CONTENTS: 1. Frist stage: 1701-1800 Sivatattvaratnakara by Keladi Basavaraja of 1709 A.D. Chapter III and IV on geology 2. Second stage: 1801-1850 Trigonometric survey, geological survey persons envolved and others 3. Third stage: 1851-1900 Geological survey of India

1419 ROY RD. Outline survey of some aspects of technology in India, 1750-1900 and its legacy. IJHS. 1982, 17; (1) : 18-27. Ref 15 Calcutta; Pre-Census Population Studies Unit, Indian Statistical Institute, 203 Barrackpore Trunk Road. 700 035

CONTENTS: 1. 1750-1800 Surveys by British to know the geography and natural resources of the conquered territory 2. 1801-1850 (a) Statistical survey of the country took place by Dr. Francis Buchanan (b) Trigonometrical and topographical surveys 3. 1851-1900 (a) Geological survey of India formed formally (b) Establishment of scientific and technical departments

1420 SINGH S. Agricultural science and technology in the Punjab in the nineteenth century. IJHS. 1982, 17; (2) : 191-204. Ref 76 Amritsar; Research Fellow. Department of History. Guru Nanak Dev University

CONTENTS: 1. History of Punjab area - Rulers 2. Modes of irrigation, crops, techniques of cultivation, water-logging, live stock breeding etc. with illustrative cases

1421 SINGH RD. Development of mining technology during the nineteenth century in India. IJHS. 1982, 17; (2) : 205-22. Ref 20 Dhanbad; Department of Mining Engineering, Indian School of Mines. 826 004 KEYWORD: TECHNOLOGY - MINING

CONTENTS: 1. Growth of mining industry in the nineteenth century -coal mining, gold mining, copper mining, lead and zinc mining, iron ore mining, mining of other minearls (manganese, mica, salt, petroleum) 2. Development in mining technology - coal mining technology, hard-rock mining technology 3. Growth of safety legislation 4. Development of mining education and training

1422 BHARDWAJ HC. Development of iron and steel technology in India during 18th and 19th centuries. IJHS. 1982, 17; (2) : 223-33. Ref 24 Varanasi; Dept. of Ancient Indian History, Culture and Archaeology. BHU. 221 005

CONTENTS: 1. Indigenous iron and steel industry during 18-19th centuries (in Assam, Bengal, M.P., A.P., Tamil Nadu, Orissa) 2. Ores, their mining and preparation 3. Furnaces and their working 4. Indigenous smelting process of iron and wootz steel 5. Modernization of Indian iron industry under British rule (in Madras Presidency, Bengal, Burdwan)

1423 FORBES EG. European astronomical tradition: Its transmission into India, and its reception by Sawai Jai Singh II. IJHS. 1982, 17; (2) : 234-43. Ref 50 UK; University of Edinburgh, Scotland KEYWORD: EUROPE; KNOWLEDGE TRANSMISSION; ASTRONOMY

CONTENTS: 1. Samarkand observatory and Ulugh Beg 2. Tycho Brahe's instrumentation- European astronomy 3. Jean Picard- French astronomer 4. Johannis Baptista Carbone 1723 - telescopic observations. Hindu Traditions 5. Jai Singh II- Translation work of Ptolemy's Almagest, improved many European tables, astronomical observatory and instruments

1424 SHARMA ML. Jagannath Samrat's outstanding contribution to Indian astronomy in eighteenth century AD. IJHS. 1982, 17; (2) : 244-51. Ref 21 Varanasi; Dept. of Jyotish, Sampurnanand Sanskrit University

CONTENTS: 1. Biography of Jagannath Samrat 2. Jagannath translated Ptolemy's 'Almagest' into sanskrit title Samrat Siddhanta, Euclid's elements from their Arabic version into sanskrit title Rekhaganita 3. Discussion of the books -chapters, sections, figures etc. 4. Siddhanta sara kaustubha is a translation of Almagest, whereas Samrat Siddhanta is an original contribution of the author 5. Examined Kamalakara and Bhaskara

1425 MEHRA AK. Solar eclipse theory and observations in the 18th century Inida. IJHS. 1982, 17; (2) : 252-59. Ref 20 Department of Physics, University of Wisconsin Green Bay. Green Bay, WI54302 KEYWORD: ECLIPSE SOLAR; ECLIPSE MOON; SOLAR ECLIPSE THEORY

CONTENTS: 1. Tables for total solar eclipses and total eclipses of the moon have been given 2. Manuscripts examined for this are- Zij Muhammad-shahi; Samrat-Siddhanta; Jaivinodsarani

1426 MADHYASTHA MN; RAHIMAN MA; KAVERIAPPA KM. History of scientific technology, research and educational progress of South Kanara, Karnataka state. IJHS. 1982, 17; (2) : 260-67. Ref 9 Mangalore; Post-graudate Dept. of Bio-Sciences, University of Mysore, Post graduate Centre, Konaje. 574 152 KEYWORD: SCIENCE AND TECHNOLOGY

CONTENTS: 1. Early period 9th century- 1300 AD. Kannada language 2. Nineteenth century -education, books, art of lithography, medicine, agricultural technology, implements, tile technology, printing technology, chemical technology

1427 NOBUO K. Characteristic feature of Japanese culture in the field of science in the eighteenth-nineteenth century. IJHS. 1982, 17; (2) : 268-78. Ref 6 Japan; Professor of Tokai University. Hiratsuka KEYWORD: CONFUCIANISM

CONTENTS: 1. Survey of Japanese culture 2. Japanese view of nature under the influence of Confucianism Chinese philosophical thought 3. Wasan: Japanese mathematics 4. Process of acceptance of European science 5. Japanese mentality

1428 BLACKMORE JT. Rise and fall of three fashionable expectations. IJHS. 1982, 17; (2) : 279-88. Ref 28 Austria; Institute fur Theoretische Physik Der Universitat Wien, Boltzmanngasse 5, A-1090 Wien KEYWORD: PHYSICAL REALITY

CONTENTS: 1. Physical reality- its electromagnetic characters electromagnetism, energeticism and phenomenalism gone out of fashion in their reductionist form

1429 KULKARNI RP. Irrigation engineering of India. IJHS. 1982, 17; (1) : 28-45. Ref 31 Nasik; Research Officer. Soil Mechanics Division No. II. Maharashtra Engineering Research Institute. 422 004 KEYWORD: DAMS

CONTENTS: 1. State of irrigation before 18th century 2. Development of irrigation during 18th-19th Century 3. Design of earth and masonry dams 4. Advances in irrigation engineeering brought by British engineers 5. Irrigation administration compared between 19th century and in Kautiliya Arthasastra (3rd-4th century B.C.)

1430 DEEPAK KUMAR. Economic compulsions and the geological survey of India. IJHS. 1982, 17; (2) : 289-300. Ref 90 Kurukshetra; Department of History, Kurukshetra University KEYWORD: GEOLOGY

CONTENTS: 1. From British rule to present geological survey of India 2. Colonial purposes of GSI- as explorer and advancing coal, Iron, gold, and manganese knowledge of geological sciences

1431 KHATUN S. Impact of the eighteenth and nineteenth century European science and technology on the life and thought of the people of Bengal. IJHS. 1982, 17; (2) : 301-12. Ref 28 Bangladesh; Institute of Education and Research, University of Dacca. KEYWORD: KNOWLEDGE TRANSMISSION

CONTENTS: 1. Britain's industrial revolution indirectly influences the European S&T in the minds of people 2. Diffusion of Western S&T 4. Impact of S&T 5. Impact of industrial revolution (Change in structure of trade and commerce, slow growth of indigenous industries, urbanization, disintegration of village life etc.) 6. Impact of thought (new methods of enquiry and new way of thinking, awareness of the new concepts of natural rights, ideals of liberty, freedom and equality, secularization, sense of unity) 7. New spirit in Bengal

1432 SINGH G; JOSHI PD. Impact of European science and technology on the development of modern Ayurveda during 19th century. IJHS. 1982, 17; (2) : 313-25. Jamnagar; Dept. of Kayachikitsa. Institute of Post-Graduate Teaching and Research. Gujarat Ayurved University. KEYWORD: DRUG RESEARCH; SUSRUTA SAMHITA; KNOWLEDGE TRANSMISSION

CONTENTS: 1. Publications of Ayurvedic texts in differnt periods 2. Susruta samhita was the 1st Ayurvedic treatise, 1st published in 1831-39 1861-1870AD 1871-1880AD 1881-1890 1890-1900 2. Preservation of Ayurvedic literature 3. Drug research 4. Ayurvedic education 5. History 6. Diagnosis and treatment 7. Pharmacy

1433 BASU AK. Perspectives of the cultural impact of European medical sciences on the development of scientific medicine in India. IJHS. 1982, 17; (2) : 326-32. Ref 22 Calcutta; Psychiatry Unit, N.R.S. Medical College Hospitals. 700 014 KEYWORD: MEDICINE; EDUCATION

CONTENTS: 1. Institutions between 1845-1900 2. Biologists in the 19th century

1434 SHARMA VN. Jai Singh, his European astronomers and the Copernican revolution. IJHS. 1982, 17; (2) : 333-44. Ref 67 USA; Associate Professor of Physics and Astronomy, University of Wisconsin-Fox Valley, Menasha, Wisconsin. 54952

CONTENTS: 1. Reasons why Jai Singh did not involved the Copernican revolution 2. Copernican revolution 3. Delegation of Europe 4. Portuguese contacts 5. Jesuits contacts with Jai Singh 6. Opposition of the Catholic Church -Bruno, Galileo, Copernicus and Kepler etc. were amongst the'Indey' of prohibited book 7. Jesuits and the Copernican revolution 8. Jai Singh used telesocpe in his studies

1435 SHARMA VN. Impact of the eighteenth century Jesuit astronomers on the astronomy of India and China. IJHS. 1982, 17; (2) : 345-52. Ref 44 USA; Dept. of Physics, University of Wisconsin-Fox Valley, Menasha, WI 54952

CONTENTS: 1. Background of Jesuits 2. Jesuit astronomers in India - had little impact 3. Jesuits in China - had great impact 4. Jesuits and copernican revolution

1436 MATTHEW KM. Botany and its technologies in Peninsular India in the eighteenth and nineteenth centuries. IJHS. 1982, 17; (2) : 352-64. Tiruchirapalli; Rapinat Herbarium, St. Joseph's College. 620 002

CONTENTS: 1. Travel and survey accounts 2. Check list of plants and preliminary accounts of vegetation 3. Floras - Burman's (1768) Flora Indic 4. Botanical illustrations (12 volumes 'Hortus Malabaricus Indicus') 5. Economic botany 6. Botanical gardens and horticulture 7. Forestry, ecology and conservations (Bombay, Madras and Mysore) 8. Pharmacology and/Museums and other education aids 9. Documentation (Bibliographies) on the subject

1437 BHATTACHARYYA PK. Beginning of modern botany in India by Dutch in 16th-18th century (basic features and characteristics). IJHS. 1982, 17; (2) : 365-76. Ref 12 Kalyani; Botany Department, Kalyani University. 741 235 Burdwan, Dept. of Botany, Burdwan University. 713 104 KEYWORD: KNOWLEDGE TRANSMISSION

CONTENTS: 1. Portuguese 2. Dutch 3. First botanic garden "Hortus Ciccus" and Floristic study- 'Flora Malabarica'

1438 JHA DP. Science and technology(Coal mining) in India in eighteenth-nineteenth century.

IJHS. 1982, 17; (2) : 377-91. Ref 21 Dhanbad; Indian School of Mines. 826 004

CONTENTS: 1. East India Company Rules/Authority for coal mines 2. 'Coal'-origin in earlier texts 3. Invention of steam engine accelerated the coal mines

1439 MEHTA MJ. Indigenous paper industry and Muslim entrepreneurship: Case study of paper technology and trade of Ahmedabad with speical reference to the 19th cenury. IJHS. 1982, 17; (1) : 46-62. Ref 76 Baroda; Department of History, M.S. University of Baroda

CONTENTS: 1. History of paper making 2. Raw materials, tools and equipments 3. Manufacturing process 4. Crisis in the latter part of the 19th century. Reasons

1440 NAQVI HK. Technology and process of some of the principal industries of eighteenth century Hindustan. IJHS. 1982, 17; (1) : 63-69. Ref 59 Delhi; Jamia Millia Islamia. 110 025 KEYWORD: INDUSTRY, TECHNOLOGY; TECHNOLOGY, INDUSTRY

CONTENTS: 1. Iron 2. Cotton textiles 3. Paper

1441 BHARDWAJ HC; JAIN KK. Indian dyes and dyeing industry during 18-19th century. IJHS. 1982, 17; (1) : 70-81. Ref 18 Varanasi; Department of Ancient Indian History Culture and Archaeology. BHU. 5. KEYWORD: INDUSTRY-DYEING

CONTENTS: 1. Extraction of dyes- animal dyes, vegetable dyes 2. Extraction of Indigo 3. Mordants in dyeing 4. Dyeing techniques-cotton dyeing, silk, wool dyeing 5. Chintz printing 6. Trade in dyes during 18-19th century 7. British impact over Indian dyes and dyeing industry 8. Reasons of dwindling of Indian dye industry

1442 BAG AK. Technology in India in the eighteenth-nineteenth century. IJHS. 1982, 17; (1) : 82-90. Ref 28 Delhi; INSA KEYWORD: INDUSTRIES

CONTENTS: 1. Textile 2. Metallic and allied works, iron, zinc soldering, carvings etc. 3. Some miscellaneous technology- paper, ink, coconut or mustard oil, soap

1443 RAO EGK. Development of banking institutions in India in the eighteenth-nineteenth century. IJHS. 1982, 17; (1) : 91-113. Ref 4 Kotekar; Anand Aram, P.O. Kotekar. 574152

1444 CHATTERJEE RB. Indian surgery - past and present. JAS. 1982, 24 : 1-4. Ref 12 KEYWORD: SURGERY

CONTENTS: 1. Vedic Indian (1300 BC) 2. Sushruta great Indian Surgeon (1000 BC) 3. Khalif of Bengal-Harun al Rashid (786-806) 4. Firuz Shah (1351-1388) 5. Greek medicine under Hippocrates (460-370 BC) 6. European development also given 7. British rule medical education and science 8. Present concept of surgery

1445 INDE AJ. Dexter Award: Silver Anniversary. ISIS. 1982, 73; 267 : 268-70.

CONTENTS: 1. The Dexter Award, given in recognition of distinguished contributions to the history of chemistry. The award is sponsored by the Dexter Chemical Corporation and administered by the History of Chemistry Division of the American Chemical Society.

1446 MERCHANT C. ISIS consciousness raised. ISIS. 1982, 73; 268 : 398-409. Ref 25

CONTENTS: 1. ISIS (Egyptian mother goddess associated with the annual flooding of Nile) 2. Ideology of objectivity 3. Role of Symbol 4. Women and Science

1447 FORBES EG. International organization of the History of Science : A personal view. ISIS. 1982, 73; 269 : 552. KEYWORD: HISTORY OF SCIENCE - INTERNATIONAL

CONTENTS: 1. Activities of the Division of History of Science (DHS) of the International Union of the History and Philosophy of Science (IUHPS) 2. IUPH is a member of the International Council of Scientific Union (ICSU) a non-governmental scientific organization created in 1931 as the immediate successor to the International Research Council 3. Propose reorganization of the DHS

1448 WHITROW M. ISIS Cumulative Bibliography, 1-90, 1913-65. ISIS. 1982, 73; 266 : 95. KEYWORD: INFORMATION SCIENCE

1449 BLOOMFIELD . South Asia and Burma Retrospective bibliography. JRAS. 1983, 1 : 83-84.

CONTENTS: 1.Describe project of compiling bibliography 2. Scope 3. Period covered

1450 GERSHENOWITZ H. Arthur Koestler's osculation with Lamarckism and Neo-Lamarckism. IJHS. 1983, 18; (1) : 1-8. Ref 16 USA; Department of Life Sciences, Glassboro State College, Glassboro, New Jersey. 08028

CONTENTS: 1. Koestler A - scientific investigation 2. Reinvestigation of Lamarckian experiments by Kammerer P 3. Land breeding midwife toads - experiment 4. Bombina orientalis - incomplete experiment

1451 RAMAKRISHNAN P. History of Powder Metallurgy. IJHS. 1983, 18; (1) : 109-14. Ref 10 Bombay; Department of Metallurgical Engineering,IIT.400076 KEYWORD: METALLURGY

CONTENTS: 1. Basic principles of process 2. Early developments from 3000 BC 3. Delhi iron pillar 4. Modern renaissance of powder metallurgy

1452 NAQVI HK. Some varieties of Indian silken stuffs in Persian sources C. 1200-1700 AD. IJHS. 1983, 18; (1) : 115-29. Ref 92 Delhi; Department of Indian History and Culture, Jamia Millia Islamia Karachi; B 214/10 Federal B,Area.38 (present address) KEYWORD: SILK

CONTENTS: 1. List of 58 varieties occurring in persian sources 1200-1700 AD (provenance, price, etc)

1453 MAHDIHASSAN S. Etymology of Kim-Purusa : a rejoinder. IJHS. 1983, 18; (1) : 130. Ref 3 Karachi; SD - 34, Block A, North Nazimabad.33

CONTENTS: 1. Kim-Purusa - golden-man, wild tribesman, how-man, etc

1454 GERSHENOWITZ H. Why did Gregory Bateson overlook some basic Lamarckian tenets ?. IJHS. 1983, 18; (2) : 137-53. Ref 43 USA; Department of Life Sciences,Glassboro State College,Glassboro,New Jersey.08028 KEYWORD: LAMARCKISM

CONTENTS: 1. Kammerer P and Bateson W and others view on Lamarckian theories 2. 'Lucky Gene' theory

1455 GERSHENOWITZ H. Georges Clemenceau - traditional Lamarckian. IJHS. 1983, 18; (1) : 15-22. Ref 26 UAS; Department of Life Sciences, Glassboro State College, Glassboro, New Jersey.08028 KEYWORD: LAMARCKISM

CONTENTS: 1. Clemenceau G - philosophy of evolution thought and Lamarckian theory

1456 GERSHENOWITZ H. Encyclopedias and Lamarck. IJHS. 1983, 18; (2) : 154-65. Ref 24 USA; Department of Life Sciences,Glassboro State college,Glassboro,New Jersey.08028

CONTENTS: 1. Analysis of nines encyclopedia articles on their treatment of Lamarck J (1744-1829 AD) 2. Four errors are repeated in seven encyclopedias - Lamarck's chemical theories were worthless, Charles Darwin considered Lamarck as one of his forerunners, Lamarck was inept, Lamarck's colleagues did not study his work:considering it unimportant

1457 SINGH RS; VYAS VD(MS). Identity and critical appraisal of the basis of nomenclature and ancient socio-cultural and geographico-historical reflections evinced with the Paninian Perfume-plant/plant-part 'Kisara'. IJHS. 1983, 18; (2) : 166-71. Ref 21 Varanasi; BHU

CONTENTS: 1. Kersara - Crocus sativus Linn location, habitat and habits 2. Kisara in Kisaradi Ganapatha of Paninian aphorism IV. IV.53 3. Ayurvedic Nighantus - Kesara-origin and plant description, perfume plant-part

1458 SINGH AN; SINGH RS. Identity of the Indo-Greek relation reflected in the plant-names and uses evinced in the Kautiliya Arthasastra with particular reference to 'Kiratatikta' of 'Katuvarga' (group of spices). IJHS. 1983, 18; (2) : 172-75. Ref 12 Varanasi; BHU KEYWORD: SPICES

CONTENTS: 1. Economic plants/plants products 2.Kiratatika - bitter taste, ancient views about plant 3. Mauryan period - kiratatika of Arthasastra

1459 SHEKAR NC. Antiquity of mining and metallurgical activities at Ambaji, Kumbaria and Deri, Gujarat and Rajasthan. IJHS. 1983, 18; (2) : 176-83. Ref 4 Nagpur; Deputy Chief Geologist. Mineral Exploration Corporation Limited,Seminary Hills.440006 KEYWORD: METALLURGY

CONTENTS: 1. Location, and radiocarbon dating 2. Ore-artifact correlation 3. Literary evidences

1460 KIRSANOV VS. Characteristic features of the development of science and technology in Europe in the 18-19th centuries. IJHS. 1983, 18; (2) : 184-99. Ref 3 USSR; Institute of History of Science and Technology,Staropansky, 1/5 Moscow A-12. KEYWORD: EUROPE - S AND T

CONTENTS: 1. Newton and Leibniz's systematization of knowledge 2. Mathematical principles of natural philosophy 3. Mechanics 4. Experimental research into electricity, thermal phenomena 5. Great French Revolution 6. France -mathematical sciences, celestial mechanics 7. German -mathematical thoughts 8. 19th century development of science 9. Theory of heat 10. Electormagnetism, chemistry, theory of evolution

1461 MAJUMDAR PK. Rationale of Bhatta Govinda's method for solving the equation ax-cby and a comparative study of the determination of 'Mati' as given by Bhaskara I and Bhatta Govinda. IJHS. 1983, 18; (2) : 200-05. Ref 6 24 Parganas; 60 Talpukur Road, P.O. Naihati.WB KEYWORD: GOVINDA BHATTA; INDETERMINATE EQUATION

CONTENTS: 1. Bhatta Govinda (c.850) sanskrit verses, its english translation with modern interpretation 2. Comparative study of determination of 'Mati' as given by Bhaskara and Bhatta Govinda

1462 CHAUDHURI M. Technique of Glass making in India. IJHS. 1983, 18; (2) : 206-19. Ref 44 Calcutta; Asiatic Society, 1 Park Street.700016

CONTENTS: 1. Archaeological excavations on glass (Sirpur, Kolhapur, Nevasa etc) 2. Technique of glass making 3. Raw materials, chemical analysis, modelling, moulding, blowing, tools and implements, method of fabrication and technique of glass-beads, technique of plain bangles etc.

1463 GUPTA RC. Spread and triumph of Indian Numerals. IJHS. 1983, 18; (1) : 23-38. Ref 59 Ranchi; Department of mathematics, Birla Institute of Technology,P.O. Mesra.835215

CONTENTS: 1. Menniger and Indian numerals 2. Mayan system of place-value and zero 3. Chinese work on Indian numerals 4. Europe and arab countries numerals in ancient and medieval time

1464 DATTA B; SINGH AN; (SHUKLA KS, REVISED BY). Hindu Trigonometry. IJHS. 1983, 18; (1) : 39-108. Lucknow; Hussainganj crossing, Behind Lakshman Bhawan

CONTENTS: 1. Trigonometrical functions - definitions 2. Trigonometrical formulae 3. Addition and subtraction theorems 4. Functions of particular angles 5. Trigonometrical tables 6. Interpolation 7. Spherical trigonometry

1465 JOSHI VK. Evolution of the concept of Astavarga. IJHS. 1983, 18; (1) : 9-14. Ref 16 Varanasi; Department of Dravyaguna, Institute of Medical Sciences, BHU.221005

CONTENTS: 1. Concept evolved on the basis of Jivaniyagana of Caraka and Kakolyadigana of Susruta 2. Concept developed in three phases as in Nighantus

1466 CHONDAR SL. Osteology of Gudusia Chapra (Pisces:Clupeidae) III. Systematic account. JAS. 1983, 25; 1-4 : 46-50. Ref 12

CONTENTS: 1. 13 characters by which Chapra resembles Ilisha and 14 characters of differences

1467 KING DA. Astronomy of the Mamluks. ISIS. 1983, 74; 274 : 531-55. KEYWORD: TIME

CONTENTS: 1. Main Cairo Corpus of Tables for timekeeping

1468 NYE MJ. Scientific Decline : Is quantitative evaluation enough?. ISIS. 1983, 75; 279 : 697-708. Ref 44 KEYWORD: SCIENTOMETRICS

CONTENTS: 1. Perceptions of national decline 2. Quantitative measures of scientific achievement 3. French Style

1469 ADAMSON PB. Anatomical and pathological terms in Akkadian : Part IV. JRAS. 1984, 1 : 3-18. Ref 105 part-III,1974,102-6;1979,2-8,1981/125-32 KEYWORD: DICTIONARY; PATHOLOGY; ANATOMY

CONTENTS: 1. Mention 13 termsZl.h a. za diptheria,USULTU/blood vessel Musaru/anus, sapulu/between legs, umatu/nipples of a pregnant woman, Terku/Tirku/Skin Tumour, Halu/blackmole, sibaru/Skin externally, Uttetu/reddishlesion of skin, Seu/over whole body skin, Kittabru tumour, Setum/heat of sun, Ummu/non-specific fever

1470 SINGH P. Varga-Prakrti the cakravala method of its solution and the regular continued-fractions. IJHS. 1984, 19; (1) : 1-17. Ref 16 Bihar; R.N. College,Hajipur (Vaishali)

CONTENTS: 1. Recursive character of process which leads to solution of Nx2 + Cy2, (C1) 2. Cakravala method - Acarya Jayadeva views, Narayana views, Brahmagupta's views

1471 GERSHENOWITZ H. Encyclopaedia Judaica's interpretaton of Neo-Lamarckism. IJHS. 1984, 19; (2) : 105-08. Ref 13 USA; Department of Life Sciences,Glassboro State College,Glassboro, New Jersey.08028

CONTENTS: 1. Publication of encyclopaedia Judaica - critical evaluation 2. Article on evolution - misinterpretation of Neo-Lamarckism

1472 GUPTA PD. Helminthology in India in 18th - 19th centuries with some remarks on its recent progress. IJHS. 1984, 19; (2) : 109-17. Ref 65 Patna; Zoological Survey of India, Gangetic Plains Regional Station.800016

CONTENTS: 1. Development of helminthology in European countries (Renaissance period and in India) 2. Effects of foreign invasions on India 3. Plea for National Laboratory for Helminthology

1473 ROY M. Concept of yantra in the Samarangana Sutradhara of Bhoj. IJHS. 1984, 19; (2) : 118-24. Ref 9 Calcutta; 80-B, Vivekananda Road.700006 KEYWORD: BIJAS

CONTENTS: 1. Elements and yantras 2. Characteristic of bijas - primary elements, material substances, elements in innovations of yantra (5 principal activities), fire-element 3. Composition of yantras

1474 BHATNAGAR MS. Concept of matter and its states in Indian literature. IJHS. 1984, 19; (2) : 125-42. Ref 50 Dehradun; 15, Negi Road.248001

CONTENTS: 1. Classification of matter 2. Types of matter - properties of matter 3. Types of properties 4. Structure of matter 5. Types of system 6. Origin of matter - radiations, gaseous state, liquid state, solid state, solution and colloidal sate, glassy state 7. Types of solutions

1475 MERCIER R. Astronomical tables of Rajah Jai Singh Sawai. IJHS. 1984, 19; (2) : 143-71. Ref 2 UK; Faculty of Mathematical Studies, University of Southampton, Southampton.S09 5NH

CONTENTS: 1. Parameters of mean motion 2. Meridian of reference 3. Mean motion tables 4. Tables of geographical coordinates 5. Catalogue of stars 6. Index to chapters and tables of Zij of Muhammad Shah

1476 PRAKASH B; IGAKI K. Ancient iron making in Bastar district. IJHS. 1984, 19; (2) : 172-85. Ref 15 KEYWORD: METALLURGY

CONTENTS: 1. Chemical analysis of ancient iron samples from Bihar and Koraput 2. Chemical analysis of raw material and products of Trials 3. Description of furnace 4. Supply of blast, collection of raw material 5. Preparation and operation of furnace 6. Chemical analysis of charcoal, slag and iron by primitive method

1477 CHHABRA JG; SHARMA SD; KHANNA M. Prediction of Pluto by V B Ketakar. IJHS. 1984, 19; (1) : 18-26. Ref 23 Patiala; Department of Physics,Punjabi University KEYWORD: KETAKAR VB

CONTENTS: 1. Ketakar's orbital elements of trans-neptunian planets 2. Pickering's planets 3. Element of Lowel's X planet 4. Orbital elements of Pluto by different astronomers 5. Predicted positions of Pluto 6. Position and other elements of Brady's object

1478 DESHPANDE V. Transmutation of base-metals into gold as described in the text Rasarnavakalpa and its comparison with the parallel chinese methods. IJHS. 1984, 19; (2) : 186-92. Ref 11 Chandigarh; SRF INSA, Panjab University KEYWORD: METAL - TRANSMUTATION; CHINA

CONTENTS: 1. Chinese alchemy 2. Transmutation processes in Rasarnavakalpa 3. Uniform substrate alloys 4. Surface layer enrichment by addition and by withdrawal 5. Sulphide film formation and tinging

1479 RAMAN VV. Jean Le Rond D'Alembert 1717-1783 AD. IJHS. 1984, 19; (3) : 201-14. Ref 45 USA; Department of Physics and Humanities, Rochester Institute of Technology,1 Lomb Memorial Drive,Rochester,New York.14623 KEYWORD: ALEMBERT JLRD

CONTENTS: 1. Biography 2. Mechanics, wave equation, hydrodynamics, mathematics, probability theory, philosophy of science

1480 BANSAL UR(MS); BANSAL BB. Industries in India during 18th and 19th centuries. IJHS. 1984, 19; (3) : 215-23. Ref 21 Varanasi; Department of History,BHU.221005 Varanasi; Department of Mechanical Engineering, Institute of Technology,BHU.221005

CONTENTS: 1. Decline of Indian industry 2. Establishment of new industries - cotton industry, plantation industries, coal mines and mining, paper industry, iron and steel

1481 SHEKHAWAT V. Standards of scientific investigation: logic and methodology of science in

Caraka Samhita. IJHS. 1984, 19; (3) : 224-52. Ref 128 Jaipur; Philosophy Department, Rajasthan University KEYWORD: CARAK SAMHITA

CONTENTS: 1. Dialectical method - logical structure of science, psychology of scientific investigation, skills of theorising and fallacies 2. Method of examination 3. Therapeutic strategy

1482 KUMAR D. Science in higher education: a study in Victorian India. IJHS. 1984, 19; (3) : 253-60. Ref 70 Delhi; National Institute of Science, Technology and Development Studies, Hillside Road.110012

CONTENTS: 1. Place of science in education 2. Industrial revolution and productive mechanism 3. European science and education and its effects in India 4. Colonization

1483 GERSHENOWITZ H. Professor Conway Zirkle's Vitriolic attack on Lamarck. IJHS. 1984, 19; (3) : 261-71. Ref 13 USA; Department of Life Sciences, Glassboro State College,Glassboro,New Jersey.08028 KEYWORD: ZIRKLE C

CONTENTS: 1. Life history and his works

1484 PARAMHANS SA. Units of measurements in medieval India and their modern equivalents. IJHS. 1984, 19; (1) : 27-36. Ref 11 Allahabad; Mehta Research Institute of Mathematics and mathematical Physics,26 Dilkusha, New Katra.211002

CONTENTS: 1. Units of length 2. Measurements of weights 3. Measurement of Time

1485 YADAV CL; CHUNEKAR KC. Wonder ayurvedic drugs Laksmana for progeny: a historical appraisal. IJHS. 1984, 19; (3) : 272-78. Ref 63 Varanasi; Department of Dravyaguna, Faculty of Indian Medicine, Institute of Medical Sciences,BHU.5 KEYWORD: DRUG

CONTENTS: 1. Review and discussions - Laksmana in vedic literature, samhita period 2. Use of plant as drug

1486 KULAICHEV AP. Sriyantra and its mathematical properties. IJHS. 1984, 19; (3) : 279-92. Ref 12 USSR; Biology Faculty of Moscow University,Moscow KEYWORD: MATHEMATICS

CONTENTS: 1. Analysis of Sriyantras - mathematics 2. 11 Sriyantra were shown (spherical figures)

1487 DEY SC. Ichthyological developments in Assam during nineteenth century. IJHS. 1984, 19; (4) : 297-313. Ref 19 Assam; Fish and Fishery Biology Laboratory, Post-graduate Dept. of Zoology, Gauhati University.781014

CONTENTS: 1. Prologue and ichthyological research and developments 2. Systematic account of the ichthyospecies (list of 123 species)

1488 AGARWAL DK; SHUKLA SC. Waherman and washing materials in ancient India. IJHS. 1984, 19; (4) : 314-22. Ref 13 Lucknow; Department of Chemistry,Lucknow University.226001

CONTENTS: 1. Purification - washing materials

1489 BHATNAGAR MS. Atom from Veda to date. IJHS. 1984, 19; (4) : 323-28. Ref 44 Dehradun; 15 Negi Road.248001

CONTENTS: 1. Life, matter and universe 2. Atomic structure

1490 NAQVI HK. Cultivation under the sultans of Delhi c. 1206-1555. IJHS. 1984, 19; (4) : 329-40. Ref 66 New Delhi; B5/76 Safdarjang Enclave.110029 KEYWORD: CROPS CULTIVATION

CONTENTS: 1. Principal crops in the sultanat c.1200-1550

1491 MAJUMDAR M. Risala dar falahat. IJHS. 1984, 19; (4) : 341-60. Calcutta; SRF, INSA,75/18 SN Roy Road.700038 KEYWORD: AGRICULTURE; HORTICULTURE

CONTENTS: 1. 12 chapters of manuscript were described 2. Saving plantations from insects and animals

1492 CHATTOPADHYAY PK. Archaeometallurgical studies in Indian subcontinent: a survey on metallography of iron objects. IJHS. 1984, 19; (4) : 361-65. Ref 22 Durgapur; Training Department,Alloy Steel Plant.713208 KEYWORD: METALLURGY

CONTENTS: 1. Iron pillar of Delhi 2. Specimens from 1200 BC

1493 JOSHI BC. Neurology in ancient India : some evidences. IJHS. 1984, 19; (4) : 366-96. Ref 23 Jhansi; Department of Anaesthesia,MLB Medical College.284128 KEYWORD: YOGA

CONTENTS: 1. Available literature 2. Ancient texts (18 plates) 3. Rise and fall of yoga-vidya in India 3. Communication gap 4. Yogic approach to neurology 5. Impact of neurology on Ayurveda 6. Development of neurology in west (chronology)

1494 BHATLA N; MUKHERJEE T; SINGH G. Plants:traditional worshipping. IJHS. 1984, 19; (1) : 37-42. Ref 5 Delhi; Publications and Information Directorate (CSIR),Hillside Road.110012 KEYWORD: BOTANY

CONTENTS: 1. 11 Plants worshipped as deities - customs

1495 BAG AK; SHEN KS. Kuttaka and Qiuyishu. IJHS. 1984, 19; (4) : 397-405. Ref 11 New Delhi; INSA. China; Hangzhou University KEYWORD: INDETERMINATE EQUATION

CONTENTS: 1. Kuttaka in India 2. Equation by-axc 3. Qiuyishu in China 4. Indeterminate equation

1496 KRISHNAMURTHY KH; MOULI GC. Siddha System of medicine:a historical appraisal. IJHS. 1984, 19; (1) : 43-53. Ref 15 Pondicherry; Department of Biology, Jawaharlal Institute of Post-graduate Medical Education and Research.605006

CONTENTS: 1. Historical evaluation 2. Evidence from inscriptions 3. Comparison of two systems 4. Note on specialities of Siddha system 5. Some points for future study

1497 GERSHENOWITZ H. Robert Broom's misinterpretation of Lamarck and Darwin. IJHS. 1984, 19; (1) : 54-63. Ref 27 USA; Department of Life Science,Glassboro State College,Glassboro,New Jersey.08028

CONTENTS: 1. Renaissance man - Broom R :fossils misinterpreted Lamarckian and Darwinian theories

1498 ISMAIL A; KHAN AB. Surgery in the medieval muslim world. IJHS. 1984, 19; (1) : 64-70. Ref 16 Aligarh; Department of Unani Tib and Surgery,Ajmal Khan Tibbiya College,AMU KEYWORD: SURGERY - MUSLIM

CONTENTS: 1. Greek physicians in medieval world - Aetus Amidenus, Alexander of Tralles, Caliph Haroun-al-Raschid 2. Greek medical literature into Arabic 3. Hunain on opthalmology 4. Al-Majusi - tubercular glands, treatment of fracture of lower jaw 5. Albucaris 6. Avicenna etc.

1499 MAHDIHASSAN S. Chinese origin of the sanskrit word for Wheat. IJHS. 1984, 19; (1) : 71-73. Ref 6 Pakistan; Karachi, SD 34,Block A, North Nazimabad.33 KEYWORD: CHINA

CONTENTS: 1. History of wheat plant 2. Atharvaveda, rgveda and chinese work

1500 MAHDIHASSAN S. Tridosa doctrine and the constituents of chinese humorology. IJHS. 1984, 19;

(1) : 74-76. Ref 4 Pakistan; Karachi,SD-34,Block A,North Nazimabad.33 KEYWORD: CHINA

CONTENTS: 1. Indian and chinese systems of humorology 2. Elements of humorology and modern biochemistry

1501 RIZVI SAH. Trisection of an angle leading to the derivation of a cube equation and computation of value of sine. IJHS. 1984, 19; (1) : 77-85. Ref 10 Cochin; Department of Mathematics and Statistics,University of Cochin KEYWORD: ANGLE - TRISECTION; EQUATION - CUBIC

CONTENTS: 1. Neusis method (insertion of a static line) - trisection of an angle 2. Derivation of a cubic equation equivalent 3. Computation of value of sines

1502 DATTA B; SINGH AN; (SHUKLA KS REVISED BY). Use of calculus in Hindu mathematics. IJHS. 1984, 19; (2) : 95-104. Ref 23 Lucknow; Argara,Hussainganj Crossing.226001

CONTENTS: 1. Problems in astronomy 2. Differential formula and its proof 3. Method of infinitesimal-integration, surface of sphere, volume of sphere

1503 . Vedanga Jyotisa of Lagadha. IJHS. 1984, 19; (4) : B. Ref 21 KEYWORD: TIME; ASTERISM; ASTRONOMY

CONTENTS: 1. Benediction and valediction of astronomy 2. Measures of time, asterisms 3. Fundamental and derived yuga constants 4. Tithi, Naksatra etc. of certain special days 5. Daily tithi, Naksatra etc.

1504 KESHAVARZ F. Horoscope of Iskandar Sultan. JRAS. 1984, 2 : 197-208. KEYWORD: ASTROLOGY

CONTENTS: 1. Description of ms. of Horoscope and date of Iskandar Sultan 2. Position of planets and their movements

1505 CHONDAR SL. Raciation of Gudusia Chapra between ganga river and Keetham reservoir by osteological characters. JAS. 1984, 26; 1-4 : 16-22. Ref 18 KEYWORD: OSTEOLOHY; FISH

CONTENTS: 1. Racial separation of Ganga from Keetham Chapra by D2 analysis of three skull bones measurements

1506 BISWAS UN. Coal and Indigo interests of Tagores : as reflected in some unpublished records and manuscripts. JAS. 1984, 26; 1-4 : 54-62. Ref 1

CONTENTS: 1. Historical perspective 2. 12 records and some photos of area given 3. Reaction of coal industry to new technology 4. Navigability of Damodar and behavious of Boatmen 5. Commercial rivalry and local administrative characters 6. Indigo plantation 7. Lease indenture between Prince Dwarka Nath Tagore and Maharaja of Burdwan 8. First mercantile coal company in India

1507 ELLIOTT CA. Joint Committee on Archives of science and technology (JCAST): Summary from the final report. ISIS. 1984, 75; 276 : 58-62. KEYWORD: INFORMATION SCIENCE; DOCUMENTATION RESEARCH

CONTENTS: 1. Six chapters 2. Science and Technology in the US: its organisation and documentation 3. S & T studies documentary requirements 4. Characteristics of scientific and technological activity: implications for archivists 5. Unique documentation of S&T 6. Documentating S & T : Actions for reconstruction 7. Research in Documentation : Needs and opportunities

1508 MUTHURAJ JG. Survey of Tamil Christian Writings in European libraries. JRAS. 1985, 2 : 187-90. Ref 2 KEYWORD: KNOWLEDGE TRANSMISSION; EUROPE

CONTENTS: 1. Mentioned Tamil mss in England, Denmark, Netherlands,France,Germeny,Spain,and Portugal etc. 2. Mentioned name of libs holding Tamil mss. and appearance in foreign writings

1509 SARMA KV. Survey of source materials [Astronomy]. IJHS. 1985, 20; (1-4) : 1-20. Ref 45 Panjab; Formerly Director, Visvesvarananda Vedic Research Institute, Hoshiarpur KEYWORD: ASTRONOMY - SOURCE MATERIALS

CONTENTS: 1. Archaeological sources 2. Vedical literature 3. Jaina literature 4. Astronomical siddhantas 5. Arabic and persian sources 6. Astrolabes and observatories

1510 BAG AK. Astronomy in Indus Civilization and during vedic times.. IJHS. 1985, 20; (1-4) : 122-30. Ref 58 New Delhi; Indian National Science Academy KEYWORD: TIME; SEASONS

CONTENTS: 1. New Year Festival 2. Lunar Mansion and Lunar Months 3. Unit of Time 4. Lunar Months, Seasons & Solar Months 5. Solstices (Ayanas) and Equinoxes

1511 SHARMA SD. Post - Vedic Astronomy. IJHS. 1985, 20; (1-4) : 131-44. Ref 27

CONTENTS: 1. Planetary Kinematics 2. Jaina Astronomical Tradition 3. Three Schools of Siddhantic Astronomy 4. Karanagranthas and Saranis 5. Karanagranthas based on three principal Siddhantas

1512 SOMAYAJI DA. Yuga System and the Computations of mean and true Planetary Longitudes. IJHS. 1985, 20; (1-4) : 145-87. Ref 9 KEYWORD: LONGITUDES - PLANETARY

CONTENTS: 1. Yuga Systems 2. Rationale of Yuga Systems 3. Computation of mean longitudes of planets 4. Method finding length of the Day

1513 SHARMA SD. Eclipses, Parallax and Precession of Equinoxes. IJHS. 1985, 20; (1-4) : 188-211. Ref 73

CONTENTS: 1. Eclipses - Algorithm for computing eclipses in present recension of Surya - Siddhanta 2. Parallax (Lambana) - Brahmagupta's method of parallax, Makaranda Sarani, Surya Siddhanta 3. Precession - Length of Solar year, sidereal or tropical, Hypothesis of trepidation of equinoxes, Zero of Hindu Zodiac & Schools of Ayanamsa

1514 GHORI SAK. Development of Zij Literature in India. [Astronomy]. IJHS. 1985, 20; (1-4) : 21-48. Ref 138 Allahabad; Formerly Registrar of Arabic and Persian Madrassas of U.P. Directorate of Education KEYWORD: ASTRONOMY - SOURCE MATERIALS

CONTENTS: 1. Graeco - Arabic Astronomy 2. Zijes prepared in west-central Asia 3. West-Central Asian Zijes in India 4. Zijes compiled in Medieval India, Pre-Mughal India, Great Mughals, Later Mughals 5. List of 86 Zijes in Ain-I Akbari 6. List of 26 Zijes by Mulla Farid in Zij-i-Shahjahani

1515 SHUKLA KS. Phases of the Moon, Rising and setting of planets and Stars and their conjunctions. IJHS. 1985, 20; (1-4) : 212-51. Ref 141 KEYWORD: PLANETS - RISING AND SETTING

CONTENTS: 1. Phase and sita of moon 2. Geographical representation of Sita 3. Visible moon 4. Altitudes of Sun & Moon 5. Base & upright of Sun & Moon-Lalla 6. Rising and setting of planets & stars - Heliacal of planets and stars, Lalla of planets & stars 7. Diurnal rising & setting of planets & stars 8. Time-interval from rising to setting 9. Conjunction of planets & stars - Union & Encounter 10. Conjunction of planet & star

1516 CHATTERJEE SK; CHAKRAVARTY AK. Indian Calendar from Post-vedic period to AD 1900. IJHS. 1985, 20; (1-4) : 252-307. Ref 54 CONTENTS: 1. Vedanga Jyotisa 2. Suryaprajnapti-Jaina astronomical treatise 3. Transition period from Vedanga Jyotisa to Siddhanta Jyotisa 4. Siddhantic astronomy: Surya Siddhanta 5. Motion of Moon 6. Rapid change of

Moon's orbit around earth - Pancanga or Indian Almanac

1517 RAI RN. Astronomical Instruments. IJHS. 1985, 20; (1-4) : 308-36.

CONTENTS: 1. Nadivalaya-Yantra 2. Pitha-Yantra and Chatra-Yantra 3. Karttari-Yantra 4. Gola-Yantra 5. Ghati-Yantra 6. Gnomon 7. Yasti or Staff for Time measurement from Sun-rise 8. Cakra-Yantra 9. Dhanur Yantra - Turiya-yantra or the Quadrant 10. Phalaka-Yantra 11. Kapala-Yantra

1518 SHARMA SD. Astronomical Observatories. IJHS. 1985, 20; (1-4) : 337-62. Ref 55

CONTENTS: 1. Development of astronomical observatories 2. Observatories of Sawai Jai Singh - Ujjain Observatory, Delhi Observatory, Jaipur Observatory, Mathura Observatory 4. Other Observatories

1519 ANSARI SMR. Modern Western Astronomy in India during 18-19 centuries. IJHS. 1985, 20; (1-4) : 363-402. Ref 389

CONTENTS: 1. Early Indian observatories 2. Jai Singh's reception of European astronomy 3. European astronomy in India during 17-19 centuries - Observatories in India 4. Control by astronomers Royal 5. Conditions of scientific work

1520 BHATTACHARYA JC; VAGISWARI A. Astronomy in India in the 20 century. IJHS. 1985, 20; (1-4) : 403-435. Ref 97

CONTENTS: 1. Astronomical studies in India 2. Astronomy during first half of present century 3. Post-independence astronomy - Optical astronomy, Radio astronomy, X-ray astronomy, Cosmic ray research, Infrared astronomy, Gamma ray astronomy, UV astronomy, Theoretical astrophysics

1521 SEN SN. Survey of Studies in European Languages. IJHS. 1985, 20; (1-4) : 49-121. Ref 140

CONTENTS: 1. Seventeenth Century beginning 2. Progress in eighteenth century: Le Gentil to Hunter 3. Hindu Lunar Mansions, Zodiacal Sign 4. John Bentley on Surya - Siddhanta & its dating 5. Astronomical Observatories 6. Indian Astronomy in Nineteenth Century - Colebrooke on Zodiac, Precession of Equinoxes, Chronology of Astronomers 7. John Warren on Tamil reckoning of time 8. Progress in Twentieth Century

1522 MATILAL BK. Theory of number and paryapti in Navya-Nyaya. JAS. 1985, 27; 4 : 13-21. Ref 8 KEYWORD: NUMBER - THEORY

CONTENTS: 1. Chain relation of number 2. Philosophy of number-logical 3. Svasamanavrttikatvam 4. Nyaya-Vaisesika school notion of number

1523 REDDY VR. Indian dental anthropology in retrospect and prospect. JAS. 1985, 27; 3 : 5-34. Ref 94 KEYWORD: ANTHROPOLOGY - DENTAL

CONTENTS: 1. Dental eruption, dental morphology 2. Dental caries, supernumerary teeth, carabelli's cusp, shovel-shaped incisors, diastema, crowding, cingulum or lingula cusp, occlusion

1524 RIDDLE JM. Ancient and medieval chemotherapy for cancer. ISIS. 1985, 76; 283 : 319-30. Ref 43 KEYWORD: CANCER - DRUGS

CONTENTS: 1. Cancer in antiquity and the middle ages 2. Methodology cancer 3. Ancient and medieval cancer drugs

1526 RAHMAN A. Science and technology in Indian culture. ISIS. 1985, 76; 281 : 71. KEYWORD: SCIENCE AND TECHNOLOGY

CONTENTS: 1. National Institute of Science Technology and Development studies has initiated a major programme's in Indian Culture 2. Programme is divided into the following periods

-Science and technology in ancient India (from the earliest times to A.D. 1200), -Science and technology in medieval India (A.D. 1200-1800), -Science and the Indian response in the colonial era, -Science and technology in India since independence

1527 KRAHL R. Export porcelain fit for the chinese emperor. Early chinese blue-and-white in the Topkapi Saray Museum, Istanbul (Turkey). JRAS. 1986, 1 : 68-92. Ref 27 KEYWORD: POTTERY; DYE

CONTENTS: 1. 14 Plates of Porcelain given 2. History of export for Turkey 3.Chinese porcelain-history export 4. Chinese porcelain works, places, shapes etc

1528 WENZEL M. Manuscript sources for some motifs in early Islamic glass painting. JRAS. 1986, 2 : 214-27. Ref 19

CONTENTS: 1. Glass decoration and the naturalistic tradition 2. Possible manuscript model 3. Manuscripts model for more stylized glass decoration and a note on vertical kuficinscriptions

1529 RAMANATHAN AS. Contribution to 'Weather Science in Ancient India' II-Indra in Rgveda. IJHS. 1986, 20; 1 : 15-21. Ref 8 Madras; Kuppuswami Sastri Research Institute-600 004 KEYWORD: METEOROLOGY

CONTENTS: 1. Weather disturbances by Aryan settlers 2. Indra principle 3. Discussion of 32nd sukta of first mandala

1530 MAHDIHASSAN S. Elphedra as Soma meaning hemp fibres with soma later misidentified as the hemp plant itself. IJHS. 1986, 21; 1 : 1-6. Ref 19 Pakistan; SD-34, Block-A, North Nazimabad-33, Karachi

CONTENTS: 1. Ephedra Sinica, fire-yellow hemp fibre, hao-ma 2. Epherdra-hemp yellow, ma-huang 3. Hum-epherda 4. Connabis-huang-ma 'yellow hemp'

1531 KUNITZSCH P. Star Catalogues and star tables in medieval oriental and European astronomy. IJHS. 1986, 21; 2 : 113-22. Ref 37 West Germany; Insititute for semitic studies, University of Munich KEYWORD: TIME

CONTENTS: 1. Fixed Stars 2. Greek, Islamic & Hellenistic Times 3. Almagest 4. Ulugh Beg 5. Alfonsine Tables 6. Zijes

1532 SINGH P. Narayana's treatment of magic squares. IJHS. 1986, 21; 2 : 123-30. Ref 17 Bihar; Vaishali, R.N. College, Hajipur-844 101

CONTENTS: 1. Rule for first term & Horizontal difference of magic square whose total of a row and number of terms are separately given 2. Narayana's rule for finding vertical difference 3. Alternative rule for finding vertical difference for even magic squares

1533 GUPTA RC. Madhavacandra's and other octagonal derivations of the Jaina value Pie . IJHS. 1986, 21; (2) : 131-39. Ref 15 Ranchi; Department of Mathematics, Birla Institute of Technology, Mesra-835 215 KEYWORD: PIE - OCTAGONAL DERIVATIONS

CONTENTS: 1. Madhavacandra's derivation 2. G.Chakravarti's computations 3. Aryika Visuddhamati's exposition 4. Applying the process of averaging

1534 RIZVI SAH. Arithmetical ratio of diameter to its circumference of a circle with special reference to Jame-i-Bahadur Khani. IJHS. 1986, 21; (2) : 140-47. Ref 23 Lucknow; National Insititute of Science, Technology and Development Studies, M-7 Janpath, Chandralok, Aliganj-226 020 KEYWORD: PIE

CONTENTS: 1. Ghulam Husain Jaunpur (1790-1862 AD) - Method of ratio of Diameter to circumference of circle 2. Pie value

1535 JOSHI BC. Neurology in ancient India: Muladhara cakra - a physiological reality. IJHS. 1986, 21; (2) : 148-78. Ref 18 Jhansi; Department of Anaesthesia, MLB Medical College-284 128 KEYWORD: SEX; REPRODUCTION

CONTENTS: 1. Review of Literature 2. Source material and method 3. Plates for muladhara cakra are given (Root-Base Nerve Station) 4. Nerve impulses 5. Sex & reproduction

1536 UPADHYAY OP. Evolution of Kusta. IJHS. 1986, 21; (2) : 179-86. Ref 19 Varanasi; Department of Dravyaguna, Institute of Medical Sciences, BHU-221 005 KEYWORD: DRUG

CONTENTS: 1. Saussurea Lappa Clarke (Kusta) drug - Botanical details of Plant, Roots of Plant - Ecological state 2. Adulteration in Kusta

1537 MAHDIHASSAN S. Lac and its decolourization by Orpiment as traced to Babylon. IJHS. 1986, 21; (2) : 187-92. Ref 16 Pakistan; SD-34, Block-A, North Nazimabad, Karachi-33

CONTENTS: 1. Lac insect - Yellow pigment erythrolaccin, red dye, laccaic acid 2. Laksa as a drug

1538 RAHMAN A. Relevance of Ibn Sina Today. IJHS. 1986, 21 ; (3) : 207-36. Ref 17 Delhi; NISTADS-110 012

CONTENTS: 1. Scientific tradition and remembering scientists of Bygone days 2. European attitude to asia scientific and technological tradition 3. New look at old knowledge

1539 AINTABI MF. Ibn Sina Genius of Arab-Islamic Civilization. IJHS. 1986, 21; (3) : 217-19. Ref 3 Syria; Aleppo

CONTENTS: 1. View on Ibn Sina - his life & philosophy

1540 ROY RD. Great Trigonometrical Survey of India in a historical perspective. IJHS. 1986, 21; 1 : 22-32. Ref 16 Calcutta; Pre-Census Population Studies Unit, Indian Statistical Institute, 203-Barrackpore Trunk Road-700 035 KEYWORD: TRIGONOMETRIC SURVEY

CONTENTS: 1. Traditional knowledge & institutions in India & interface of society with knowledge 2. Introduction of European knowledge & British system 3. Impact of new knowledge and interaction between old & new

1541 ASIMON M. Life and Teachings of Ibn Sina. IJHS. 1986, 21; (3) : 220-43. Ref 61 USSR; Corresponding Member of the USSR Academy of Sciences, Moscow

CONTENTS: 1. Landmarks in Ibn-Sina & Life and creative works 2. Philosophic doctrine 3. Contribution to the development of medical science

1542 KHAIRULLAYEV MM. Treatises and Epistles of Ibn Sina. IJHS. 1986, 21; (3) : 244-50. USSR; Institute of Oriental Studies. Uzbek Academy of Science, UzbekSSR

CONTENTS: 1. Lists treatises - Treatise on terms, treatise of different branches of rational sciences etc., Epistles - Brethren of Purity etc. 2. Classification of Mathematical Sciences etc. 3. 'Comments' to philosophy of Aristotle & ancient Greek thinkers

1543 PANDIT KN. Central Asian Society in Ibn Sina's Time. IJHS. 1986, 21; (3) : 251-56. Srinagar; Centre of Central Asian Studies, University of Kashmir

CONTENTS: 1. Samanid Kingdom, Saffarids of Siestan of Bukhara 2. Buwayhinds of Isfahan, Ghaznavids of Ghazna - description of societies

1544 ASIMOV M. Poetic and Socio-ethic views of Ibn Sina. IJHS. 1986, 21; (3) : 257-260. USSR; Academy of Sciences of Tajik SSR, Dushanbe

CONTENTS: 1. Khaiyam regarded Ibn Sina as his teacher 2. Man-nature's best creation

1545 SAID HM. Ibn Sina as a Scientist. IJHS. 1986, 21; (3) : 261-69. Pakistan; Hamdard National Foundation, Karachi

CONTENTS: 1. Music 2. Geology 3. Meteorology 4. Mathematics & Astronomy 5. Chemistry 6. Physics 7. Medicine - Al-Qanun

1546 BAG AK. Ibn Sina and Indian Science. IJHS. 1986, 21; (3) : 270-75. Ref 17 Delhi; INSA

CONTENTS: 1. Education 2. His Works 3. Contact with Indian Science 4. Glimpses from Ibn Sina's Works - Medical knowledge, Alchemy, Physical concepts, Method of longitude

1547 RZVI SAA. Ibn Sina's Impact on the Rational and Scientific Movements in India. IJHS. 1986, 21; (3) : 276-84. Ref 32 Australia; Department of Asian History and Civilizations, Australian National University, Canberra

CONTENTS: 1. Ghazali (450-505/1058-1111) 2. Muhammad bin Munawwar 3. Sultan Muhammad bin Tughluq (725-752/1325-1351) 4. Sultan Sikandar Lodi (894-923/1489-1517) 5. Hakim Fathullah Shirazi 6. Shah Abdal-Aziz 7. Mawlana Fazl-i Haqq

1548 ABDI WH. Ibn Sina's critique of Mutakallimin's Atomic Theory. IJHS. 1986, 21; (3) : 285-92. Ref 5 Cochin; Dept. of Mathematics and Statistics, University of Cochin-682 022

CONTENTS: 1. Three proves (proofs) are given 2. Logicality of Proofs 3. Non-existence of Khala 4. Motion 5. Metaphysics, the real science 6. Ibn Sina's Ilahiyat (Metaphysics)

1549 PERWAZ SRA. Ibn Sina's medical works. IJHS. 1986, 21; (4) : 297-314. Ref 10 Delhi; Institute of History of Medicine and Medical Research-110 062 KEYWORD: MEDICINE - SOURCE MATERIALS

CONTENTS: 1. List of 40 books written by Ibn Sina is given 2. Abstract of 3 books & their translations, commentories and abridgements are given - Al-Qanun Fi al tibb, Al-Adwiyat Al-Qalbiyah, Arjuzas 3. Other books locations are given

1550 KHAN MS. Section on Cardiac diseases and their treatment in Qanun of Ibn Sina. IJHS. 1986, 21; (4) : 315-25. Ref 58 Delhi; Visiting Scientist, NISTADS KEYWORD: DISEASES - CARDIAC; AL-QANUN

CONTENTS: 1. Different sections of the "Qanun" of Ibn Sina are described 2. Symptoms, causes, effects & treatment of heart diseases - inflammation due to swelling, Palpitation - hot, cold

1551 SIDDIQUI T. Ibn Sina on Materia Medica. IJHS. 1986, 21 ; (4) : 326-57. Ref 3 Delhi; Institute of History of Medicine and Medical Research, Hamdard Nagar-110 062

CONTENTS: 1. English Translation of Al-Qanun Book II: Materia Medica, Sec.I - Ibn Sina of temperaments of simple drugs, Sec.II - Determination of degree of temperament of simple drugs through experiment, Sec.III - Determination of temperaments of drugs through analogy (smells, tastes & colours of simple drugs), SecIV - Determination of properties & actions of drugs, Sec.V - Rules applying to drugs extrinsic properties of drugs, Sec.VI - Collection & storage of simple drugs 2. 12 Tables are described 3. 7 drugs are described in details.

1552 DANIED RR. Method of Science in Astronomy. IJHS. 1986, 21; 1 : 33-50. Ref 9 Bombay; Tata Institute of Fundamental Research,

Homi Babha Road-400 005 KEYWORD: SCIENCE - METHOD

CONTENTS: 1. Deducitive & Inductive Methods 2. Pre-Socratic Greek Science 3. Athenian Science 4. Science during Renaissance 5. Newtonian Science 6. Einstein's theory of general relativity & geometry

1553 CHATURVEDI GN; SINGH KP; UPADHYAYA SD. Impact of Ibn Sina on pulse examination and Materia Medica of medieval period of Ayurveda. IJHS. 1986, 21; (4) : 358-62. Ref 15 Varanasi; Department of Kayachikitse, Faculty of Indian Medicine, Institute of Medical Sciences, Banaras Hindu University. Hardwar; Govt. Rishikul Ayurvedic College

CONTENTS: 1. Ibn Sina View on Pulse 2. Pulse in Ayurveda 3. Drugs in Indian Materia Medica of Ayurveda - Opium, Mercury

1554 TAIYAB M. Physiological approach of Ibn Sina towards the science of behaviour. IJHS. 1986, 21; (4) : 363-67. Ref 5 Aligarh; Ajmal Khan Tibbiya College

CONTENTS: 1. Emotions & constitution of human boday - Drugs in use 2. Tonics - exhilarant, Cardiac, nerve

1555 PUSHP PN. Ibn Sina on Speech articulation. IJHS. 1986, 21; (4) : 368-73. Ref 3 Kashmir; 33-Gogji Bagh, Srinagar

CONTENTS: 1. Treatment of phonological aspects 2. Speech phenomenon 3. Sound - non-metaphysical grounds 4. Role of muscles in speech

1556 VIRK HS. Ibn Sina's approach to Physics. IJHS. 1986, 21; (4) : 374-78. Amritsar; Dept. of Physics, Guru Nanak Dev University

CONTENTS: 1. Ibn Sina's Cosmology - Aristotle & Ibn Sina

1557 SHAFI M. Contributions of Ibn Sina to geographical knowledge. IJHS. 1986, 21; (2) : 379-82. Ref 7 Aligarh; Dept. of Geography, Aligarh Muslim University KEYWORD: GEOGRAPHY; MOUNTAINS - ORIGIN

CONTENTS: 1. Origin of mountains 2. Mechanism of mountain-buildings 3. Origin & formation of fossils 4. Hydrology 5. Climatology 6. Climatic Changes

1558 GUPTA RC. Some Equalization problems from the Bakhshali Manuscript. IJHS. 1986, 21; 1 : 51-61. Ref 15 Ranchi; Department of Mathematics, Birla Insititute of Technology, P.O. Mesra-835 215 KEYWORD: EQUALIZATION - GROWTH

CONTENTS: 1. Rule for solving simple equalization problems 2. Samadhana problems involving simultaneous equations 3. Equalization of uniform & accelerated growth

1559 KAK S. Computational aspects of the aryabhata algorithm. IJHS. 1986, 21; 1 : 62-71. Ref 12 USA; Department of Electrical and Computer Engineering, Lousiana State University, Baton Rouge, LA 70803

CONTENTS: 1. Algorithm 2. Application of Algorithm 3. Complexity of Algorithm 4. Chinese Remainder (CR) algorithm

1560 RAMANATHAN AS. Contribution to 'Weather Science in Ancient India' I-The Yajna concept of the Aryans. IJHS. 1986, 21; 1 : 7-14. Ref 21 Madras; Kuppuswami Sastri Research Insititute-600 004 KEYWORD: METEOROLOGY

CONTENTS: 1. Yajnas (sacrifices) in vedic Brahmanic literature, scientists point of view 2. Rainfall - Yajnas

1561 SZALEK BZ. Decipherment and interpretation of the proto-Indian Mohenjo-daro and Harappa inscriptions. IJHS. 1986, 21; 1 : 72-76.

Ref 2 Poland; 70-74 Szczecin Ul. Mazurska 20 M.7 KEYWORD: MOHENJO-DARO HARAPPA

CONTENTS: 1. Writing Systems 2. Grammar of inscriptions 3. Interpretation of proto-Indian Inscriptions

1562 SHEKHAWAT V. Art of theory construction in caraka samhita. IJHS. 1986, 21; 2 : 99-112. Ref 4 Jaipur; Department of Philosophy, University of Rajasthan-302 004 KEYWORD: CARAK SAMHITA

CONTENTS: 1. Construction of specific theory in Medicine 2. Technique of theory construction in science 3. Structure of ground theory 4. Theory of scientific therapy

1563 CHONDAR SL. Blood grouping in raciation of Gudusia Chapra (Pisces: Clupeidae). JAS. 1986, 28; 2 : 173-75. Ref 6 KEYWORD: BLOOD GROUPS

CONTENTS: 1. Variations in blood group characteristics - Gudusia Chapra 2. 32 specimens each from Keetham reservoir (Agra) and Ganga river sy stem in Allahabad and around

1564 CHONDAR SL. Biology of fishes III. Length-weight relationship of chanos chanos (Pisces:chanidase). JAS. 1986, 28; 1 : 77-83. Ref 10 KEYWORD: FISH

CONTENTS: 1. Materials and methods length and weight 280, 490mm & 140, 750gm respectively 2. Regression analysis of log length log weight relationship in fish 3. 75 observations given in tabular form of length and weight

1565 DWIVEDI RC. Concept of the sastra. JAS. 1986, 28; 2 : 51-56. Ref 19

CONTENTS: 1. Types of sastra, nature of sastra 2. Development of ancient literature

1566 LEVILL SH. Yantras from Sri Lanka. JAS. 1986, 28; 2 : 67-89. Ref 22

CONTENTS: 1. Yantras or mystic diagrams are thought to be quite effective as Mantras, doctor treatment etc. 2. William Crooke 1896, Carl Gustav Diehl (1956) views on Yantras 3. Yantras how to perform and 14 figures are given

1567 SHARMA PV. Medicinal plants in the yogasamgraha (Rajamartanda) of Bhoja. JAS. 1986, 28; 1 : 84-109. Ref 11 KEYWORD: DRUG

CONTENTS: 1. 209 list of medicinal plants used in Bhoja's Yogasamgraha in table giving sanskrit name, synonyms, botanical name, uses with references 2. Describe biography of Bhoja King and detail chapters of Yogasamgraha - drugs and diseases, veterinary medicine, use of minerals and inorganic substances, use of animal products, use of animal products, use of vegetable drugs etc.

1568 WEITZMAN M. Statistical Patterns in Hebrew & Arabic roots. JRAS. 1987, 1 : 15-22. Ref 20

CONTENTS: 1. Frequency Tabulation (Tables are given) 2. Consonants & Correlation coefficient 3. Graphical presentation is given

1569 KAPP DB. Palu Kurumba mantra for curing head-ache. JRAS. 1987, 1 : 48-52. Ref 20 KEYWORD: MANTRA

CONTENTS: 1. Mantra from Nilgiri Hills, South India 1. Mantra,Text,Translation and notes 2. Word Index is given for words used in mantra

# HISTORY OF SCIENCE

# AUTHOR INDEX

# HISTORY OF SCIENCE

## TITLE INDEX

# HISTORY OF SCIENCE

# KEYWORD INDEX

# HISTORY OF SCIENCE

# JOURNAL INDEX

603, 634, 639, 644, 648, 660, 787, 789, 795, 797, 803, 804, 814, 817, 823, 825, 826, 837, 849, 874, 893, 897, 900, 901, 902, 909, 910, 911

JASBSC

563, 564, 565, 566, 582, 583, 584, 585, 586, 587, 588, 589, 602, 609, 610, 611, 612, 613, 614, 616, 627, 628, 635, 636, 637, 638, 640, 641, 642, 643, 645, 646, 647, 657, 658, 659, 661, 673, 674, 675, 676, 677, 678, 679, 680, 681, 682, 686, 687, 688, 689, 690, 691, 692, 693, 701, 702, 703, 704, 705, 706, 707, 708, 709, 710, 711, 712, 713, 714, 715, 716, 721, 722, 723, 724, 725, 726, 727, 728, 736, 737, 738, 739, 740, 741, 742, 753, 754, 755, 756, 757, 763, 764, 765, 766, 767, 768, 769, 770, 788, 790, 791, 792, 793, 794, 796, 798, 811, 812, 813, 815, 816, 818, 819, 820, 821, 822, 824, 827, 828, 838, 839, 840, 841, 842, 843, 844, 845, 846, 847, 848, 850, 860, 861, 862, 863, 864, 865, 866, 867, 868, 869, 870, 871, 872, 873, 875, 876, 877, 883, 884, 890, 891, 892, 898, 899, 903

JPASB

1, 2, 3, 4, 5, 6, 7, 8, 9, 11, 12, 13, 14, 15, 16, 17, 18, 19, 20, 21, 22, 23, 24, 25, 26, 27, 28, 29, 30, 31, 32, 33, 34, 35, 36, 37, 38, 39, 40, 41, 42, 43, 44, 45, 46, 47, 48, 49, 50, 51, 52, 53, 54, 55, 56, 57, 58, 59, 60, 61, 62, 63, 64, 65, 66, 67, 68, 69, 70, 71, 72, 73, 74, 75, 76, 77, 78, 79, 80, 81, 82, 83, 84, 85, 86, 87, 88, 89, 90, 91, 92, 93, 94, 95, 96, 97, 98, 99, 100, 101, 102, 103, 104, 105, 106, 107, 108, 109, 110, 111, 112, 113, 114, 115, 116, 117, 118, 119, 120, 121, 122, 123, 124, 125, 126, 127, 128, 129, 130, 131, 132, 133, 134, 135, 136, 137, 138, 139, 140, 141, 142, 143, 144, 145, 146, 147, 148, 149, 150, 151, 152, 153, 154, 155, 156, 157, 158, 159, 160, 161, 162, 163, 165, 166, 167, 168, 169, 170, 171, 172, 173, 174, 175, 176, 177, 178, 179, 180, 181, 182, 183, 185, 186, 187, 188, 189, 190, 191, 192, 193, 194, 195, 196, 197, 198, 199, 200, 201, 202, 203, 204, 205, 206, 207, 208, 209, 210, 211, 212, 213, 214, 215, 216, 217, 218, 219, 220, 221, 222, 223, 224, 225, 226, 227, 228, 229, 230, 231, 232, 233, 234, 235, 236, 237, 238, 239, 240, 241, 242, 243, 244, 245, 246, 247, 248, 249, 250, 251, 252, 253, 254, 255, 256, 257, 258, 259, 260, 261, 262, 263, 264, 265, 266, 267, 268, 269, 270, 271, 272, 273, 274, 275, 276, 277, 278, 279, 280, 281, 282, 283, 284, 285, 286, 287, 288, 289, 290, 291, 294, 295, 296, 297, 298, 299, 300, 301, 302, 303, 304, 305, 306, 308, 309, 310, 311, 312, 313, 314, 315, 316, 317, 318, 319, 320, 321, 322, 323, 324, 325, 326, 327, 328, 329, 330, 331, 332, 333, 334, 335, 336, 337, 338, 339, 340, 341, 342, 343, 344, 345, 346, 347, 348, 349, 350, 351, 352, 353, 354, 355, 356, 357, 358, 359, 360, 361, 362, 363, 364, 365, 366, 367, 368, 369, 370, 371, 376, 377, 382, 383, 384, 385, 386, 387, 388, 389, 390, 391, 392, 393, 394, 395, 396, 397, 398, 399, 400, 402, 403, 404, 405, 406, 407, 408, 409, 410, 411, 412, 413, 414, 415, 425, 427, 428, 429, 430, 431, 432, 439, 440, 441, 442, 443, 444, 445, 455, 456, 457, 458, 459, 460, 461, 462, 463, 464, 465, 474, 475, 476, 477, 478, 479, 480, 481, 482, 483, 484, 485, 499, 501, 502, 503, 504, 505, 527, 528, 529, 530, 531, 532, 533, 534, 535, 536, 537, 538, 539, 560, 561, 562, 779, 780, 781, 782, 783, 784, 785, 786

JPASBL

10

JRAS

401, 471, 472, 473, 486, 487, 488, 506, 507, 508, 522, 523, 524, 525, 526, 540, 541, 606, 615, 617, 625, 626, 649, 650, 656, 662, 663, 664, 684, 685, 699, 700, 717, 720, 729, 735, 743, 751, 752, 758, 771, 778, 802, 809, 810, 858, 859, 887, 888, 896, 908, 919, 929, 930, 937, 942, 952, 953, 971, 972, 973, 979, 983, 987, 1019, 1034, 1074, 1075, 1076, 1115, 1118, 1134, 1135, 1145, 1171, 1238, 1298, 1300, 1301, 1325, 1349, 1374, 1375, 1409, 1449, 1469, 1504, 1508, 1527, 1528, 1568, 1569

VIJ

968, 969, 970, 980, 981, 982, 988, 989, 1010, 1011, 1012, 1013, 1029, 1045, 1069, 1070, 1071, 1111, 1112, 1113, 1114, 1136, 1137, 1138, 1139, 1158, 1159, 1160, 1161, 1172, 1173, 1174, 1175, 1176, 1177, 1227, 1228, 1229, 1230, 1231, 1232, 1254, 1255, 1256, 1257, 1258, 1288, 1289, 1290, 1291, 1292, 1293, 1294, 1320, 1321, 1322, 1323, 1324, 1342, 1343, 1344, 1345, 1369, 1370, 1371, 1372, 1373